T0310448

OPERATIONAL SAFETY ECONOMICS

OPERATIONAL SAFETY ECONOMICS

A PRACTICAL APPROACH FOCUSED ON THE CHEMICAL AND PROCESS INDUSTRIES

Genserik L.L. Reniers and H.R. Noël Van Erp

Safety and Security Science Group, Delft University of Technology, The Netherlands

Registered office
John Wiley & Sons Ltd, The Atrium, Southern Gate, Chichester, West Sussex, PO19 8SQ, United Kingdom

For details of our global editorial offices, for customer services and for information about how to apply for permission to reuse the copyright material in this book please see our website at www.wiley.com.

Library of Congress Cataloging-in-Publication Data

Names: Reniers, Genserik L.L. | Van Erp, H.R. Noël.
Title: Operational safety economics : a practical approach focused on the
    chemical and process industries / Genserik L.L. Reniers, H.R. Noël Van Erp.
Description: Chichester, West Sussex : John Wiley & Sons, Inc., 2016. |
    Includes bibliographical references and index.
Identifiers: LCCN 2016010341| ISBN 9781118871126 (cloth) | ISBN 9781118871515
    (epub)
Subjects: LCSH: Chemical industry–Risk assessment. | Chemical
    industry–Safety measures.
Classification: LCC TP200 .R435 2016 | DDC 660–dc23 LC record available at
https://lccn.loc.gov/2016010341

A catalogue record for this book is available from the British Library.

Set in 10/12pt, TimesLTStd by SPi Global, Chennai, India.
Printed and bound in Malaysia by Vivar Printing Sdn Bhd

1   2016

Contents

Preface

With this book, it is our intention to fill the existing gap between the academic literature on operational safety economics within organizations, on the one hand, and the industrial situation and needs regarding the topic, on the other. The gap is wide and the bridge is difficult to construct due to the complexity and broadness of the topic and the variety of different viewpoints, perceptions and stakeholders.

The economic concepts, models and theories are explained in this book as simply as possible, but – of course – no simpler than necessary. Nevertheless, it was often a challenge to strip down the existing academic insights into clearly understandable and user-friendly practical know-how. We have provided straightforward theoretical examples and exercises and illustrated with industrial usable and credible examples wherever possible.

The book is written from the perspective of microeconomics, i.e., the single company wishing to bring more economic-related knowledge into the company's decision-making process with respect to safety. The objective is to improve decision-making based on economic approaches, models and information. Risks are considered relative, and decisions need to be made to decrease risks or certain aspects of risks, relative to other risks, or aspects of risks. Hence, this book is intended to guide the user into how risk decision-making can be improved from a single organization's viewpoint. Even if a company's safety figures are already very good, there is very often leeway for further improvement, i.e., more efficiency with the same effectiveness, or vice versa. In brief, excellence needs to be strived for. To achieve this, adequate company-specific economic considerations are required.

This book is thus not intended as a macroeconomic work. Topics such as wage differentials, inter-country or inter-company macroeconomic aspects of risks, societal cost–benefit analyses, psychometric studies of risks and so on are not discussed. The book follows the observation that in this age of technology, communication and need for respect among people, new products are being ever more cleverly engineered to accomplish incredible feats of precision and economy, but the methods and approaches used to produce them often remain stuck in an old-school, mechanistic age of production. However, the safety needs of employees, like those of customers, also require innovation and adaptation to the twenty-first century. An important way to further improve safety within many organizations is for safety managers to use economic analyses more effectively. Economic analyses, if carried out correctly, almost always show that safety investments are a no-brainer (i.e., they should be carried out), and that investing in prevention and avoiding accident costs actually is a business strategy leading to long-term profitability and to sustainable and intrinsically healthy organizations.

In summary, the purpose of *Operational Safety Economics: A Practical Approach Focused on the Chemical and Process Industries* is to investigate the complexity of operational risk with respect to economic issues and considerations from a single organization perspective, to provide dimensions and definitions that encompass and describe topics in operational safety economic topics. A variety of theories and methods for dealing with economic analyses in an organizational environment are thus addressed. To accomplish this purpose, previous work in the field is revisited, studied and discussed, and new ideas and innovative theories are conceptualized and debated. In the end, the objective is to clarify the economics surrounding operational safety and, where possible, to provide techniques useful for addressing operational safety decision-making while considering economic issues. There are no final answers, only some clues and paths to follow to where and how such answers might be obtained by a company's safety manager.

We wish the reader of this book an interesting read, as well as economic innovative thinking with respect to operational safety.

Prof. Dr. Ir. Genserik L.L. Reniers

Drs. H.R. Noël Van Erp

Disclaimer

The authors of this work take no responsibility whatsoever for incidents directly or indirectly, alleged or not, related to any form of application or use of methods, models, approaches, information, or any other knowledge presented and discussed in the book.

Acknowledgements

This work would not have been realized without the help of many people along the way in terms of writing and rewriting it, and in the publishing process. Therefore, I would like to dedicate the book to all those people who contributed in some way to the book.

Specifically, I would like to mention **Tom Brijs**, **Katheleyn Van den Driessche**, **Jeremie Achille** and **Karolien Van Nunen**, four excellent and very hard-working students whom I was privileged to guide.

A word of sincere thanks also goes to my colleagues **Dr. Nima Khakzad**, **Dr. Luca Talarico**, and **Dr. Koen Ponnet**, all of whom I am very happy and proud to also call my friends.

Furthermore, I cannot thank enough my colleague **An Kempeneers** for her very efficient and excellent help with the figures, the index and the list of acronyms of the book. Thank you, An!

I must also acknowledge my parents, **Eddy and Iona**, who are always helping when and where they can during the most hectic period of my life.

There are many other people who have contributed in some way to the realization of this work, for instance the Wiley production team who I have not explicitly mentioned, but I would also like to thank them.

Finally, I would like to dedicate the book to my wonderful wife **Carmen** who remained patient and helpful during the – sometimes difficult and demanding – writing process, and also to my ever enthusiastic, happy and very vivid daughter **Kari**. Thank you both for making me happy!

Once again, thank you all so very much!

Genserik L.L. Reniers

List of Acronyms

ALARA	As Low As Reasonable Achievable
ALARP	As Low As Reasonably Practicable
AZF	Azote Fertilisants
BACT	Best Available Control Technology
BAT	Best Available Techniques
BATNEEC	Best Available Technology Not Entailing Excessive Costs
BEST	Break-Even Safety Target
BN	Bayesian Network
CATNIP	Cheapest Available Technology Not Invoking Prosecution
CBA	Cost-Benefit Analysis
CEO	Chief Executive Officer
CER	Cost Effectiveness Ratio
CPT	Conditional Probability Table
D&O	Directors and Officers
DF*	Disproportion Factor where moral principles are considered
DF	Disproportion Factor
DF°	Disproportion Factor where the Net Present Value is equal to zero
DNA	Deoxyribonucleic acid
ERR	External Rate of Return
EV	Expected Value
FN	Frequency, Number of fatalities
FR	Failure Rate
HILP	High-Impact Low-Probability
HOFS	Human and Organizational Factors of Safety
HRO	High Reliability Organisation
HSE	Health and Safety Executive
ICER	Incremental Cost-Effectiveness Ratio
IRR	Internal Rate of Return
ISO	International Standardisation Organisation
LAC	Long-term Average Cost of production
LIHP	Low-Impact High-Probability
LIMID	Limited Memory Influence Diagram
LTI	Lost Time Injury
LTIFR	Lost Time Injury Frequency Rate

LTIIR	Lost Time Injury Incident Rate
LTISR	Lost Time Injury Severity Rate
MC	Markov Chain
MCA	Multi-Criteria Assessment
MGT	management
MJS	Maximum Justifiable Spend
MTIFR	Medical Treatment Injury Frequency Rate
NPV	Net Present Value
PBP	Pay Back Period
PF	Proportion Factor
PV	Present Value
QAAP	Quality-Adjusted Accident Probability
QALY	Quality-Adjusted Life Years
QRA	Quantitative Risk Assessment
RM	Risk Management
RSSB	Rail Safety and Standards Board
SAR	Social Acceptability of Risk
SFAIRP	So Far As Is Reasonably Practicable
SIC	Safety Investment Cost
SIO	Safety Investment Option
SIP	Safety Investment Project
SMART	Specific, Measurable, Achievable, Relevant, Time bound
STOP	Strategic, Technology, Organisational, Personal
TEAM	The Egg Aggregated Model
TOP	Technology, Organisational, Personal
TRIFR	Total Recordable Injury Frequency Rate
VaR	Value at Risk
VoL	Value of Life
VoSL	Value of Statistical Life
WF	Weight Factor
WTA	Willingness To Accept
WTP	Willingness To Pay

1

Introduction

1.1 The "Why" of Operational Safety

In this book, safety within organizations, or safety linked with the operations of an organization (i.e., goods, services, installations, equipment, employees, and so on), is termed "operational safety." The term "operational safety," instead of "organizational safety," is employed to make it very clear that there is a distinction between, for instance, operational organizational safety, and finance-related organizational safety, health-related organizational safety, or public safety. Operational safety, for example, includes making strategic decisions on safety, or using tactical tools to deal with safety. The term "operational" merely indicates the relationship with the operations (all operations) of an organization, nothing more, nothing less.

Operational safety, or the lack thereof, is the result of a series of choices, great and small, within organizations. These choices are extremely complex and depend on a variety of factors within every organization. Important factors are legislation, available technology, socioeconomic aspects, ethical considerations, to name a few. Trade-offs often need to be made and, importantly, a diversity of assumptions need to be made and agreed upon within an organization prior to the safety-related choices. Uncertainties are involved and preferences may differ hugely between people making the decisions. Nevertheless, the goal is always the same: avoid losses! The idea is that by avoiding losses, non-tangible (because hypothetical) gains are realized. Gains can obviously be very small as well as very high, depending on the avoided losses. Determining the avoided losses quantitatively is often not simple, and qualitative aspects sometimes need to be considered when doing so. The ideas and mental models about how to achieve the end objective (i.e., to avoid losses) can thus be quite different, but the goal itself remains the same. In any case, operational safety is thus very much related to the fields of economics, management, and business, and it is therefore very important to be able to grasp and assess the prevention costs together with the level of avoided losses, preferences of decision-makers, assumptions to be made in relation to certain economic models, moral aspects of safety, and so on.

The father of industrial safety, H.W. Heinrich, in his seminal book *Industrial Accident Prevention*, the first edition published in 1931, starts the book by explaining why operational safety (and accident prevention) is important for company management and why he is in favor of a more scientific approach to this important phenomenon with substantial business impact.

Operational Safety Economics: A Practical Approach Focused on the Chemical and Process Industries,
First Edition. Genserik L.L. Reniers and H.R. Noël Van Erp.
© 2016 John Wiley & Sons, Ltd. Published 2016 by John Wiley & Sons, Ltd.

He does so by giving an example of a conversation at a conference involving the CEO, the production manager, the treasurer, and the insurance manager of a large manufacturing company. The first five and a half pages of the book are devoted to them talking about money and how they would be able to cut a huge amount of their costs simply by being safer [1]. Remember, this is a time when safety was really in its infancy and the job/function of "safety manager" simply did not exist. Safety management in those days was a synonym to insurance management. Nonetheless, the direct and obvious reason for adding more importance to operational safety in any organization was clear, even in that era, at the very beginning of industrial safety: economic considerations and the profitability of a company.

Things have changed dramatically with respect to the "how" question of operational safety – the safety regulations and procedures, prevention management, techniques and technology available, and so on. However, things have stayed exactly the same regarding the "why" question of operational safety, the answer being to avoid losses to be more profitable as an organization and to be able to "stay in business." Of course, the benefits for people and society are also very welcome. Nonetheless, for example, in the seminal work by Lees on *Loss Prevention in the Process Industries* [2, 3], a book of over 3600 print pages on process safety and loss prevention in the chemical and process industries, only a mere 20 pages cover the topic of "economics and insurance." The literature in other industries is similarly lacking.

It is thus remarkable to note that economic issues have been important from the beginning of industrial safety, but the focus has been on technology (e.g., new risk assessment techniques and innovative ways of prevention), organizational issues (e.g., compliance, new procedures and safety management systems) and, most recently, psychological/sociological human factors issues (e.g., leadership, training and collaboration within groups, safety climate and culture). However, all these advancements should be linked, in some way, to economic assessments. At the end of the day, safety choices are made to be profitable (in some way, not necessarily according to a strict interpretation of the term), not "just to be safe." This assumes adequate economic assessments within an organization.

There have been attempts to bring more economics into the operational safety decision-making process, but these have not (or hardly) been successful. The reason for their limited success is that they focus on one aspect of economic assessment (e.g., a cost-benefit analysis) but fail to develop the "big picture" of operational safety economics, where all aspects are considered by an integrated economic assessment. Moreover, most economic attempts have focused on macroeconomic issues and have tried to depict operational safety within a macro environment. Such a macroeconomic picture and theory are not interesting or applicable to concrete and microeconomic industrial practice and operational safety decision-making.

Furthermore, even in the present era there is still too much unproductive competition between "objective" and "subjective" as labels to attach to beta science and technology activities on the one hand, and social science activities on the other. However, one should recognize that people, in essence, only wish to distinguish between what is experimentally reproducible within certain limits of uncertainty, and what is either unknown or unpredictable. Looking at the objective–subjective debate in this way leads to the insight that the difference between exact sciences and social sciences is actually rather small, as some risk calculations (of so-called type II risks – see Chapter 2) as well as their rational assessment (using the best risk assessment techniques available to date) are not experimentally reproducible within certain limits of uncertainty. Hence, it is not a question of "or", but rather one of "and": the use of "objective" and "subjective" as pejorative terms is counter-productive, and risk experts should understand that risks need to be considered by all kinds of disciplines to improve the decision-process. These disciplines then need to work together based on, among other things, technological, economic, and moral aspects to inform the decision process.

1.2   Back to the Future: the Economics of Operational Safety

One theory proposes that the word "risk" is derived from the Greek word "*riza*," which means, amongst others, "cliff" [4]. In Ancient Greece, most transactions were done via shipping. If a ship sank after running into a cliff, it was lost. For an individual shipowner this was obviously a disaster, but a number of shipowners agreed that the possible misfortune should be shared among them, and that an individual shipowner should be compensated for his lost ship by the joint budget of the shipowners. In this way, the future became less uncertain for individual shipowners due to an increased confidence in doing business. This ancient version of insurance is possibly the root cause of the propensity to take risks, and the willingness to loan more for commercial undertakings, as a result of an increased confidence in the future. Hence, risk has, since its origins, been linked with economics.

Furthermore, as Bernstein put it [5], probability theory seems to be a subject made to order for the Ancient Greeks, given their zest for gambling, their skill as mathematicians, their mastery of logic, and their obsession with proof. Yet, although the Greeks were the most civilized of all the ancients, they never ventured into that fascinating world. Only in the Renaissance period, some thousands of years later in the seventeenth and eighteenth centuries, were the laws of probability conceptualized and developed in contemporary Europe. All the great scientists of this era were in some way involved in the development of probability theory. Gambling and insurance were the fields in which the laws of probability were derived, and where they were used and applied.

A new insight into risk science came in 1738 with the St. Petersburg paper [6]. The author was Daniel Bernoulli, from the famous family of mathematicians. From the late 1600s to the late 1700s, eight Bernoullis had been recognized as celebrated mathematicians. The founding father of this remarkable tribe was Nicolaus Bernoulli of Basel, a wealthy merchant whose Protestant forebears had fled from Catholic-dominated Antwerp, nowadays Belgium, around 1585. The St. Petersburg paper is truly important on the subject of risk as well as on human behavior, because it establishes that people ascribe different values to risk and introduces a pivotal idea: "Utility resulting from any small increase in wealth will be inversely proportionate to the quantity of goods previously possessed." In the paper, Bernoulli thus converts the process of calculating probabilities into a procedure for introducing subjective considerations into decisions that have uncertain outcomes, and for the first time in history measurement was applied to something that could not previously be counted. In this paper, the intellectual groundwork was laid for many aspects of microeconomic theory and decision theory.

Many improvements have been made since Bernoulli, with the introduction of operational risk management, including risk analysis and risk assessment methods, in the mid-twentieth century, mainly in the aftermath of the Second World War and the beginning of the atomic era. Due to many consistency problems with risk analysis, utility theory, and risk acceptability in general, it has been difficult to link existing theories and methods in a practical way. Nonetheless, as expressed by Reith [7], the utility of the notion of "risk" lies not necessarily in its ability to correctly predict future outcomes (at which, on the level of the individual, it is currently not particularly successful), but rather in its ability to provide a basis for decision-making.

The mandate of the economics of operational safety can be interpreted in two different ways. Instead of predicting the exact future for every individual or for an organization as a whole, the economics of operational safety should be seen as an instrument to make decisions that are as good as possible, or optimal, for the organization. Alternatively, the economics of operational safety should aim to resolve the consistency problems with risk analysis, utility theory, and risk acceptability in general, so that it may serve as a predictive instrument for the future for every individual or for an organization. In this book, both views of the role of the economics of operational safety are discussed. The current state of the art, discussed in Chapter 4, sees the economics of operational safety as a non-normative decision support tool. The advantage is that the approach is user-friendly for organizations, while the downside is that the approach gives the perception of accuracy, although it only has a limited predictive resolution, depending on the information available and on the assumptions made. The Bayesian decision theory, a neo-Bernoullian utility theory which is introduced in Chapters 7 and 8, goes beyond the state of the art and sees the economics of operational safety as a powerful normative decision support tool with considerable predictive resolution. However, the disadvantage of this approach is that its user-friendliness, in terms of easy to use and readily available software modules, needs to be improved for it to be widely employed by companies. In any case, organizations may opt to use either or both approaches to adequately deal with operational safety economics. Most importantly, the approaches need to be used correctly. This book aims to provide the background knowledge and know-how for interested readers to be able to use economic approaches with respect to operational safety.

Economic considerations have led to the development of risk theory and indirectly to operational safety, and they will also lead to its further advancement. Until some decades ago, operational safety was a field in which autonomous action was taken and certain objectives were pursued without considering the economic impact on the company. In recent years, interest in economic considerations and the financial impact of operational safety decisions has been growing. Hence, you could say that the story of the economics of safety is one of going back to the future.

1.3   Difficulties in Operational Safety Economics

The process of making optimal economic assessments with respect to operational safety, and the use of these assessments as a background for company policy-making, offers a large number of difficult challenges. For instance, no widely accepted method is available to give a monetary value to certain aspects of health, or to human life. Also, the result of any economic assessment method is only as good as the data available. As many costs are either unknown or incalculable, this can pose a serious problem. Furthermore, operational safety is a productive factor, but enhanced productivity is difficult to include in cost-benefit analyses of operational safety. Moral and ethical aspects need to be taken into account when carrying out economic assessments related to operational safety. For example, those people who bear the risks are not necessarily the same as the ones taking the risks, posing equity problems. Another problem is the fact that decision-makers often have a difficult choice to make between short-term real (safety investment) costs and long-term hypothetical (related to avoided accidents) benefits. As many decision-makers are responsible (and accountable) only for a brief period for safety budget allocations, and the accident probability of possible disaster scenarios is often extremely low ("once in a million years"), depending on a myriad factors such as the macroeconomic circumstances, they sometimes will opt for budget cuts and safety downsizing, regardless of the outcome of economic assessments for operational safety. Psychological biases present within humans (such as the principle of loss aversion; see Chapter 4) may also lead to difficulties in adequate operational safety economics. As an example, probability neglect among laypeople in general leads to a remarkable disparity between the views of the public and the views of experts, as explained by Sunstein [8], leading to consistency problems.

Fischhoff *et al.* [9] further remark that the reduction of risk typically entails the reduction of benefits, thus posing dilemmas for society. At first glance, this is of no interest in this book focusing on risk and economics only at the level of the organization. However, the question at the root of this remark is the same for society and for any company: is this product, activity, technology, etc., acceptably safe? Classic methods such as "revealed preference" or "expressed preference" (see also Chapter 4) are difficult to implement in a single organization, and, if applied, would only allow one to get an idea of the perceived risk and benefit. For operational safety decision-making to be successful, more is needed than merely perception. Based on company information available, and all other information that can be found in the academic and professional/industrial literature, and on company preferences within the wider context of societal aspects and the regulatory framework in which the company is operating, decisions need to be made about safety investments to avert operational losses. Besides the need for optimization of decisions through the use of economic analyses (where perception and moral aspects are also considered), if something goes wrong, decisions need to be defensible vis-à-vis top management, shareholders, citizens, politicians, and – sometimes forgotten – the company's employees. This seems a nearly impossible task, but the theories and concepts explained in this book, if applied correctly, may help towards achieving this task.

1.4 The Field of Operational Safety within the Profitability of an Organization

Any organization tries to be profitable, and to be healthy in the long term. Adequate managerial decisions should be taken about uncertain outcomes to achieve this goal. Hence, uncertainties, whatever they are, should be well managed within organizations. In this regard, it is important that managers realize that the results of decisions are double-edged: they may have a positive outcome, and they may have a negative outcome. Furthermore, there are essentially two fields of uncertainty that are important for organizations, and decisions are taken in these fields to make a company profitable or to continue its profitability: financial uncertainties and operational uncertainties. Figure 1.1 provides an idea of the different fields in which adequate or optimal managerial decisions matter for the long-term viability of any organization.

As can be seen from Figure 1.1, operational risks should be seen as an essential domain of any organization, to increase profitability. Similar to the fact that good decisions should be taken in the fields of financial uncertainties and profit-related operational uncertainties (e.g., production investments and innovation), good decisions should be taken to avoid operational losses. In other words, this book takes the viewpoint that operational safety is actually a domain fully contributing to the profitability of a company by generating hypothetical benefits (through the avoidance of real losses via the use of adequate operational safety). Organizations, and managers within organizations, should genuinely look at operational safety in this way. Operational safety is not part of the cost structure of a company; on the contrary, it is a

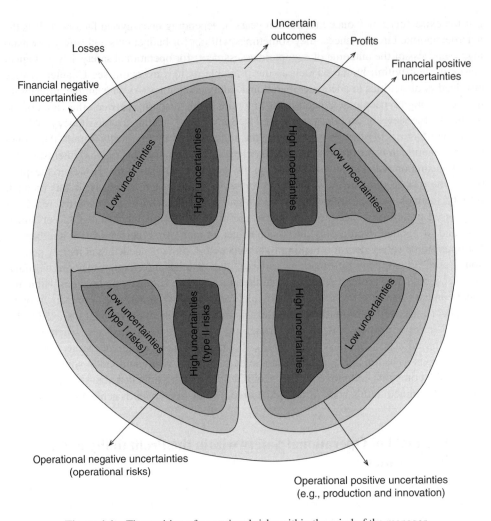

Figure 1.1 The position of operational risks within the mind of the manager.

domain that truly leads to making profits (in both the short and long term). Similar to financial uncertainties, one needs to be careful with high-uncertainty decisions (related to so-called type II risks; see later in this book), and one needs to focus on making profits via low-uncertainty decisions (related to so-called type I risks). This book will treat the risk phenomena that are fundamentally different due to the highly diverging levels of uncertainty (type I and type II) separately wherever needed.

1.5 Conclusions

Traditionally, economists are not much occupied with operational safety. Microeconomics is usually focused on well-known research fields, such as production and technology, profit and cost, utility and consumer preference, demand and supply, choice theory and game theory, to name but a few. Operational safety is thus not at all a common field of study within the economic sciences, and, in the rare cases where it is studied at universities, most academics take a macroeconomic viewpoint. This is a regretful observation, which it is not easy to explain, as operational safety is an important aspect of production and can lead to economic optimization within organizations and also within society. It is also highly relevant at all levels of society and within all subgroups and professions. Newspapers are filled with news related to operational safety and to microeconomics; it is merely the combination of the two that is lacking attention. Based on previous observations, the overlap of operational safety and microeconomics is clearly important and of interest to universities and research institutions. The field of operational safety economics within organizations therefore deserves much greater attention, from academics as well as practitioners.

References

[1] Heinrich, H.W. (1931). *Industrial Accident Prevention. A Scientific Approach*. McGraw-Hill Publishing Co., London.
[2] Lees, F.P. (1996). *Loss Prevention in the Process Industries*. 2nd edn. Butterworth-Heinemann, Oxford.
[3] Mannan, S. (2004). *Lees' Loss Prevention in the Process Industries*. 4th edn. Butterworth-Heinemann, Oxford.
[4] Drayer, E., Gude, R. (2005). *Leven in de Risicosamenleving*. Amsterdam University Press, Amsterdam.
[5] Bernstein, P.L. (1998). *Against the Gods. The Remarkable Story of Risk*. John Wiley & Sons, Inc., New York.
[6] Bernoulli, D. (1738). Exposition of a new theory on the measurement of risk. [Translated from Latin into English by dr. Louise Sommer from 'Specimen novae de mensura sortis', Commentarii academiae scientiarum imperialis petropolitanas]. *Tomus*, **V**, 175–991.
[7] Reith, G. (2009). Uncertain times: the notion of 'risk' and the development of modernity. In: *The Earthscan Reader on Risk* (eds Lofstedt, R. & Boholm, A.). Earthscan, London.
[8] Sunstein, C.R. (2005). *Laws of Fear. Beyond the Precautionary Principle*. Cambridge University Press, Cambridge.
[9] Fischhoff, B., Slovic, P., Lichtenstein, S., Read, S., Combs, B. (2009). How safe is safe enough? A psychometric study of attitudes towards technological risks and benefits. In: *The Earthscan Reader on Risk* (eds Lofstedt, R. & Boholm, A.). Earthscan, London.

2

Operational Risk, Operational Safety, and Economics

2.1 Defining the Concept of Operational Risk

A "risk" is defined by ISO 31000:2009 as "the effect of uncertainties on (achieving) objectives" [1]. Our world can indeed not be perfectly predicted and life and businesses are always exposed to uncertainties, which have an influence on whether objectives will be reached or not. Risks are double-sided: we call them negative risks if the outcome is negative, and positive risks if the outcome is positive. It is straightforward that organizations should manage risks in a way that minimizes the negative outcomes and maximizes the positive outcomes. Such management is called risk management RM) and contains, among other things, a process of risk identification, analysis, evaluation, prioritization, handling, and monitoring (see, e.g., Meyer and Reniers [2]), aimed at controlling all existing risks, whether known or not, and whether they are positive or negative. In this book, to make it workable, "operational risks" are assumed to arise from involuntary undesirable events within an organizational context. The rest of the book will thus be concerned with taking decisions regarding the management of these undesirable events and thereby considering economics-related issues.

The adoption of consistent risk management processes within a comprehensive framework can help to ensure that all types and amounts of risk are managed effectively, efficiently, and coherently across an organization. As mentioned, the economics of operational risks are focused upon. Operational risks imply unwanted events with possible negative consequences resulting from industrial operations. The economics implies approaches (concepts, models, theories, etc.) linked to financial considerations, whatever they are and in whatever form they occur. Evidently, economic considerations are very important while dealing with operational risk. Managing risks always demands making choices and allocating available budgets in the best possible way. This is not an easy task; on the contrary, it can be extremely difficult.

These days, companies and their safety managers are usually overwhelmed with tasks concerning the operational safety policy of a company. The number of tasks is huge, as are the responsibilities accompanying the decisions and choices that have to be made. Economic considerations are only one part of the larger domain of risk management. Other elements

Operational Safety Economics: A Practical Approach Focused on the Chemical and Process Industries,
First Edition. Genserik L.L. Reniers and H.R. Noël Van Erp.
© 2016 John Wiley & Sons, Ltd. Published 2016 by John Wiley & Sons, Ltd.

Figure 2.1 The operational risk management set. (Source: Meyer and Reniers [2]. Reproduced with permission from De Gruyter.)

that form part of risk management, and which are, in a way, also related to economic considerations, include safety training and education, on-the-job training, management by walking around, emergency response, business continuity planning, risk communication, risk perception, psycho-social aspects of risk, emergency planning, and risk governance. Meyer and Reniers [2] define operational risk management as "the systematic application of management policies, procedures, and practices to the tasks of identifying, analyzing, evaluating, treating, and monitoring risks." Figure 2.1 illustrates the operational risk management set.

Although economic issues of risk may only be one part of the risk management set, as can be seen in Figure 2.1, it is a very important part, being interconnected with all other parts of the risk management set, affecting the effectiveness of a company's safety policy as a whole, and, by extension, of a company's profitability in the long term. Therefore, this domain deserves to be well elaborated, both in theory and in practice. This book provides practitioners as well as the academic community new insights into this very interesting and challenging research domain, and offers practitioners concrete approaches and models to improve their risk management practice from an economic perspective.

2.2 Dealing with Operational Risks

As defined by the Center for Chemical Process Safety [3], operational risk can be seen as an index of potential economic loss, human injury, or environmental damage, which is measured in terms of both the incident probability and the magnitude of the loss, injury, or damage. The operational risk associated with a specific unwanted event can thus be expressed as the product of two factors: the likelihood that the event will occur (L_{event}) and its consequences (C_{event}). Therefore, such an operational risk index, as calculated according to Eq. (2.1), represents the "expected consequence" of the undesired event (see also Chapters 4 and 7):

$$R_{event} = L_{event} \times C_{event}. \tag{2.1}$$

However, the risk estimation always refers to specific scenarios in which the perception and the attitude to the consequences of the decision-maker may also differ in an important way. For example, most people judge a high-impact, low-probability HILP) event as more undesirable than a low-impact, high-probability LIHP) event, even if the expected consequence of the two events is exactly the same (e.g., a fatality). By introducing a risk preference parameter, the previously formulated risk index, taking into account decision-makers' preferences, can be re-formulated into:

$$R_{event} = L_{event} \times (C_{event})^a, \tag{2.2}$$

where the parameter a represents the attitude of the decision-maker to the consequences. If a decision-maker is consequence-averse (also called "risk-averse"), $a > 1$; if risk-neutral, $a = 1$, and if risk-seeking, $a < 1$. It is obvious from this that the way a risk is calculated, which depends on preferences of people, has an influence on the resulting index outcomes. The risk index as calculated according to Eq. (2.2), should thus be seen as the "calculated perception of risk reality" by a person or a group of persons using a certain calculation method that they agreed upon. In any case, the index outcomes allow us to distinguish between different types of risks.

Hence, in general, a "risk" calculation includes four terms: likelihood, consequences, risk aversion, and what can go wrong, in terms of "the event" (sometimes also called "the scenario"). To have an idea of the accumulated risk in an organization, the risks of different events (or scenarios) thus need to be summed.

Furthermore, as companies face many risks, especially when operating in high-risk environments, but also in "low-risk" environments, operational risks are usually classified into the following three categories:

- very small risks where no further investments in risk reduction are necessary;
- very large risks with an outcome so unacceptable that these risks need to be reduced immediately;
- risks that fall between the previous two risk categories.

For each of these categories of risks, discussion is possible; for example, what does the company consider to be a "very small risk" or what is considered a "very large risk"? The definitions of these categories are usually decided upon and discussed within the organization and they can differ from company to company. The risks, using their likelihood and consequences, are usually displayed on a so-called (likelihood, consequences) risk assessment decision matrix (often shortened to risk matrix), and the need for further reduction of risk (or not) is usually determined by the position of the risk within the risk matrix. The reader wishing more information on the concept of a risk matrix and on how to use it adequately is referred to Meyer and Reniers [2]. Depending on their position in the matrix, the risks should be simply monitored (negligible risk region), reduced immediately (unacceptable risk region), or reduced to the lowest level practicable [tolerable risk region or as low as reasonably practicable ALARP) region – see also Chapter 4], bearing in mind the benefits of further risk reduction and taking into account the costs of that risk reduction. A company usually has two choices when the risk is located in the ALARP region: either take further risk reduction measures or show that additional risk reduction is not reasonably practicable. "Not reasonably practicable" usually means that the risk reduction costs are disproportionally higher than the accompanying benefits [4].

However, such an approach as described above implies that the company has sufficient knowledge of the risk to assign it to a certain risk matrix cell. This is the first problem: companies often do not possess adequate information on all risks to be able to assign them unambiguously to one risk matrix cell. Unknown risks will never occur in risk matrices, for example. The more information there is about a risk, the easier it is to assign it to one risk matrix cell. Another disadvantage is that the risk matrix does not distinguish between different types of risks and thus that all risks are treated in the same way. This is very dangerous, and it has led to "blindness for disaster" in the past (cf. the BP Texas City disaster in 2005 [5]). Operational risks of different types should thus not be mixed when dealing with them and when making safety investment decisions for them. But how can the different types of operational risk be distinguished?

2.3 Types of Operational Risk

"Risk" means different things to different people at different times. However, as already mentioned, one element characterizing risk is the notion of uncertainty. Unexpected things happen and cause unexpected events. The level of uncertainty can, however, be very different from event to event.

In this book there is only a focus on operational risks, composed of the elements belonging to the negative risk triangle, i.e., "hazards – exposure to hazards – losses." If one of these elements is removed from this triangle, there is no operational risk. Hence, the economic aspects of operational risk management, discussed in this book, are all about what decisions to take, considering economic features, to diminish, decrease, or soften in an optimal way one of the three elements of the risk triangle (hazards, exposure, or losses), or a combination thereof. In any case, every risk triangle element is accompanied by uncertainty: indeed, not all hazards are known, not everything is known about the recognized hazards, not all information is available about possible exposures, certain losses are simply not known or considered, and there is also substantial uncertainty about all potential losses being taken into account.

Roughly, three types of uncertainty can be distinguished: uncertainties where a lot of historical data are available (type I), uncertainties where little or extremely few historical data are available (type II), and uncertainties where no historical data are available (type III).

Whereas type I negative risks usually lead to LIHP events (e.g., most work-related accidents, such as falling, small fires, slipping), type II negative risks can result in catastrophes with major consequences and often multiple fatalities, so-called HILP events. Type II accidents do occur on a (semi-)regular basis in a worldwide perspective, and large fires, large releases, explosions, toxic clouds, and so on belong to this class of accidents. Type III negative risks may turn into "true disasters" in terms of the loss of life and/or economic devastation. These accidents often become part of the collective memory of humankind. Examples include disasters such as Seveso (Italy, 1976), Bhopal (India, 1985), Chernobyl (USSR, 1986), Piper Alpha (North Sea, 1988), the 9/11 terrorist attacks (USA, 2001), and more recently Deepwater Horizon (Gulf of Mexico, 2010) and Fukushima (Japan, 2011). It should be noted that once type III risks have turned from the theoretical phase into reality, they become type II risks.

To prevent type I risks from turning into accidents, risk management techniques and practices are widely available. Statistical and mathematical models based on past accidents can be used to predict possible future type I accidents, indicating the prevention measures that need to be taken to prevent such accidents. Type II uncertainties and related risks and accidents are much more difficult to predict. They are extremely difficult to forecast via commonly used mathematical models, as the frequency with which these events happen is very low within one organization and the available information is therefore not enough to be investigated using, for example, regular statistics. The errors of probability estimates are very large and one should thus be extremely careful while using such probabilities. Hence, managing such risks is based on the scarce data that are available within the organization, and, more generally, on a global scale, and on extrapolations, assumptions, and expert opinions. Such risks are also investigated via available risk management techniques and practices, but these techniques should be used with much more caution, as the uncertainties are much higher for these types of risks than for type I risks. A lot of risks (and latent causes) are present which never turn into large-scale accidents due to adequate risk management, but very few risks that are present turn into accidents with huge consequences. The third type of uncertainties are extremely high, and their related accidents are simply impossible to predict. No information is available on them and they only happen extremely rarely. They cannot be predicted by past events in any way; they can only be predicted or conceived by imagination. Such accidents are also called "black swan accidents" [6]. Such events can truly only be described as "the unthinkable" – which does not mean that they cannot be thought of, but merely that people are not always capable of appreciating (or mentally ready to appreciate) that such an event may really take place.

Figure 2.2 illustrates in a qualitative way the three uncertainty types of events as a function of their frequency.

As mentioned before, type I unwanted events from Figure 2.2 can be regarded as "occupational accidents" (e.g., accidents resulting in an inability to work for several days, accidents requiring first aid). Type II and III accidents can both be categorized as "major accidents" (e.g., multiple fatality accidents, accidents with huge economic losses). Type III events are not considered further in this book, due to the fact that it is simply impossible to carry out economic analyses for such events. They are so rare that precaution and application of the high reliability organization (HRO – see later) principles are the only ways to rationally deal with them. In fact, they can also be considered as an extremum of type II events.

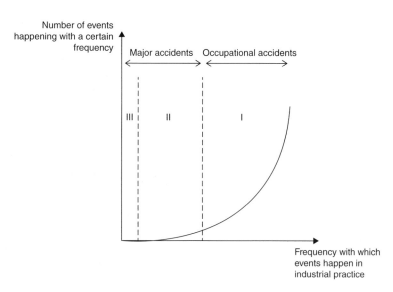

Figure 2.2 Number of events as a function of the events' frequencies (qualitative figure). (Source: Meyer and Reniers [2]. Reproduced with permission from De Gruyter.)

	Limited	Extensive
High	A (Type II)	C (Type I)
Low	B (Type II)	D (Type I)

Figure 2.3 Risk type matrix based on variability and information availability.

For each of these types of event, different economic considerations should be made and different kinds of economic analysis carried out, as will be explained and elaborated upon later on in this book. It is thus obvious that a different kind of matrix should be used before an adequate economic analysis can be carried out. An organization should therefore be able to distinguish between the different types of risk in the most objective way possible. A matrix that can be used to this end is the (information, variability) risk type matrix (see Figure 2.3).

The matrix in Figure 2.3 can be set up and employed by a company to determine the risk type, and thus the approach that is to be followed to tackle the risk from an economic viewpoint. Area A in the center of the matrix, as displayed in Figure 2.3, will be the most difficult to deal with and thereby to take economic considerations into account, while area D will be the easiest for decision-making. The importance of the factor "variability" in cost-benefit decisions can be illustrated using the following reasoning. Assume that a decision about prevention investment

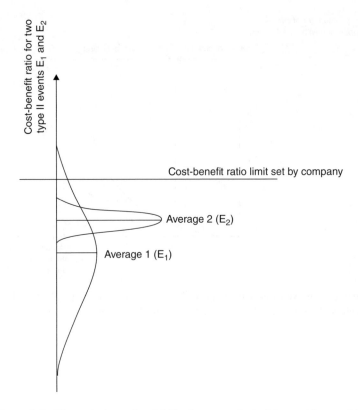

Figure 2.4   Uncertainties and variability in economic decision-making on risks.

has to be made for two type II event scenarios. A cost-benefit ratio of the required prevention investments can then be calculated in both cases. Assume that one event (E_1) is characterized with a high level of variability, and the other is characterized with a low level of variability (E_2). If there were a cost-benefit ratio limit set by the company, it would be possible that, by looking only at the average ratio position, a different decision would be made than if one were to also consider the variability. Figure 2.4 illustrates this reasoning.

Figure 2.4 illustrates that without the variability, Average 1 of E_1 would be the best choice as it is situated further away from the company limit. Average 2 of E_2 is closer to the limit, and thus the average cost-benefit ratio is higher. Because the company would prefer a cost-benefit ratio that is as low as possible (defining cost-benefit ratio as the costs divided by the benefits), prevention investments would be chosen for E_1. However, since E_1 is characterized with a higher level of variability, the tail of the distribution of E_1 exceeds the company's cost-benefit limit and thus a small possibility still exists that the ratio for prevention investment related to E_1 is worse than the absolute boundary set by the company. Conversely, the ratio of prevention investments with respect to E_2 never exceeds the company limit, thanks to the low level of variability of E_2. Hence, although on average the cost-benefit ratio is worse for E_2 than for E_1, investing in prevention for E_2 is preferred in this case due to the variability differences between the two events.

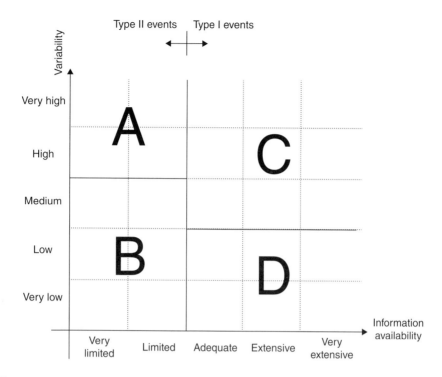

Figure 2.5   Illustrative example of matrix for determining the operational risk type and the area.

In reality, any company can choose its own means of developing such a (information, variability) matrix. For example, a concrete risk type matrix for an organization may look like the one in Figure 2.5.

The company can then further elaborate and define the qualitative parameters of the matrix from Figure 2.5 ("very low," "low," "very limited," "limited," "adequate," etc.) and use the matrix to distinguish between the different areas A–D, to determine which economic analysis technique(s) should be employed to make objective safety investment decisions with respect to the risks present within the company. In Chapter 8, indications and suggestions are given to use a certain decision-making technique for the domains A–D.

2.4   The Importance of Operational Safety Economics for a Company

As already mentioned, operational safety economics is extremely important for the profitability of a company in the long term. Figure 2.6 provides an overview of the economic effects and advantages resulting from safety investments.

The financial consequences of accidents cannot be underestimated, and avoiding accidents leads to a double positive effect. On the one hand, as a result of accidents, real financial as well as opportunity costs emerge (see also Chapter 5); hence, by avoiding such accident costs via adequate operational safety, health and safety performance is enhanced and operational negative uncertainties (see Figure 1.1) decrease (see also Figure 2.6). On the other hand,

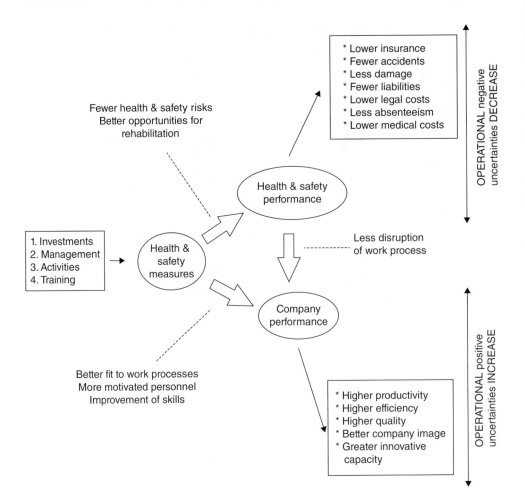

Figure 2.6 Economic consequences of health and safety investments. (Source: Fernández-Muñiz *et al.* [7]. Reproduced with permission from Elsevier.)

productivity decreases due to accidents happening, both quantitatively and qualitatively (due to the influence of factors such as reputation and the "happiness" of working in the company). Therefore, by avoiding accidents, operational positive uncertainties (see Figure 1.1) increase (see Figure 2.6).

It should be clear that occupational accidents (and not only major accidents) can be a type of negative and harmful publicity for a company, possibly leading to repercussions such as excellent employees leaving the company (and having to replace them), and small or large customers not becoming (or remaining) interested in the company . These are all possible societal consequences of accidents, in a worst-case scenario leading to bankruptcy, companies losing shareholder value on a massive scale, or companies losing a competitive market position. Chapters 5–8 elaborate further on all the direct and indirect costs of accidents for organizations, and how they contribute to and can be used in the decision-making process regarding safety investments.

It should be obvious by now that companies invest in safety. Safety investments not only lead to fewer costs of accidents, but also to an increasing company performance and competitiveness. But why is it then sometimes so difficult to convince managers to further increase safety budgets/investments or to make the necessary safety investments for HILP risks (in other words, type II risks)? There are many answers to this question, among them psychological, emotional, and economic ones. At first sight, it seems evident that risk management and safety management are essential in any manager's decision. However, Perrow [8] indicates that there are indeed reasons why managers and decision-makers would not put safety first. One very important reason for focusing on production over safety (and not on top of safety) is that the harm and the consequences are not evenly distributed: the latency period may be longer than any decision-maker's career. Few managers are punished for not putting safety first even after an accident, but they will be punished quickly for not putting profits, market share, or prestige first. But in the long term, this approach is obviously not the best management solution for any organization.

The economic issues of risk, playing a crucial role in the decision-making process with respect to operational safety management, as well as safety budgets and budget constraints, are explained in this book. One question increasingly being asked by a lot of corporate senior executives concerns the risk–opportunity trade-offs of investing in operational safety. Accidents and illness at work are matters of health and operational safety, but they are also indirectly matters of company profitability.

Operational safety is not without cost. Providing it absorbs scarce resources that could have alternative uses; they constitute the visible cost of safety. The question of whether it would be worth investing in stock options the money intended to be invested in operational safety (so-called "opportunity costs"; see also Chapter 5) can always be posed. At the same time, one should realize that the fact that safety (or prevention) has a cost does not mean that it does not have a benefit – on the contrary. However, the benefit is much harder to acknowledge by managers, as it has a hypothetical and uncertain nature. This is due to the very nature of safety. One of the definitions of safety is that it is a "dynamic non-event" [9], consistent with the understanding of safety as "the freedom from unacceptable risk." If safety is regarded as a dynamic non-event, the question of how to count or detect the non-events in time arises, as this is actually what safety represents or can be regarded as a proxy for. A non-event is, by definition, something that has not happened or will not happen, and is therefore quite difficult to measure. At the end of every working day, employees from a company may come home safely and may ask themselves, "How many times was I not injured at work today?" or "How many accidents did I not encounter today?" or "How many cyclists or cars did I not hit when I drove to work this morning or when I drove home from work this evening?" These are all legitimate questions, but they are very hard to answer. Nonetheless, this is what companies pay for: for dynamic non-events, or in other words, for events not to happen. However, statistics of non-events within companies do not exist. There are no statistical data or there is no information on non-events. Therefore, it is obviously very difficult to prove how costly, or perhaps how cost-efficient, safety really is. It is only possible to use non-safety information, such as accidents, incidents, and the like, to verify the cost and benefit of non-safety. One of the problems with this is that non-safety information can only be collected easily for type I events, and not – or it is much more difficult to do so – for type II events, as it is evidently not possible to have information based on a number of disasters that happened within the company. A disaster usually only strikes once. Hence, as can be seen from the earlier discussion, the economics of operational safety is not an easy subject.

More about cost, benefit, uncertainty, and other economic topics around safety will be explained in this book in the chapters to come. It is easy to understand that a minimum safety level is needed (it is even required by legislation) within a company. Without the minimum operational safety level, there would be unacceptable losses to the company, but also to victims and to society as a whole. But there also exists a maximum safety level, which can be seen as some kind of economic constraint. This is much harder to explain, as it depends on the type of risk and it varies from organization to organization. The problem lies at the very essence of risk: risk is uncertain and dynamic. Even the perception of what is an acceptable risk varies over time within society. Thirty years ago it was common practice not to wear a safety belt in western Europe, for example, while nowadays this is unacceptable to western European societies. The acceptability of risk ("How safe is safe enough?") is thus a very difficult question and one that will be discussed in Chapter 4.

2.5   Balancing between Productivity and Safety

To provide a tentative answer to the question of the acceptability of risk, the classic approach is the view that productivity is the enemy of safety. However, this is a false premise and a meaningless discussion. The comparison can be made with the philosophical discussion of the chicken and the egg: "Which came first?" This is a silly question; they are both equally important, and one simply cannot exist without the other. If productivity were the nitrogen of air, then safety would be the oxygen: together, they make it possible for life as we know it to exist. Analogously, the combination of productivity and safety and the balance between them make it possible for an organization to exist and to be profitable in the long run.

With the knowledge that both productivity and safety are very important for the profitability of an organization, there is indeed an "optimal situation" which can be represented by an equilibrium state. On the one hand, there are "safety means for the zero-accident (*ideal*) situation," and on the other, there are "safety means for the *as is* situation." Both of these means should be aligned as much as possible, and the difference between both should be well considered to achieve an equilibrium situation (e.g., HRO safety; see later in this section, and also the next section).

The safety means in both cases are composed of all known safety features, e.g., to a greater or lesser extent, safety management system, business continuity plans, available technology, reliability and maintenance expertise, training, competences, staffing levels, compliance, and so on. This concept of an equilibrium situation can be represented as in Figure 2.7.

A distinction can thus be made between "absolute safety" and "*as is* safety," and every company can be situated somewhere on the continuum between both extreme situations. The absolute safety situation comprises all safety measures and actions that should be present in the organization to achieve the mythical zero-accident situation. Hence, it can be seen as a theoretical situation that can never be reached, unless there are infinite safety resources. The *as is* safety situation represents all safety measures and actions as they exist at present in the organization. This situation is therefore the safety result based on current practice.

If there were a disproportionate focus on safety over production in an organization, this would lead to the "absolute safety situation" and to an economically suboptimal situation. Conversely, if there were a disproportionate focus on production over safety, the consequence would be a hazardous situation within the organization. Therefore, it is important that

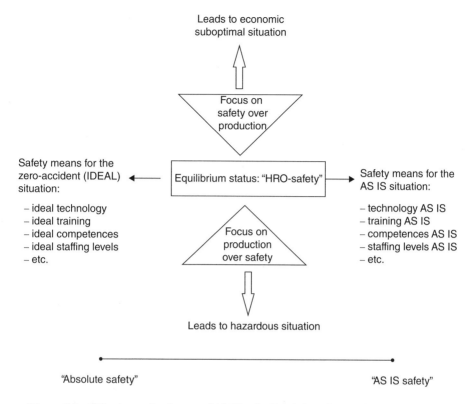

Figure 2.7   "Absolute safety" versus "AS IS safety" and the safety equilibrium situation.

operational safety economics, and all costs and benefits of safety and non-safety, are well elaborated and well managed in any organization.

One of the premises of good safety economics is to make a distinction between the different existing types of accidents. For the different risk types, a conceptual figure (see Figure 2.8) can be drawn displaying the different levels of safety and the fluctuation of the real safety level, trying achieve an optimum situation for the company. Regretfully, in many companies the actual fluctuating safety level curve is not situated around the equilibrium situation, but rather below it.

The reader may have noticed that the equilibrium situation is also referred to as "HRO safety." But what is HRO safety? The principles that apply in HRO safety are mainly aimed at type II risks, but type I risks also benefit just as much from the practices and the mindset of such environments – hence the suggestion to consider it as the safety equilibrium situation. HRO safety is explained more in detail in the next section.

2.6   The Safety Equilibrium Situation or "HRO Safety"

Organizations capable of gaining and sustaining high reliability levels are called "high reliability organizations" (HROs). Despite the fact that HROs operate hazardous activities within a high-risk environment, they succeed in achieving excellent health and safety figures. Hence,

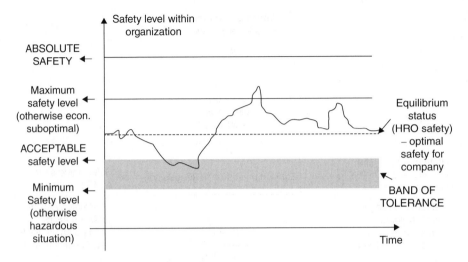

Figure 2.8 Company fluctuating safety level (to be drawn for type I and type II risks separately).

they identify and correct risks very efficiently and effectively. A typical characteristic of HROs is collective mindfulness. Hopkins [10] also indicates that HROs organize themselves in such a way that they are better able to notice the unexpected in the making and halt its development. Hence, collective mindfulness in HROs implies a certain approach in the way they organize themselves. Five key principles are used by HROs to achieve such a mindful and reliable organization, as discussed in the following (see also Weick and Sutcliffe [11]).

The first three principles mainly relate to anticipation, or the ability of organizations to cope with unexpected events. Anticipation concerns disruptions, simplifications, and execution, and requires means of detecting small clues and indications with the potential to result in large, disruptive events. Of course, such organizations should also be able to decrease, diminish, or stop the consequences of (a chain of) unwanted events. Anticipation implies the ability to imagine new, uncontrollable situations, which are based on small differences with well-known and controllable situations. HROs take this into account through principles 1–3. Whereas the first three principles relate to proaction, the fourth and fifth principles focus on reaction. It is evident that if unexpected events happen despite all the precautions taken, the consequences of these events need to be mitigated. HROs take this into account via principles 4 and 5.

2.6.1 HRO Principle 1: Targeted at Disturbances

This principle points out that HROs are actively and in a proactive manner looking for failures, disturbances, deviations, inconsistencies, and the like, because they realize that these phenomena can escalate into larger problems and system failures. They achieve this goal by urging all employees to report (without a blame culture) mistakes, errors, failures, near-misses, and so on. HROs are also very much aware that a long period of time without any incidents or accidents may lead to complacency among the employees of an organization, and may thus further lead to less risk awareness and less collective mindfulness, eventually leading to accidents. Hence, HROs rigorously see to it that such complacency is avoided at all times.

2.6.2   HRO Principle 2: Reluctant for Simplification

When people – or organizations – receive information or data, there is a natural tendency to simplify or reduce it. Parts of the information considered as unimportant or irrelevant are – almost automatically – omitted. Evidently, information which may be perceived as irrelevant might in fact be very relevant in terms of avoiding incidents or accidents, especially those of type II. HROs will therefore question the knowledge they possess from different perspectives and at all times. This way, the organizations try to discover "blind spots" or phenomena that are hard to perceive. To this end, extra personnel (as a type of human redundancy) can, for example, be used to gather information.

2.6.3   HRO Principle 3: Sensitive toward Implementation

High reliability organizations strive for continuous attention toward real-time information. All employees (from frontline workers to top management) should be very well informed about all organizational processes, and not only about the process or task for which they are responsible. They should also be informed about the way that organizational processes can fail and how to control or repair such failures. To this end, an organizational culture of trust among all employees is an absolute must. A working environment in which employees are afraid to provide certain information (e.g., to report incidents) will result in an organization that is information-poor, and one in which efficient working is impossible. A so-called "engineering culture," in which quantitative data/information are much more appreciated than qualitative knowledge/information, should also be avoided. HROs do not distinguish between qualitative and quantitative information.

High reliability organizations are also sensitive toward routines and routine-wise handling. Routines can be dangerous if they lead to absent-mindedness and distraction. By instituting job rotation and/or task rotation in an intelligent way, HROs try to prevent such routine-wise handling.

Furthermore, HROs view near-misses and incidents as opportunities to learn. The failures that go hand in hand with the near-misses always reveal potential (or otherwise hidden) hazards, and hence such failures serve as an opportunity to avoid future similarly caused incidents.

2.6.4   HRO Principle 4: Devoted to Resiliency

High reliability organizations define resiliency as the capacity of a system to retain its function and structure, regardless of internal and external changes. The system's flexibility allows it to keep on functioning, even when certain system parts no longer function as required. An approach to ensure this is for employees to organize themselves into ad hoc networks when unexpected events happen. Ad hoc networks can be regarded as temporary informal networks capable of supplying the required expertise to solve the problems. When the problems have disappeared or are solved, the network ceases to exist.

2.6.5   HRO Principle 5: Respectful for Expertise

Most organizations are characterized by a hierarchical structure with a hierarchical power structure, at least to some degree. This is also the case for HROs. However, in HROs, the power structure is no longer valid in unexpected situations in which certain expertise is required. The decision process and the power are transferred from those highest up in the hierarchy (in normal situations) to those with the most expertise regarding certain topics (in exceptional situations).

But how to achieve "HRO safety" and the correct equilibrium between productivity and safety? The first concept to consider, in this respect, is an adequate organizational safety culture. This can be reached by using performance management science in combination with The Egg Aggregated Model (TEAM) for safety culture. The second concept that should be taken into account is that of "safety futures." The TEAM model and safety futures are explained in the following sections.

2.7   The Egg Aggregated Model (TEAM) of Safety Culture

A lot of research has been carried out on the subject of safety culture. This research has been carried out by a variety of scientific disciplines, e.g., engineering, sociology, psychology, safety scientists, and others. There has never been an integrated and holistic overview of what constitutes a safety culture, and a vivid debate among scientists on this topic can be observed. However, Vierendeels *et al.* [12] recently developed a unifying model of safety culture, taking all aspects of safety science within any organization into consideration in the model and explaining their position toward each other. Figure 2.9 illustrates TEAM of safety culture.

A safety culture is like a relationship: it needs constant attention and constant labor to ensure its success. An (internal or external) audit only provides an idea of the safety climate at a certain point in time and thus does not give a true indication of the safety culture or of the "safety DNA" of an organization. To obtain a more accurate picture, several research methods should be combined and safety performance management should be established within the organization to make sure that there is continuous improvement over time.

The different research methods that need to be used are: document and quantitative analyses for assessing the "observable factors" domain of the TEAM model in the company; surveys and questionnaires for assessing the "perceptual factors" domain of TEAM (or the so-called "safety climate"); and in-depth interviewing with individuals and with groups of individuals to assess the "personal psychological factors" domain of the TEAM model within the company. Hence, both quantitative and qualitative research techniques should be used to obtain a good idea of an organization's safety culture.

Furthermore, performance management science should be employed to make sure that the company's safety culture is constantly monitored and improved where needed. To this end, there should be clear and unambiguous performance indicators linked to objectives to be able to evaluate the company safety culture parts.

Indicators should be "SMART," which is an acronym for:

- Specific and clearly defined;
- Measurable so that one can check on a regular basis how the indicator is performing;
- Achievable so that each indicator provides a target that is challenging but not so extreme that it is no longer motivational (the indicator needs to have sufficient support);

- **R**elevant to the organization and what it is aiming to achieve;
- **T**ime-bound in terms of (realistic) deadlines or timing regarding when each indicator will be achieved.

Objectives can be formulated in different ways: as an absolute number (target numbers), as a percentage (e.g., decrease of $x\%$, satisfy $x\%$ of criteria, satisfy $x\%$ of a checklist), or as a relative position to a benchmark (e.g., higher than the national mean, lower than the mean of the industrial sector, lower than one's own performances of the past x years).

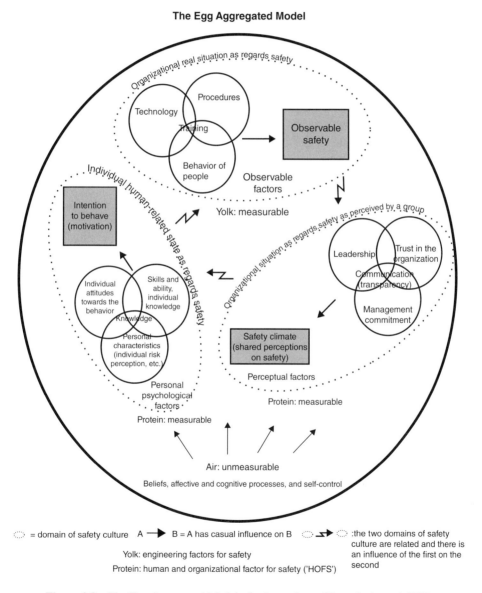

Figure 2.9 The Egg Aggregated Model of safety culture (Vierendeels *et al.* [12]).

Moreover, there are different types and levels of indicators. First, indicators should be identified for the two realistic types of risk (type I and type II). Second, different decision levels require different indicators: management, process, and result indicators. Management indicators establish whether the conditions are present for achieving certain predefined goals. They answer the question, "With what means can the goal be reached?" Process indicators provide an idea as to whether a predefined goal is achievable and whether the (different stages of the) efforts that are prefaced to achieve this goal are executed. They provide information on the working processes within the organization and answer the question, "How can the goal be reached?" Such indicators are very important, as they allow one to gain a systems view on the operational safety of the organization. Next to management and process indicators, result indicators give an indication as to what has been achieved, and whether a predefined goal has been reached or not. They answer the question, "What goals have been reached?" Another distinction is the position of the indicator: measuring "before" the result (proactive or leading indicator) or "after" the result (reactive or lagging indicator). Management and process indicators are usually leading, whereas result indicators are lagging. It should be obvious that some theoretical indicators will sometimes be extremely hard to realize in practice and others will be difficult to think of, or will simply not exist.

Linking the different parts of the TEAM safety culture model with the different possible types of indicator, practitioners should focus on elaborating process indicators and objectives for the different state variables of the model (i.e., the Venn diagrams and their intersections), whereas mainly management and result indicators and objectives need to be worked out for the gray rectangles (i.e., the aggregated results of the state variables). It should be stressed that performance management science is not an easy task and, depending on the organization, a trial-and-error approach will most probably have to be employed to eventually achieve a good performance management system and policy to monitor and continuously improve the company's safety culture.

2.8 Safety Futures

A "safety future" could be seen as an agreement between various parties to have achieved a specified level of safety at some agreed point in time in the future. This could, for example, involve senior management and the safety managers working in an organization. Preventive investments will be required to achieve this goal. This example also immediately further demonstrates the link between safety and economic issues. All companies deal in "safety futures," even if they don't usually look at it this way explicitly.

To do this properly, one should realize that the different types of risk deserve attention and that both type I and type II risks give rise to two types of safety futures, which should not be confused. Furthermore, the credo that says "you cannot put a price on safety" is wrong. A price can be put on safety, but in many cases that estimated price should be much higher than is currently assumed by many practitioners. Safety, or the avoidance of accidents, should simply be seen as part of the business of making a profit or benefit, as also explained in Chapter 1. In the case of operational safety, the profit is hypothetical, because the accidents that are being postulated have not actually happened; but nevertheless these profits/benefits can be calculated, and the sums at stake are many times higher than is generally believed. The costs and hypothetical benefits of accidents and safety are elaborated in depth in Chapter 5 on costs and benefits. In Section 8.12 the hypothetical benefit, in the form/terminology of "maximum investment willingness", is derived by way of the Bayesian decision theory.

Much criticism is also directed toward any method of calculating the cost of safety in economic terms, because so many assumptions have to be made, and choices have to be made in order to arrive at a result. The result is therefore surrounded by a great deal of uncertainty. But it is not because the calculations are imperfect, and the assumptions are many, that an issue should just be ignored. Rather, calculations need to be made more accurate. A tool that has a high degree of reliability and the validity that is required should be developed. The next sections discuss the controversy surrounding economic analyses and the requirements for adequate economic assessment analyses in greater depth.

2.9 The Controversy of Economic Analyses

The usefulness of economic analyses in operational safety has been questioned in many ways and by many people. Although economic analyses can support normative risk control decisions, they should not be used to determine the efficiency and effectiveness of prevention measures. They cannot prove that one prevention measure is intrinsically better than another. Economic analyses as regards operational safety should provide appropriate information on economic and financial aspects of safety to decision-makers, and this information should be easy to understand and interpret.

Unfortunately, economic analyses require debatable information, e.g., the price of a fatality, the price of a lost finger, the question of who pays which costs, the question of who receives which benefits, and other potentially controversial information. Using such data (or not), choices have to be made with respect to safety and prevention measures, constrained by the available safety budget of an organization. A well-known example of a possibly difficult decision, as already mentioned, is having to choose between investing in safety measures and HILP (type II) risks, or LIHP (type I) risks. How to deal with such a question is explained in this book, amongst others. Based on rigorous economic analyses, evaluative best-value-for-money risk reduction measures should obviously be sought.

As mentioned earlier, many critiques have been formulated regarding the concept of economic approaches for safety decisions. Economic assessments can only be based on the "best estimates" available. Such estimates are obtained by using models, data, and information accompanied by many uncertainties and assumptions. Hence, the accuracy of economic analyses is often limited, and thus decision-makers should be careful when using the results.

One should also be careful that an economic approach that is employed to back up safety decisions is not misused and does not merely serve the purpose of giving an organization an aura of being scientific about prevention measures taken. The complexity of an economic analysis often leads to non-experts having difficulties in understanding the premises and assumptions made. Economic analyses can indeed be misused by those who desire to do so, as there is plenty of room for the assumptions and methods to be adjusted to arrive at certain recommendations. This is possible for every type of risk. Furthermore, economic approaches and processes allow organizations to hide behind "rationality" and "objectivity" if the recommendations following an economic assessment are followed without thinking them through. In principle, economic assessments are not necessarily carried out to measure the financial aspects of safety investments. The focus of economic-based recommendations should be placed on selecting the most optimal safety investments. The aim is to improve the operational safety of an organization in the most optimal way, thereby taking financial aspects into consideration. Carrying out an economic assessment with respect to safety is not about figures, although managers are sometimes blinded by the figures obtained. However, the figures are always relative and should always be checked for their meaningfulness, and they should be explained and interpreted.

The economic assessments can be based on selective information, sometimes arbitrary assumptions, and small or large uncertainties. Nevertheless, the assessments may lead to recommendations providing input for decision-makers to select certain safety investments and to prefer one option over others. Hence, the objectivity of such assessments can, in certain cases, be contested and one should be aware of simplistic and unrealistic claims and/or recommendations. The disguised subjectivity of economic analyses is thus potentially dangerous and open to abuse, if it is not recognized.

However, there is no alternative to rigorous economic assessments, unless one is being naïve about the financial aspects of operational safety in an organization. Such naivety can lead to imbalance in two directions: either one has an unclear view of the possible losses due to lack of safety, and safety investments are inadequate (which leads to an undesired and dangerous situation), or one invests much more in safety than would be recommended from a rational, economic point of view (which leads to an economically suboptimal situation).

To support and continuously improve decision-making about prevention and safety measures, economic assessments need to be made. The right way forward is therefore not to reject the economic approach in safety decision-making, but to improve the methods, data, information, concepts, and their use.

2.10 Scientific Requirements for Adequate Economic Assessment Techniques

Aven and Heide [13] indicate that a scientific method, such as, for example, an economic assessment, should be characterized by the following requirements:

1. The scientific work shall be in compliance with all rules, assumptions, limitations or constraints introduced, and the basis for all choices, judgments, and so on, given shall be clear, and finally the principles, methods, and models shall be subjected to order and system, to ensure that critique can be raised and that it is comprehensible.
2. The analysis is relevant and useful – it contributes to a development within the disciplines it concerns, and it is useful with a view to solving the "problem(s)" it concerns or with a view to further development in order to solve the "problem(s)" it concerns.
3. The analysis of the results are reliable and valid.

As Aven [14] mentions, the first two requirements are based on standard requirements for scientific work. Economic assessments provide decision support by systematization of financial aspects of safety-related choices. As there is no general consensus about the how, when, and why of performing an economic analysis in relation to operational safety, there are many possible principles and methods that are available and that can be employed at any time for a variety of purposes. The third requirement is therefore important to ensure that the results (and recommendations based on them) of an economic assessment are reliable and valid. The consistency of the "measuring instrument" (analysts, experts, methods, procedures) is expressed by its reliability. The success at "measuring" what was set out to be "measured" in the analysis is determined by an analysis's validity. The following definitions are proposed by Aven [14]:

reliability is the extent to which the analysis yields the same results when repeated, and validity can be seen as the degree to which the analysis describes the specific concepts that one is attempting to describe. Aven and Heide [13] formulated more specific criteria for both concepts (in relation to risk analysis).

In the case of reliability, the criteria, applied to economic assessments, are as follows: the degree to which the economic analysis methods produce the same results at reruns of these methods; the degree to which the economic analysis produces identical results when conducted by different analysis teams, but using the same methods and data; and the degree to which the economic analysis produces identical results when conducted by different analysis teams with the same analysis scope and objectives, but with no restrictions on methods and data.

The validity criteria of an economic assessment are as follows: the degree to which the economic/financial numbers produced are accurate compared with the underlying true number; the degree to which the assigned probabilities adequately describe the assessor's uncertainties of the unknown quantities considered; the degree to which the epistemic uncertainty assessments are complete; and the degree to which the economic analysis addresses the right quantities.

Some recommendations can thus be made for decision-makers who decide to use an economic analysis to help with safety investment decisions. An economic analysis is as accurate as its input information, and it is often easier to obtain data on costs than on potential (hypothetical) benefits. Indirect and invisible financial information can play an important role in the (lack of) accuracy of an economic assessment. People carrying out the economic analysis should be objective and open-minded, such that their perception regarding safety and risks within the organization becomes as close to reality as feasible. After all, it should be kept in mind that while an economic analysis creates an image of precision, (mostly) it is not precise.

2.11 Four Categories of Data

Based upon the characteristics of the numbers, data can be classified into four categories [15]: *ratio*, *interval*, *ordinal*, and *categorical*. As certain economic models and mathematical approaches can only be used with certain data, it is important for decision-makers to understand this. Ratio data, for example, can be used with all statistical approaches, while categorical data can only be used with statistical tools designed specifically for such data. Hence, identifying the format of the data prior to deciding about the economic approach to be employed is essential, as the approach that the data can be used with is dependent upon this format.

Ratio data are continuous data and it is the only data scale in which it is possible to make comparisons between the values of the scale. Hence, magnitude between values on the scale exists for this type of data. This means that if, for example, one safety investment leads to the avoidance of four accidents of a certain type, and another investment avoids eight accidents of the same type, it is correct to say that the second investment leads to twice as many avoided accidents as the first.

Interval data are a form of continuous data, but less strict than the ratio data, i.e., there is no magnitude between values on the scale. An interval scale is divided into equal measurements and should be used like this. For example, the difference between 10 and 20 units of the scale is the same as that between 20 and 30 of the scale. However, it is not accurate to say that 20 units of the scale is twice as much as 10 units of the scale. Utility values, for instance, could be designed and determined in a way that corresponds to interval data.

Ordinal data are rank-order data. The term "ordinal" implies that the data are ordered in some way. For example, rankings from "very low" to "very high," worst to best, "strongly disagree" to "strongly agree," or "bad, medium, good," belong to the ordinal data category. It is important to realize that it is not possible to make comparisons between the values if ordinal data are used. For example, if a scale from "'strongly disagree", "disagree", "undecided", "agree", "strongly agree"' is assigned the values 1–5, and decision-maker A disagrees with an item (hence value 2 in his perception), and decision-maker B agrees with the same item (hence value 4 in his perception), this does not indicate at all that decision-maker B agrees twice as strongly as decision-maker A. The only conclusion that can be drawn from such ordinal scale information, is that decision-maker B is more agreeable to the item than decision-maker A.

Categorical data are sometimes also called "discrete data" and represent categories. The values assigned to categorical data only serve to differentiate between memberships in the groups. Category data examples are the types of risk (type I or type II), the departments of an organization, the categories male–female, and so on. Magnitude does not exist between category values. For instance, it would be absurd to say that a category numbered "1" is half as large as a category numbered "2."

2.12   Improving Decision-making Processes for Investing in Safety

In general, the literature indicates that company management often has difficulties with the decision-making process for operational safety investments. An important reason for this observation is that managers within organizations have a general lack of knowledge concerning the costs of accidents. Because of this lack of understanding, most of the costs related to an accident are believed to be insured and thus are believed not to play an essential part in the financial situation of the company. In addition, costs are assumed to be limited to the direct accident costs, although indirect accident costs also need to be included. Therefore company managers often believe that there is no valid reason to spend significant capital and time on the complex decision-making process of investing in safety [16].

Another reason why companies tend not to be able to see the importance of a transparent and extensive decision-making process of operational safety investments relates to the measurement difficulties of costs and benefits of prevention. An accurate calculation of many of the required data in the economic analysis is a complex and highly time-consuming process, being costly in itself.

Furthermore, the common assumption of many managers (certainly in the past, and sometimes still in the present) is that accident costs are inevitable and thus represent sunk costs. Of course, such reasoning is very wrong indeed and leads to companies performing badly or even going bankrupt. Moreover, managers may consider investments in safety and accident prevention merely as marketing or reputation expenses, to enhance the company's, or their own, image. Therefore there will be neither time nor money for an extensive decision-making process of investments in prevention measures. In reality, taking into account all the benefits of accident prevention and mitigation while deciding on prevention investments truly leads to companies not having blind spots with regard to a very important area for the profitability of any organization, i.e., prevention investment decision-making, and lowering the operational negative uncertainties of the company (using the terminology of Figure 1.1). Chapters 5 and 6 look in more depth at cost-benefit and cost-effectiveness analyses.

There is also a psychological bias, known as the "loss aversion" principle, that prevents many managers from making the correct safety investment decisions. Indeed, due to the psychological principle of loss aversion [17], the fact that people hate to lose, safety investments to manage and control all types of accidents, but especially precautionary investments to deal with highly unlikely events, are not at all evident. Top managers, risk managers, and the like, being human beings like all other people, may also let their decision judgment be influenced by this psychological principle.

To have a clear understanding of loss aversion, consider the following example. Suppose you are offered two options: (i) you receive €5000 from me (with certainty); or (ii) we toss a coin. You receive €10 000 from me if it is heads, otherwise (if it is tails), you receive nothing. What will you choose? Although the expected outcome in both cases is identical, by far most of the people will choose option (i). They go for the certainty, and prefer to take €5000 for certain rather than gamble and receive nothing if the coin turns up tails.

Let's now consider two different options: (iii) you have to pay me €5000 (with certainty); or (iv) we toss a coin. You have to pay me €10 000 if the coin turns up heads, otherwise (in case of tails) you pay me nothing. Which option would you prefer this time? Notice that the expected outcome is still identical. By far most people in this case would prefer option (iv). Hence, they go for the gamble, and risk paying €10 000 with a level of uncertainty (there is a 50% probability that they will not have to pay anything) instead of paying €5000 for certain.

From this example, it is clear that people hate to lose and love certain gains. People are more inclined to take risks to avoid certain losses than they are to take risks to gain uncertain gains.

Translating this psychological principle into safety terminology, it is clear that company management would be more inclined to invest in production ("certain gains") than to invest in prevention ("uncertain gains"). Also, management is more inclined to risk highly improbable accidents ("uncertain losses") than to make large investments ("certain losses") to prevent such accidents.

Therefore, management should be well aware of this basic human psychological principle, and when making prevention investment decisions, managers should take this into account in their decision. The fact that human beings are prejudiced and that some predetermined preferences are present in the human mind, should thus be consciously considered in the decision-making process of risk managers.

2.13 Conclusions

Operational safety and accident risk are not adequately incorporated into the economic planning and decision processes in organizations. The business incentives for investing in operational safety are unclear. There is a need to demonstrate that safety measures have an essential value in an economic sense. A valuable question is, "To what extent is it true that businesses would not invest in higher operational safety if such values cannot be demonstrated"? An over-investment in safety measures is very likely if, for instance, the fact that there is access to an insurance market is ignored, while an under-investment in safety measures is very likely if insurance is purchased without paying attention to the fact that the probability and consequences can be reduced by safety measures.

Abrhamsen and Asche [18] stated that the final decision regarding how much company resources should be spent on operational safety measures and insurance may be very different depending on what kinds of risks are considered. It makes a difference, for instance, if the risks are of a voluntary nature or if they are involuntary and imposed by others. Clearly, there is more reason for society to enforce standards in the latter case. However, the decision criterion itself is independent of the kind of risk: an expected utility maximization should combine insurance, invest in safety measures, and take the direct costs of an accident, such that the marginal utility of the different actions are the same (see also Chapters 3, 4, and 7).

The fact that decision-makers have an in-built psychological preference to avoid losses should be consciously considered by these decision-makers when making precaution investment decisions.

Now that the importance of economics and economic considerations with respect to industrial safety has been explained, more basic information is needed about economic principles and how they can be applied to operational safety. The next chapter deals with the foundations of economics, and applies them to the field of operational safety.

References

[1] ISO NEN-ISO 31000:2009 (2009). *Risk Management – Principles and Guidelines*. NEN, Delft.

[2] Meyer, T., Reniers, G. (2013). *Engineering Risk Management*. De Gruyter, Berlin.

[3] Center for Chemical Process Safety (2008). *Guidelines for Chemical Transportation Safety, Security and Management*. American Institute of Chemical Engineers, Hoboken, NJ.

[4] Rushton, A. (2006). *CBA, ALARP and Industrial Safety in the United Kingdom*.

[5] Hopkins, A. (2010). *Failure to Learn. The BP Texas City Refinery Disaster*. CCH Australia Limited, Sydney.

[6] Taleb, N.N. (2007). *The Black Swan. The Impact of the Highly Improbable*. Random House, New York.

[7] Fernández-Muñiz, B., Manuel Montes-Peón, J., Vázquez-Ordás, C.J. (2009). Relation between occupational safety management and firm performance. *Safety Science*, **47**, 980–991.

[8] Perrow, Ch. (2006). The limits of safety: the enhancement of a theory of accidents. In: *Key Readings in Risk Management. Systems and Structure for Prevention and Recovery* (eds Smith, D. & Elliott, D.). Routledge, Abingdon.

[9] Hollnagel, E. (2014). *Safety-I and Safety-II. The Past and Future of Safety Management*. Ashgate, Burlington, VT.

[10] Hopkins, A. (2005). *Safety, Culture and Risk. The Organizational Causes of Disasters*. CCH Australia Limited, Sydney.

[11] Weick, K.E., Sutcliffe, K.M. (2007). *Managing the Unexpected. Resilient Performance in An Age of Uncertainty*, 2nd edn. Jossey-Bass, San Francisco, CA.

[12] Vierendeels, G., Reniers, G.L.L., Van Nunen, K., Ponnet, K. (2016) *An integrative conceptual framework for safety culture: The Egg Aggregated Model (TEAM) of safety culture*, forthcoming.

[13] Aven, T., Heide, B. (2009). Reliability and validity of risk analysis. *Reliability Engineering & System Safety*, **94**, 1862–1868.

[14] Aven, T. (2011). *Quantitative Risk Assessment. The Scientific Platform*. Cambridge University Press, Cambridge.

[15] Janicak, C.A. (2010). *Safety Metrics. Tools and Techniques for Measuring Safety Performance*, 2nd edn. The Scarecrow Press Inc., Lanham, MD.

[16] Gavious, A., Mizrahi, S., Shani, Y., Minchuk, Y. (2009). The cost of industrial accidents for the organization: developing methods and tools for evaluation and cost-benefit analysis of investment in safety. *Journal of Loss Prevention in the Process Industries*, **22**(4), 434–438.

[17] Tversky, A., Kahneman, D. (2004). Loss aversion in riskless choice: a reference-dependent model. In: *Preference, Belief, and Similarity: Selected Writings* (ed. Shafir, E.). MIT Press, Cambridge, MA.

[18] Abrhamsen, E.B., Asche, F. (2010). The insurance market's influence on investments in safety measures. *Safety Science*, **48**, 1279–1285.

3

Economic Foundations

3.1 Macroeconomics and Microeconomics

In economics, a distinction is made between microeconomics and macroeconomics. Whereas macroeconomics deals with aggregate economic quantities such as national output, unemployment, and inflation issues, microeconomics concerns the behavior of individual economic units, such as consumers, workers, and individual organizations. More explicitly, the field of microeconomics studies and explains the decision-making of these individual economic units from an economic and rational viewpoint. Hence, when investigating operational safety decision-making in an organizational context, microeconomics is the research field of primary interest.

Microeconomics may help us to understand, for example, safety indifference curves, safety utilities, safety investments, safety budgeting, equilibrium safety levels, and the like. It should be mentioned that microeconomics deals with both positive and normative questions. When discussing safety investment decisions, in particular, it is important to differentiate between the two approaches. Positive economics refers to the objective scientific descriptions, explanations, and predictions based on certain facts and to the explanation of how an organization takes decisions. It involves the development of economic theories, models, and facts and it can be regarded as the "science" aspect of economics, which aims to explain, optimize, and predict decisions of individual economic units. It will not, however, consider aspects such as how an individual firm distributes its bonuses among its managers, or how safety benefits should be spent among the workers. Conversely, normative economics refers to the policy decisions and views that individual economic units have about a particular issue, and therefore relates to subjective judgments or preferences [1]. There are no "right" or "wrong" answers in relation to normative economic decisions, as they are determined solely by an individual's (or a group of individuals') views.

It is important, therefore, when discussing operational safety economics and the financial implications of decisions, to distinguish between the views expressed by a safety economist who is presenting information based on evidence and those of a safety economist who is merely presenting a personal opinion. This book is mainly concerned with the position of operational safety in microeconomics from a "positive" viewpoint, and with how principles, concepts, and theories used in microeconomics may be applied to safety, in an objective way. Nevertheless,

Operational Safety Economics: A Practical Approach Focused on the Chemical and Process Industries,
First Edition. Genserik L.L. Reniers and H.R. Noël Van Erp.
© 2016 John Wiley & Sons, Ltd. Published 2016 by John Wiley & Sons, Ltd.

this book also discusses some normative perspectives, because operational safety within single organizations always also involves normative parameters in the decision-making process. In fact, Chapter 7 elaborates further on normative decision-making with respect to operational economics, as a beyond-the-state-of-the-art approach.

3.2 Safety Demand and Long-term Average Cost of Production

3.2.1 Safety Demand

An essential concept in economic theory is that of "demand." Demand is the quantity of goods and services that consumers may want to purchase at a specific price. Due to the usefulness of a certain good or service, some people are interested in purchasing it; it has a value for these consumers, and thus a certain price follows. The demand will then change depending on the price. In the case of operational safety economics, one might reason that the good or service represents safety or, more specific, safety measures: the demand curve of a safety good or service (e.g., a safety measure or a portfolio of safety measures) depends on the price of the safety good or service. Even if all goods and services (in this case, operational safety) were free to the consumer, there would be a limit on how much consumers could use. Therefore, there would be a demand figure even at this price. In case of safety, this restriction would, for example, be imposed by production and physical limitations of implementing the safety measures. The reason is simple: every production, no matter how small, goes hand in hand with risks, even if operational safety were free of costs.

In general, in economics, the demand curve defines the purchase characteristics of consumers at all conceivable prices, whereas the quantity demanded defines the purchase characteristics of consumers at a particular price. Demand depends on a number of parameters. In case of safety as a good or service, it first depends on the preferences of safety managers and their safety utility curves (see Section 3.5.1). Second, it depends on the available safety budget of an organization. In general, the higher the budget, the higher the demand. Third, it depends on the number of organizations needing this specific kind of safety good or service. The more companies that need it, the higher the demand. Fourth, the demand will be influenced by the industrial sector where the safety good or service is used. The demand of safety can differ drastically from sector to sector, due to the industrial activities and agreements with unions, legislation, profit, and income specificities within the sectors. Fifth, in case of some safety goods or services, a substitution effect may be observed, for example, if the price rises. In this case, the price relative to the prices of other safety goods or services is important, not the absolute price of the good or service.

To investigate the influence of each of the parameters on the safety demand, every parameter can be changed or fluctuated while the other factors are kept constant (according to the well-known "*ceteris paribus*" – other things being equal – principle).

Thus far, safety was looked upon from the perspective of "safety measures," and the price of safety measures influenced by the possible market demand. This reasoning assumes that safety is a regular good comparable with all other goods and/or services. However, this is not the case and a regular demand curve for safety such as for other goods and services is not applicable. Some safety measures, almost no matter how expensive, will be purchased anyway, and others might only be purchased for the happy few who have a company culture strongly tending toward high reliability organization (HRO) safety and who are also able to afford them. In

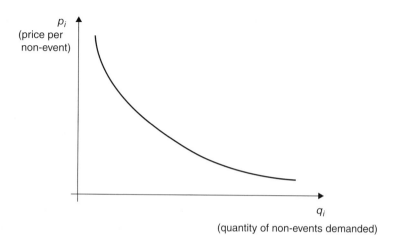

p_i
(price per
non-event)

q_i
(quantity of non-events demanded)

Figure 3.1   An illustrative safety demand curve.

other words, the purchase of safety measures within an organization strongly depends on a lot of other factors besides the market mechanism. The classic microeconomic demand function therefore does not fully apply to safety.

Safety is thus not a regular good or service, such as a book or a football match . Safety is actually, as already mentioned in this book, a "dynamic non-event." As such, in economic terms, companies purchase a certain level of non-events over a certain period of time. The higher the price of non-events, the harder it is to obtain a lot of them and the lower the number of non-events that are purchased. In other words, if safety per unit costs more, there will be less safety demand per unit. This is illustrated in Figure 3.1. Notice that this figure only holds for one type of risk at a time. Hence, the figure should be seen in terms of either type I risks and avoiding type I accident scenarios, or type II risks and avoiding type II accident scenarios.

In standard works on microeconomy, the next step is to elaborate and show a supply curve, in order to finally explain the equilibrium mechanism and the market equilibrium, based on demand and supply curves. In the case of safety, however, it makes no sense to develop a supply curve. As "supply" can be defined as the quantity of goods and services that producers may want to sell at a specified price, in safety terms this would mean the number of non-events that a company would want to sell at a certain price. Of course, the company does not want to sell safety, or non-events, it just needs a certain level of safety itself. Therefore, although a demand curve of safety exists within a single company and on a microeconomic level, it is useless in this context to talk about a supply curve.

Safety goods and services can also be viewed from the perspective of the profitability of a company, and thus in that regard be linked to micro-economic theory.

3.2.2   Long-term Average Cost of Production and Safety

Looking at how safety influences the profitability of a company can be done via the so-called long-term average cost of production (LAC). Before explaining more microeconomic theories and models in depth, it is interesting to consider why safety economics and investing in safety

– or not – may indeed make the difference to an organization in terms of its long-term profitability. The profitability of any organization depends on its profitability in the long term, and this in turn depends on the LAC of the firm. The average cost of production can be defined as the total cost of production divided by the total output in goods and services. In other words, the average cost of production can be seen as a proxy for the accumulated operational negative uncertainties divided by the accumulated operational positive uncertainties within an organization. Further, assume that the type of organization considered in this book is aimed at long-term existence and making profits in a sustainable way. In this case, to ensure the long-term viability of a company, the average cost of production should be less than the price at which the goods and services can be traded.

Inputs to production comprise any goods or services that are used to produce the output product or service, including, for example, labor costs, raw materials, utilities, and equipment, but also safety and safety-related goods and services (such as risk assessments, maintenance operations, and training). The production function determines the maximum output that can be produced from a defined level of inputs. The efficiency with which an organization achieves a specific output will define its long-term average costs. The importance of safety within the average costs of production can be illustrated by a simple example.

Assume two similar competitor companies in the same industrial sector: A and B. Company A's cost structure is €39 000 non-safety-related costs, and €10 000 safety-related costs (due to safety investments as well as incidents and accidents happening), in both cases per 1000 units produced. Conversely, company B's per-1000 units cost structure adds up to €38 000 non-safety-related costs and €13 000 safety-related costs (due to safety investments as well as accidents). In this example, company A, although characterized with a higher non-safety cost structure per 1000 units produced, produces the goods and services at a lower average cost per unit (€49) than company B (€51), purely due to the safety-related costs.

The costs, in broad terms, will vary with production output. As the production output increases, the unit costs of production will go down due to fixed costs staying the same irrespective of the production output (economies of scale). However, at a certain point, costs will rise again due to more levels of management, more difficult organizational management, transport costs, marketing costs, geographical location costs, and so on (diseconomies of scale). Hence, an organization's long-term average cost curve is likely to follow more or less the shape of the curve depicted in Figure 3.2.

Figure 3.2 shows the LAC curves of companies A and B. Both companies can sell their goods or services in the market at €50 per unit. However, company A can sell at a profit if it sells quantities of the good or service between Q_1 and Q_2, whereas company B may be able to sustain production at a loss in the short term. Company B will have to lower its production costs if it desires to keep operating for a longer period of time though. Company B can, for example, lower its production costs by improving its health and safety standards and thereby suffering lower accident costs.

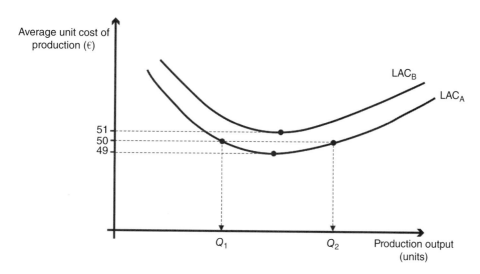

Figure 3.2   Long-term average costs of production (LAC) for the illustrative example of companies A and B.

3.3   Safety Value Function

A safety value function is a real-value mathematical function defined over an evaluation criterion that represents an option's measure of "goodness" over the levels of the criterion. If
we take "operational safety" as the criterion to be evaluated, a measure of goodness reflects a
safety manager's judged value in the performance of an option across the levels of operational
safety. In practice, the safety value function for operational safety's least preferred level (i.e.,
very low operational safety scores) takes the value 0. The safety value function of operational
safety's most preferred level (i.e., very high operational safety scores) takes the value 1.

   Notice that it is not always straightforward to assign a value to the criterion "operational
safety." As an example, the differences between the scores of type I risks (or occupational
safety) and type II risks (or safety with respect to major accidents) may be important to assess
and to take into account in the value function. Preference of one type over another type is
quite personal (cf. risk-averse/risk-neutral/risk-seeking character) and may even change over
time for an individual, depending on personal experiences and changing conditions or circumstances. There are, however, also many various preference levels even within type I or type II
operational risks.

   Ideally, a safety value function should be measurable, to be able to use it in practice. Hence,
a preference ordering and a strength of preference between operational safety measures should
be present. The safety value difference between any two measures with respect to operational
safety represents a safety manager's strength of preference between the two measures. Hence,
the vertical axis of a measurable safety value function is an ordinal scale measure of the
strength of a safety manager's preferences. An ordinal value (on the vertical axis) is thus combined with a categorical value (on the horizontal axis). For the safety manager, the values on

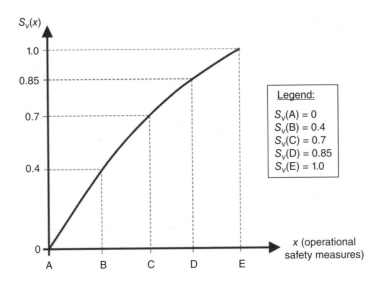

Figure 3.3   An illustrative safety value function.

the vertical axis have meaningful preference differences, while the distances between A, B, C, D, and E along the horizontal axis are not meaningful (see Figure 3.3 for an example of a safety value function).

As illustrated in Figure 3.3, the safety values not only display an ordering of preferences (ordinal scale), but also the strengths of preferences (interval scale). It is clear from the figure that there is a smaller increment in safety value between D and E than between A and B, for example. In fact, the safety value increment between A and B is approximately three times the safety value increment between D and E. The expression "safety value increment" indicates the degree to which the safety manager prefers one (higher) score over another (lower) score.

An approach to specifying a single dimensional safety value function is the direct subjective assessment of the safety value, or "direct rating." To this end, all alternatives can be ranked by a safety manager or a safety management team on a line such that the ranking and the position of the alternative reflect a numerical assessment.

The rating scale approach to value safety has its fundamentals in psychometric theory. The method concerns a single line with anchor points that represent the best possible safety situation and the worst possible safety situation. For example (see also Figure 3.4), the worst safety measure/bundle corresponds to 20% probability of a type I accident per year (this anchor point serves as the left wall or floor), whereas the best safety measure/bundle might correspond to 1% probability of a type I accident per year (this anchor point then serves as the ceiling or right wall). Safety managers (or a safety management team) can then be asked to put several alternative type I safety measures/bundles on this line, in such a way that the points they mark on the line reflect the differences between the safety measures/bundles described. Evidently, this method generates values rather than utilities (see Section 3.5), as choices between different measures/bundles are not made pairwise, and it also does not involve decision-making under uncertainty.

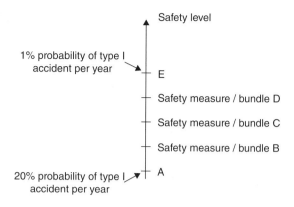

Figure 3.4   A rating scale example.

Rating scales can thus be employed to measure the responses of a safety manager or a safety management team on a continuum or in an ordered set of categories, with numerical values assigned to each point or category.

3.4   Expected Value Theory, Value at Risk, and Safety Attitude

3.4.1   Expected Value Theory

The expected value associated with a situation where uncertainty is involved is a weighted average of the payoffs or values associated with all possible outcomes, with the probabilities of each outcome used as weights. Hence, the expected value measures the average payoff. To illustrate this, assume the following possible consequences of a certain undesired event, "*ue*":

Outcome 1: a cost of c_1 with probability p_1
Outcome 2: a cost of c_2 with probability p_2
Outcome 3: a cost of c_3 with probability p_3.

Assume that no other outcomes are possible, i.e. that,

$$p_1 + p_2 + p_3 = 1$$

Then, the expected value of the undesired event is:

$$E(ue) = p_1 c_1 + p_2 c_2 + p_3 c_3$$

The decision to take preventive measures or not against this particular undesired event with three possible outcomes is in the hands of the safety manager.

The expected value also refers to the weighted average value of all costs or effectiveness outcomes in a decision analysis tree. In safety management, event trees are often used to visualize different outcomes and their consequences and/or probabilities. Such trees can thus also be used to carry out decision analyses by calculating expected values using the information in the tree. An illustrative example is provided in Figure 3.5.

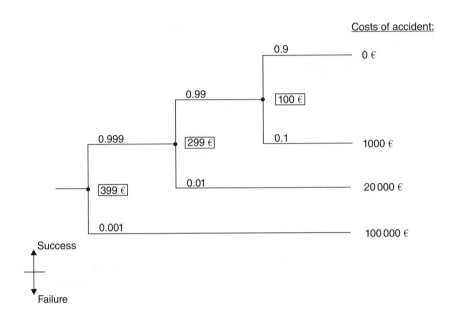

Figure 3.5   An illustrative example of a decision analysis tree for calculating the expected value.

In Figure 3.5, the costs of an accident are given at the end of the terminal nodes. The probabilities that certain situations occur are given on the branches. Working backward from each chance node, the total overall expected value for that node can be determined. For instance, the total cost of the chance node situated furthest to the right is €0 × 0.9 + €1000 × 0.1 = €100. If the same reasoning is followed for the two other chance nodes, it is possible to calculate the overall expected value of this event tree, amounting to €399.

3.4.2   Value at Risk

A term much used in financial risk management is the so-called value at risk (VaR). The question that a VaR tries to answer is simple: "What is the most that can be lost on a particular financial investment with a certain level of confidence (typically 95% or 99%)?" VaR indeed provides a predicted loss over a target time horizon within a given confidence interval. The loss is usually called "maximum loss" (or "worst-case loss") in financial risk management, but there is actually a chance that an even greater loss may occur, albeit small. In brief, a VaR measure determines an amount of money, such that there is *that* probability of the portfolio of assets not losing *that* amount of money over *that* time horizon. Hence, in statistical terms, this is called a percentile. So a 1-year 95% euro VaR is just the 95 percentile of a portfolio's yearly loss expressed in euros (see also Section 7.2.1.1).

This VaR can also be explained in terms of operational risk management. For example, if the worst-case summed consequence of accident scenarios ("VaR" in financial risk management terminology) in an organization is calculated to be €100 million/year, with a 95% confidence level, then there is only a 5% chance that the summed consequences of accident scenarios will be higher than €100 million over any given year.

In operational risk management, a VaR could be used for decision-making for type I, and also for type II, risks, as probability distributions can be used for both types of risk, and this kind of input information is needed for VaR statistic calculations. A VaR statistic indeed needs, among other things, an estimated input of loss (either in currency or in percentage terms) as a probability density.

A Monte Carlo simulation, randomly generating outcomes for a number of hypothetical trials, can be used to calculate the VaR.

The downside of using a VaR for operational risk management is that: (i) there is still a lot of epistemic uncertainty about type II risks (and thus the accuracy and probability distribution depend on the assumptions made and the scenarios considered); (ii) there is still a certain probability that a greater loss may occur; and (iii) due to (i) and (ii), avoiding a true disaster is not fully taken into account while using this VaR method for making safety investments.

3.4.3   Safety Attitude

One approach to characterizing a person's risk attitude is via the concept of the so-called "certainty equivalent." A certainty equivalent is a value V_{ce}, such that the safety manager is indifferent between the consequences of the undesired event and the value V_{ce} that he obtains for certain.

As an example, given the following possible consequences of an undesired event duirng the next year, what value would a safety manager be willing to invest in the next year with certainty that makes him indifferent between that prevention investment value and engaging in the following options of an undesired event next year?

Assume the following set, $\{(c_1, p_1), (c_2, p_2), (c_3, p_3)\}$:

Lose c_1 = €200 000 next year with probability $p_1 = 0.2$
Lose c_2 = €800 000 next year with probability $p_2 = 0.005$
Lose c_3 = €0 next year with probability $p_3 = 0.795$.

The expected value of the undesired event can then be calculated as:

$$E(ue) = 0.2 \times 200\,000 + 0.005 \times 800\,000 + 0.795 \times 0 = €44\,000$$

If the safety manager would be indifferent between investing €50 000 next year with certainty and investing nothing (i.e., engaging himself with the possible consequences), then the certainty equivalent for this undesired event is €50 000. If this were the case, the safety manager is "safety seeking," as he is willing to invest with certainty an amount of money higher than the expected amount (i.e., €44 000) that might be lost if the decision were taken not to make a prevention investment.

Hence, depending on the certainty equivalent, safety managers can be categorized as being safety-seeking, safety-neutral, or safety-averse:

Safety-seeking: Certainty equivalent of investment > expected value of undesired event
Safety-neutral: Certainty equivalent of investment = expected value of undesired event
Safety-averse: Certainty equivalent of investment < expected value of undesired event.

Looking at it from the "hypothetical benefit" point of view, safety-seeking, safety-neutral, and safety-averse can be compared to risk-seeking, risk-neutral, and risk-averse. In case of "risk," there is a possibility to win and to have a positive financial payoff, i.e., in this case, a hypothetical benefit of being safe. In the case of "safety," the gains are equal to "not losing" or "not having accidents," and hence they can be interpreted as necessary prevention investments to avoid human, social, and legal, losses. They thus represent the costs that go hand in hand with safety. Hence, in the case of safety there is no possibility of achieving a real positive financial payoff – the payoff is a hypothetical one. In this way, safety managers can be categorized into being risk-averse, risk-neutral, or risk-seeking:

Risk averse: Certainty equivalent of positive outcome < expected value of positive outcome
Risk neutral: Certainty equivalent of positive outcome = expected value of positive outcome
Risk seeking: Certainty equivalent of positive outcome > expected value of positive outcome.

A risk-averse person, for instance, is willing to accept, with certainty, an amount of money (i.e., a positive outcome) less than the expected amount that might be received if a decision were made to participate in a gamble. Risk-neutral and risk-seeking behavior can be understood in a similar manner.

3.5   Safety Utilities

3.5.1   Safety Utility Functions

A safety utility can be seen as a measure of safety management's satisfaction from a certain level of operational safety within an organization. Safety utilities can thus be employed to explain the economic behavior of a company's management in terms of attempts to increase operational safety. The "utility" concept is usually applied in economics by using indifference curves (see Section 3.7), which in operational economics represent the combination of measures that a safety management team would accept to maintain a given level of satisfaction.

A class of mathematical functions exhibits behaviors with respect to risk-seeking, risk-neutral, and risk-averse attitudes. These can be regarded as "safety utility" (S_u) functions. Figure 3.6 illustrates these functions.

A safety utility is a measure of preference that a certain safety situation or state has for a safety manager. Mathematically, it is a dimensionless number. A safety utility function is a real-value mathematical function comparable to the safety value function, the difference being that the safety value is, in this case, an abstract dimensionless "safety utility" related to a safety situation or state, and not, as in the safety value functions, a concrete safety value attached to a certain option of an operational safety measure or bundle. The approach to determine the functions is also different.

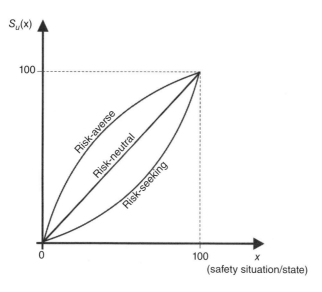

Figure 3.6 Safety utility functions (risk-averse, risk-neutral, and risk-seeking).

The vertical axis of a utility function can be any real number, but usually the axis is scaled between 0 and 100 or 0 and 1. If the latter range is chosen, the utility of the least preferred outcome is assigned "0" and the utility of the most preferred outcome is assigned "1."

3.5.2 Expected Utility and Certainty Equivalent

The expected utility of an uncertain undesired event "ue" with utilities $U(c_1), U(c_2), \ldots, U(c_n)$ of possible outcomes $c_1, c_2, \ldots, c_n$, with respective probabilities $p_1, p_2, \ldots, p_n$ is:

$$E[U(c)] = p_1\, U(c_1) + p_2\, U(c_2) + \cdots + p_n\, U(c_n).$$

Furthermore, it is possible to illustrate the relationship between the expected value $E(ue)$ of an undesired event and the expected utility $E[U(c)]$ of this undesired event. Figure 3.7 displays this relationship for a monotonically increasing risk-averse utility function.

The equation of the chord from Figure 3.7 can be written as:

$$y_{\text{chord}}(x) = \frac{S_u(c_2) - S_u(c_1)}{c_2 - c_1}(x - c_2) + S_u(c_2).$$

We also know that $E(ue) = p_1\, c_1 + (1 - p_1)\, c_2$. If we set $x = E(ue)$ in the equation of the chord, then it can algebraically be shown that (see also Garvey [2]):

$$y_{\text{chord}}[E(ue)] = p_1 S_u(c_1) + (1 - p_1)S_u(c_2) = E[S_u(x)].$$

Furthermore, the certainty equivalent (CE) is the value on the horizontal axis at which a person is indifferent between a lottery and receiving the amount CE with certainty. Hence, the utility of CE must be equal to the expected utility of a lottery; thus,

$$S_u(\text{CE}) = E[S_u(x)] \tag{3.1}$$

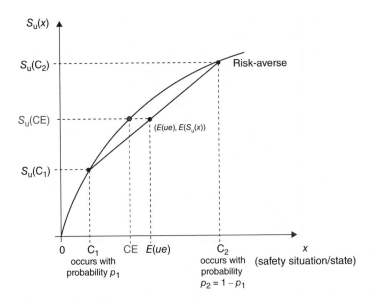

Figure 3.7 Relationship between expected value and expected utility.

The information in Figure 3.7 shows this reasoning. From Eq. (3.1), it follows that when a safety utility function of a safety management team has been drafted and is known, the certainty equivalent CE can be determined by taking the inverse of the safety utility function:

$$CE = S_u^{-1}(E[S_u(x)]).$$

As a concrete example, consider the possible undesired event scenario ue with the following possible outcomes: lose €80 000 with $p_1 = 0.6$, and lose €10 000 with $p_2 = 0.4$. We are curious about the CE for this undesired event by a safety management team whose safety utility function is given by $S_u(x) = 10x^{1/2}$. The CE provides an idea of the level of the possible safety investment required to deal with the undesired event scenario. The CE for this example turns out to be equal to €44 000 (readers are invited to verify the calculations), and hence, the cost of safety measures for this undesired event scenario should certainly not be higher than €44 000 for the management team with this safety utility function.

3.6 Measuring Safety Utility Functions

Utilities cannot be observed directly, and hence they have to be determined in another way. Economists therefore devised an approach to measure managers' relative utilities underlying their choice-making. Such an approach is also called the "revealed preference" in economics.

Measuring safety utilities involves a two-step procedure. The first step involves defining which safety states or situations are of interest. The second step involves trying to value those safety states or situations, and place them in a ranking order (first, second, third, etc.).

Straffin [3] indicates that to determine a safety manager's or a safety management team's safety utils on a cardinal (i.e., ratio or interval; see Section 2.11) scale, numbers need to be assigned to outcomes so that the ratios of differences between the numbers reflect something about the managers' or the team's preferences. As Von Neumann and Morgenstern [4] indicate, the relevant information can be obtained by asking questions about lotteries.

Suppose that the following outcomes are possible: A, B, C, D. The manager's or team's preference is the following: A, D, C, B. However, this is a ranking on an ordinal scale, and thus ordinal safety utils are given. To determine the safety utils on a cardinal scale, assign numbers to A and B, for example, assign 0 to A and 100 to B. Next, the safety manager/team should be asked the following question: "Which would you prefer: D for certain, or a lottery which would give you A with probability $\frac{1}{2}$, and B with probability $\frac{1}{2}$?" Hence, the lottery can be denoted $\frac{1}{2}$A, $\frac{1}{2}$B. If the safety manager/team indicates that he/it would prefer D to the lottery, it implies that D ranks higher than the midpoint between A and B, and thus D must be assigned a number higher than 50 (and evidently lower than 100). The next question to be asked might be: "Would you prefer D for certain, or the lottery $\frac{1}{4}$A, $\frac{3}{4}$B?" Assume that the preference this time is the lottery. Consequently, D is ranked below $\frac{3}{4}$ of the way from A to B, i.e., below 75. By continuing to change the lottery odds, the lottery can be found for which the safety manager/team is indifferent between D and this lottery. For example, that lottery may be $\frac{3}{10}$A, $\frac{7}{10}$B. Consequently, D should be assigned the number 70 (or 7/10 of the way from A to B). An analogous approach can be employed to determine the cardinal number assigned to C, e.g., the number 85. These assignments, if they are consistent, provide the representation of a safety manager's or safety management team's cardinal safety utilities for the outcomes.

These ratio safety util numbers can be used to make further calculations (instead of interval or ordinal numbers where no mathematical manipulations are allowed), but it should be stressed that safety utils are defined only for one individual or one team, and refer to how that individual or that team makes choices among alternatives. Simply put, safety utils are just a convenient numerical way of organizing information from safety managers or a safety team about their preferences, given that those preferences satisfy certain consistency conditions.

It should be mentioned, however, that in general, the particular ratio util number is often not that important. For example, although it is impossible to say that safety managers on a higher utility value, $S_u(2)$, are exactly twice as happy as they might be on a certain lower utility value, $S_u(1)$, an interval or an ordinal util ranking is often sufficient to obtain an insight into how safety managers make individual decisions.

3.7   Preferences of Safety Management – Safety Indifference Curves

Given the large number of possible choices that have to be made by safety management regarding safety investments and safety measures, for example, between type I and type II risks, but also within the different types of risks, how can safety management preferences be described in a coherent way? The economic theory of consumer behavior provides the answer.

The theory of consumer behavior begins with basic assumptions regarding people's preferences for one choice (e.g., a safety bundle) versus another. The basic assumptions can be formulated with respect to safety management as follows:

1. Preferences are complete. Safety managers are thus able to compare and rank all safety bundles in order of their preference or utility value. For any two safety bundles X and Y, managers will prefer X over Y, or Y over X, or they will be indifferent between the two and will be equally happy with either bundle. Note that these preferences ignore costs. A safety manager might prefer a technology-based safety measure to a person-based measure but would buy the person-based measure because it is cheaper.
2. Preferences are transitive. Transitivity means that if a safety manager prefers safety bundle X to safety bundle Y, and prefers Y to Z, then he also prefers X to Z. For example, if a manager prefers 6 technology measures + 10 organizational measures (bundle X) over 8 technology measures + 8 organizational measures (bundle Y) and further prefers option Y to 10 technology measures + 6 organizational measures (bundle Z), then the safety manager prefers X to Z. This transitivity assumption guarantees that the safety manager's preferences are consistent, and hence rational. Safety managers can, for example, use safety utility values to rank different bundles of safety goods and services.
3. All bundles are "desirable," so that, leaving costs aside, safety managers always prefer more than fewer safety goods and services (bundles).
4. Safety indifference curves are convex, i.e., bowed inward. The term convex indicates that the slope of the safety indifference curve becomes less negative as one moves down along the curve. Safety managers have a marginal rate of substitution for one safety commodity compared with another safety commodity. Managers will therefore demonstrate a diminishing marginal rate of substitution, because decreasing quantities or qualities of one type of safety bundle are given up to obtain equal increases in the quantity or quality of the other safety bundle.

Safety management's preferences are described by safety indifference curves (also called iso-utility curves of safety in safety commodity space – see also the next paragraph), which identify those bundles of safety goods and services that provide the same utility to the individual safety manager or to the safety management team. A safety indifference curve thus represents all combinations of safety bundles that provide the same level of satisfaction to a safety manager or a team of safety managers. The actual shapes of the indifference curves will depend on the personal preferences of the manager or the team.

When indifference curves are discussed in microeconomic theory, the assumption is often made that the consumer has only two goods to choose from. Let S_1 be a variable which represents the number of safety goods "1" purchased by the safety management team, and let S_2 be a variable representing the safety management team's purchases of safety goods "2." The couple (S_1, S_2) represents a choice of a number for both safety goods and is called a "safety commodity bundle." Assuming further that S_1 and S_2 should be non-negative numbers, then all possible safety commodity bundles can be represented geometrically in what can be called the "safety commodity space." Safety managers have preferences about safety commodity bundles in safety commodity space: given any two safety commodity bundles, safety management either prefers one bundle over another or is indifferent between the two. If the safety management team's preferences satisfy the basic consistency assumptions from above, they

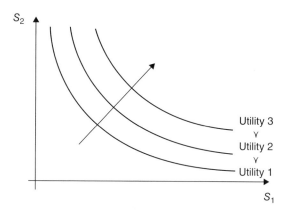

Figure 3.8 Indifference curves or iso-utility curves in a safety commodity space.

are represented by so-called iso-utility functions in safety commodity space. A utility function assigns a real value to each safety commodity bundle. If safety management prefers commodity bundle (S_1, S_2) to bundle (S_3, S_4), then a higher number is assigned to (S_1, S_2) than to (S_3, S_4). The iso-utility function U_s assigns the same number to all bundles on any given indifference curve. Figure 3.8 shows the indifference curves or iso-utility curves in the safety commodity space. The arrow in Figure 3.8 indicates the direction of preference. Safety commodity bundles on indifference curves far from the origin are preferred to those on indifference curves near the origin.

Indifference curves are often expressed, especially as a proxy, as functions of the Cobb-Douglas type. These functions are used because they represent nicely convex curves. The function then looks as follows:

$$U_s(S_1, S_2) = S_1^a S_2^{1-a},$$

where a is a parameter such that $0 < a < 1$.

Obviously, other indifference curve functions are possible and feasible as well.

3.8 Measuring Safety Indifference Curves

Determining safety indifference curves can be carried out by direct or indirect measurement. Direct measurement can be performed in a number of ways, one of which, the "time trade-off," will be discussed further. Indirect measurement concerns using questionnaires that are given to safety managers or a safety management team. Questionnaires can be used for drafting indifference curves for type I risks, while time trade-off safety utilities can be used for drafting safety indifference curves for type II risks.

3.8.1 Questionnaire-based Type I Safety Indifference Curves

Investigating the preferences of an individual safety decision-maker or of a safety management team is possible through the development and use of questionnaires. Using human preferences derived from a well-constructed survey, an indifference curve can be drafted. In an organizational context, it is assumed that the individual decision-maker or the team represents the organization and makes decisions in its interest, and not in their own interest.

With respect to safety management, "revealed preference with discrete choice" (mentioned earlier in this book) is a very interesting and straightforward method (used within consumption theory) to obtain preferences. It is an approach to determine which bundle of goods and/or services is preferred over another bundle of goods and/or services. The method can be explained as follows. When a person (e.g., a safety manager) chooses a bundle A from a series of affordable alternatives, he thus reveals his preference for this bundle A over a series of other bundles he could have chosen. All other bundles within his budget that this person did not choose, represent a lower utility for him. Hence, if it were possible to determine *all* bundles having a higher utility and a lower utility than bundle A, it would also be possible to pinpoint all bundles having the same utility as bundle A (i.e., those bundles that were left over at the end of the exercise). Graphically, these leftover bundles together form the so-called indifference curve [5].

The purpose of the questionnaire is to determine such indifference curves. This can be achieved by letting a safety manager or a safety management team choose between two safety commodity bundles. If it is indicated that bundle 1 is preferred over bundle 2, the utility of bundle 2 will be increased, or the utility of bundle 1 will be decreased, until the situation occurs that the manager or management team displays an indifference between both safety commodity bundles. By repeating this procedure, it is possible to also determine other safety commodity bundles showing an equal utility. The more this procedure is repeated, the more accurate the resulting indifference curve will be.

In the following paragraphs, a concrete working procedure is described to reveal the preferences of safety management with respect to safety measure X in contrast to all other possible safety measures. It may be interesting to position the most important safety measure, according to the opinion of company safety management, in relation to all other safety measures, and in this way to determine the allocation of the safety budget.

The initial questions serve to obtain the required background information:

1. What is the company's safety budget?
2. What are all possible safety measures and what are their investment costs?
3. What is, according to you, the most important safety measure for the company?

The most important measure is then displayed in a graph versus the aggregate of all other possible measures. To determine the indifference curve, the respondent needs to make a number of choices using a number of closed questions, whereby every choice is between two safety commodity bundles.

A possible closed question can be:

Bundle 1 contains four measures of type X and five other measures.
Bundle 2 contains two measures of type X and seven other measures.

Make your choice:

Bundle 1 is preferred over bundle 2.
Bundle 2 is preferred over bundle 1.
Indifference between bundle 1 and bundle 2.

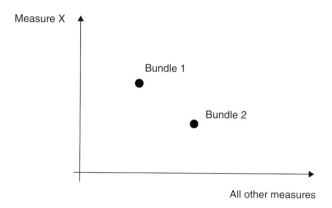

Figure 3.9 Choice between two safety commodity bundles.

It is then possible to develop an indifference curve in safety commodity space in the following manner. Figure 3.9 displays the initial choice between the two safety commodity bundles.

If safety commodity bundle 1 is chosen over safety commodity bundle 2, bundle 1 obviously represents a higher utility for the respondent than bundle 2. The next step is to lower the utility of bundle 1 by decreasing the number of measures of type X by one unit. Hence, bundle 1', positioned perpendicular beneath bundle 1, is used to formulate the following closed question:

> Bundle 1' contains three measures of type X and five other measures.
> Bundle 2 contains two measures of type X and seven other measures.

Make your choice:

> Bundle 1' is preferred over bundle 2.
> Bundle 2 is preferred over bundle 1'.
> Indifference between bundle 1' and bundle 2.

In an analogous way, the utility of commodity bundle 1 can also be decreased by lowering the number of other measures, and hence by choosing a commodity bundle 1' situated to the left of bundle 1 (see Figure 3.10a). This procedure is repeated until the respondent is indifferent between the two bundles. At that point, the two bundles have an equal utility and they then belong to the same indifference curve. Another analogous procedure is to increase the utility of bundle 2, and in this way determine the indifference curve (see Figure 3.10b).

The procedure described allows us to identify two points belonging to one indifference curve. To further determine other points belonging to this curve, bundles need to be compared two-by-two with a bundle belonging to the curve. Theoretically, this procedure needs to be repeated infinitely. Evidently, applying a procedure infinitely is impossible, and therefore in practice a limited number of points will be satisfactory to establish a proxy for the indifference curve of an individual safety decision-maker or a safety management team.

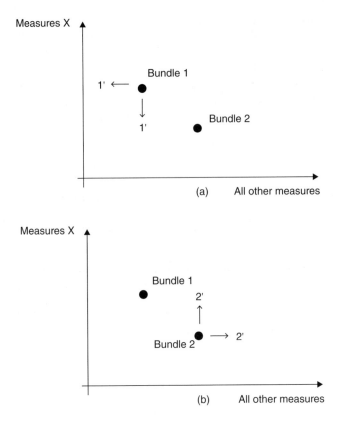

Figure 3.10   Determining the points belonging to an indifference curve versus measure X: (a) method 1; (b) method 2.

3.8.2   Problems with Determining an Indifference Curve

In trying to determine the preferences of safety management by using the procedure described in the previous section and the closed questions, a number of problems may be encountered that can influence the quality of the result. The main problem stems from the fact that humans do not always take decisions rationally, i.e., in a way that maximizes their utility. Often, decisions depend on rather arbitrary parameters, such as mood, temper, or the weather, leading to not following the condition of transitivity of preferences. One possible approach to deal with such variations is to add an extra term to take stochastic influences into consideration. However, introducing such a term would probably make the procedure too complex to be usable by most companies.

3.8.3   Time Trade-off-based Safety Utilities for Type II Safety Indifference Curves

As type II risks are characterized with much lower probabilities and much more severe outcomes, the method described in Section 3.8.1 would not be suitable for determining an indifference curve for such risks. It may be possible, however, to use the time trade-off approach in this case.

The time trade-off method asks people to consider relative amounts of time (such as the number of years a certain safety situation exists) that they would be willing to give up in order to achieve another safety situation. Consider the following two alternatives of one investment where a safety team may choose from:

Safety situation A (e.g., 7% probability of an accident) during time period t;
Safety situation B (e.g., 0.1% probability of an accident) during time period x.

Then, time period x is varied until the safety team is indifferent between alternative A and alternative B. In this case, a point p belonging to the indifference curve is obtained: $p = x/t$.

As an example, 7% probability of an accident over 8000 years (remember we are looking at type II risks) for a safety team may be similar to 0.1% probability of an accident over 2000 years. Hence, if an indifference curve for 0.1% probability is drafted, 7% probability of an accident corresponds to 2000/8000 = 0.25. This figure, 0.25, can be called the "quality-adjusted accident probability," or QAAP (referencing the well-known QALY, which stands for quality-adjusted life years in healthcare; see also page 121). In a similar way, other accident probabilities can be assigned a QAAP.

The following illustrative table may, for example, be obtained.

Table 3.1 Quality-adjusted accident probabilities (QAAPs) for 0.1% probability over 2000 years

Accident probability (%)	QAAP
70	0.05
50	0.1
20	0.12
10	0.20
7	0.25
5	0.35
1	0.4
0.5	0.5

Quality-adjusted accident probabilities can be used to make decisions concerning prevention investments for type II risks more objective. As an illustrative example, from Table 3.1, if an installation I decreases in accident probability, for example, from 10% to 5%, and hence there is an increase in QAAP from 0.2 to 0.35, over 2000 years, and also (via other safety investments) from 10% to 7%, hence from 0.2 to 0.25, in the next 2000 years, then the total equivalent safety benefit for installation I is

$$(0.35 - 0.2) \, \text{QAAPs} + (0.25 - 0.2) \, \text{QAAPs} = 0.20 \, \text{QAAPs}.$$

If installations J and Q, by a similar approach of calculations, received 0.35 and 0.40 QAAPs respectively, the total safety benefit would be $(0.20 + 0.35 + 0.40)$ QAAPs = 0.95 QAAPs. Notice that this safety investment would be similar to (for example) the choice for a prevention investment leading to 0.95 QAAPs only for installation I, and no safety changes/improvements/investments for installations J and Q. In summary, this number (i.e., 0.95 QAAPs) represents a numerical estimate of a company's valuation of the safety investments for installations I, J, and Q, and this can be compared with other possible safety investments within the company. This approach can be employed for very low probability risks.

Using QAAPs, different safety investments leading to changing accident probabilities at an industrial site can be compared with one another. Any investment leading to a decrease in accident probability leads to an increase in safety utility (expressed in QAAPs). This utility measure can then be used for comparing various safety investments.

A QAAP can thus be interpreted as a utility number expressing the value that a safety team assigns to a specific time period in a particular safety (accident probability) situation relative to the value of a reference safety situation (e.g., the 0.1% accident probability situation in the example above).

In short, time trade-off is a technique for generating a safety management team's or a safety manager's preferences by offering a hypothetical decision between a shorter time period of a higher safety situation (lower accident probability), on the one hand, or a longer time period of a lower safety situation (higher accident probability) on the other. The time trade-off scores can then be used, in turn, to calculate QAAPs. QAAPs may be particularly useful in the case of prevention with respect to extremely low probability type accidents.

3.9   Budget Constraint and n-Dimensional Maximization Problem Formulation

Looking at the iso-utility curves in the safety commodity space shown in Figure 3.8, it is obvious that safety management would like to purchase safety commodity bundles as far to the top-right as feasible. The only restriction in doing so is the safety budget at the disposal to safety management. Hence, a budget constraint curve needs to be determined to be able to make an optimal – and realistic – choice regarding the safety commodity bundle. Associated with each commodity is a price. Assume that S_1 comes at a price q_1 and S_2 has a price q_2. Safety management has a budget of B euros to divide among the two safety commodities. The cost of the safety commodity bundle (S_1, S_2) is $q_1 S_1 + q_2 S_2$. All budget sets that safety management could conceivably face can then be expressed as $q_1 S_1 + q_2 S_2 \leq B$.

The budget constraint is easy to visualize. Safety management will choose from the budget set so as to be on as high an indifference curve as possible (see Figure 3.11).

In Figure 3.11 the safety commodity bundle c^* is the most preferred bundle within reach of the safety budget. This is the safety management's maximization problem for a two-dimensional situation and it is further explained in Section 3.10. The optimal bundle c^* can be graphically characterized by the fact that the indifference curve U_s, of which c^* is a

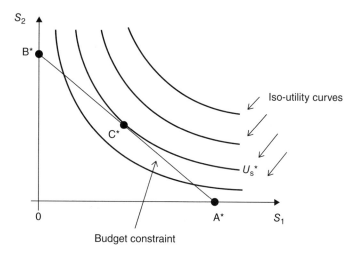

Figure 3.11   Budget constraint and indifference curves.

member, lies completely outside the budget set except at point c*, where it is tangential to the budget constraint line. In the microeconomic theory literature, this is usually stated as: "At c*, the marginal rate of substitution (which is the slope of the indifference curve through c*) equals the price ratio (which represents the slope of the budget line)."

This two-dimensional representation allows us to easily carry out and visualize some thought experiments. For example, Figure 3.12(a) shows the result of a budget decrease (if the budget goes down, c* will move toward c**), whereas Figure 3.12(b) shows what happens if the price q_1 of one of the safety goods, S_1, were to increase (again, c* would move toward c**).

Now assume that there are n safety goods for investing the safety budget in, instead of only two. Safety commodity bundles are then lists $(S_1, S_2, \dots, S_n)$, and a utility function assigns a number $U_s(S_1, S_2, \dots, S_n)$ to each such list. The safety manager or safety management team's n-dimensional maximization problem can then be stated in the following way:

$$\max[U_s(S_1, S_2, .., S_n)]$$

$$s.t.$$

$$q_1 S_1 + q_2 S_2 + \dots q_n S_n \leq B$$

$$S_1 \geq 0; S_2 \geq 0; \dots; S_n \geq 0.$$

In order to solve such problems, mathematical techniques such as multivariate calculus and matrix algebra are needed. This falls outside the scope of this book, but the two-dimensional case, which is only a special case of the n-dimensional problem, is discussed further in the next section.

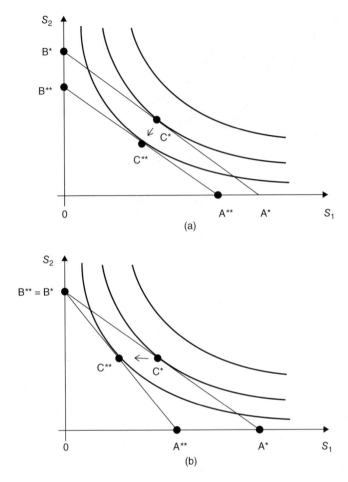

Figure 3.12 (a) Effect of decreasing the safety budget on the maximization problem. (b) Effect of increasing q_1 (price of S_1) on the maximization problem.

3.10   Determining Optimal Safety Management Preferences within the Budget Constraint for a Two-dimensional Problem

Assume a two-commodity space: safety commodity S_1 and safety commodity S_2. For instance, S_1 represents one specific safety measure and S_2 represents "all other safety measures." The incline of the budget line is determined by the respective prices of S_1 and S_2. Assume, for instance, that the price of S_2 is the average price of all the measures composing the parameter. The intersection points with the x-axis and the y-axis, respectively, represent the quantities that safety management could purchase of either S_1 or S_2 if the entire budget were spent on this measure. To obtain an optimum point, the indifference curve is shifted in a parallel way until it touches the budget line. This tangent is the optimum c* (see Figure 3.13). The coordinates

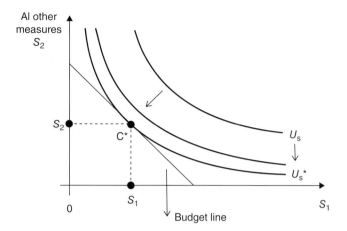

Figure 3.13 Determining the optimum safety commodity bundle within the budget constraint.

s_1 and s_2 belonging to this tangent represent the number of S_1 and S_2, respectively. It might also be interesting to calculate from these numbers the proportion of S_1 and S_2 with respect to the budget. This is possible by simply multiplying by the price per unit and dividing by the total budget.

A simple numerical example can be given to illustrate the theory. Assume that the safety utility function is given by the following function of the Cobb-Douglas type:

$$U_s(S_1, S_2) = (S_1)^{0.2}(S_2)^{0.8}$$

Assume furthermore that safety measure S_1 has a price of €1000, and that S_2 (being all other measures) has a price of €20 000. The budget is set to be €100 000. To solve the problem, the Lagrange function is formulated (for information on using the Langrangian method to solve maximization problems, see, e.g., De Borger *et al.* [6]):

$$L = (S_1)^{0.2}(S_2)^{0.8} + \lambda(100\,000 - 1000\,S_1 - 20\,000\,S_2)$$

The three first-order conditions look as follows:

$$0.2(S_1)^{-0.8}(S_2)^{0.8} = 1000\lambda$$

$$0.8(S_1)^{0.2}(S_2)^{-0.2} = 20\,000\lambda$$

$$100\,000 - 1000 S_1 - 20\,000 S_2 = 0$$

Solving this system of equations gives $S_1 = 20$ and $S_2 = 4$. Hence, safety management optimally buys 20 safety measures S_1, and spends the rest of its budget, i.e., €80 000, on other safety measures.

3.11 Conclusions

In this chapter, the foundations of microeconomic theory are applied to the safety research field. It is important to realize that economic principles such as the value function, expected value, utility function, expected utility, and indifference curves are all applicable within the field of safety. Safety utility functions can be drafted, as well as iso-utility curves in the safety commodity space. The safety budget constraint together with the safety iso-utility curves can further be used to solve the safety utility maximization problem, to decide on the optimal choice of a bundle of safety measures within an organization.

References

[1] Fuller, C.W., Vassie, L.H. (2004). *Health and Safety Management: Principles and Best Practice*. Prentice Hall, Essex.
[2] Garvey, P.R. (2009). *Analytical Methods for Risk Management: A Systems Engineering Perspective*. Taylor & Francis Group, Boca Raton, FL.
[3] Straffin, P.D. (2003). *Game Theory and Strategy*. The Mathematical Association of America, Washington, DC.
[4] Von Neumann, J., Morgenstern, O. (1967). *Theory of Games and Economic Behaviour*. John Wiley & Sons, Inc., New York.
[5] Samuelson, P.A. (1948). Consumption theory in terms of revealed preference. *Economica*, **15**(60), 243–253.
[6] De Borger, B., Van Poeck, A., Bouckaert, J., De Graeve, D. (2013). *Algemene Economie* 8th edn. De Boeck, Antwerpen.

4

Operational Safety Decision-making and Economics

4.1 Economic Theories and Safety Decisions

4.1.1 Introduction

Making a decision is often difficult because of the uncertainties involved and the risks these uncertainties entail. A balance needs to be struck between different alternatives and the corresponding consequences. The decision therefore depends in part on how a choice problem is defined and the value that is attached to the pros and cons of an alternative. However, this involves a lot of uncertainty, as people can never be completely sure of the exact consequences of the chosen course of action [1].

Much research on decision-making has already been carried out in different disciplines. Various models and theories can be found in the literature that describe decision-making under conditions of uncertainty and which attempt to explain the phenomena observed. The first theory, developed as early as the seventeenth century, is the "expected value theory" (see also Chapter 2). According to this theory, people take decisions by maximizing the expected value.

However, the St. Petersburg paradox – which is a paradox related to probability and decision theory – made it clear that the expected value theory would not hold in real situations. The paradox is based on a particular (theoretical) lottery game that leads to a random variable with an infinite expected value (i.e., infinite expected payoff) but that nevertheless seems to be worth only a very small amount to the participants. The St. Petersburg paradox is a situation where a naïve decision criterion, which takes only the expected value into account, predicts a course of action that no actual person would be willing to take (see also Section 4.1.2). The St. Petersburg paradox therefore made a more accurate model necessary, and Bernoulli developed a new theory, the "expected utility theory." This theory states that people attempt to maximize not the expected value, but the expected utility, when making decisions (see also Chapters 3, 7, and 8). People assess the utility of each outcome based on probabilities and then choose the alternative with the highest sum of weighted outcomes.

Operational Safety Economics: A Practical Approach Focused on the Chemical and Process Industries,
First Edition. Genserik L.L. Reniers and H.R. Noël Van Erp.
© 2016 John Wiley & Sons, Ltd. Published 2016 by John Wiley & Sons, Ltd.

Kahneman and Tversky then developed a theory that took into account the human factor: the "prospect theory." This theory was very successful as it included the psychological aspects of human behavior. In this theory, a choice problem is defined relative to a reference point. Depending on coding in terms of gains and losses, people will show more risk-averse or risk-seeking behavior [2–5].

People are constantly taking decisions, consciously or unconsciously. It is therefore not surprising that the subject of decision-making is studied in various disciplines, from mathematics and statistics to economy and political science, to sociology and psychology.

4.1.2   Expected Utility Theory

As described in the introduction, the first universally applicable decision-making theory was the "expected value theory," developed as early as the seventeenth century. According to this theory, people take decisions based on maximization of the expected (monetary) value. However, this theory was repudiated by the St. Petersburg paradox. In the St. Petersburg paradox, a coin is tossed until "heads" appears. If "heads" appears on the first toss, the casino pays €1, otherwise it doubles the payout each time "heads" appears. What price would you be prepared to pay to enter this game? This paradox describes a gamble with an infinite expected value. Even so, most people intuitively feel that they should not pay a large price to take part in the game [3, 6].

To solve the St. Petersburg paradox, Bernoulli developed a new theory. He suggested that people want to maximize not the expected value, but the expected utility. As a result, the utility of wealth is not linear, but concave. After all, an increase in wealth of €1000 is worth more if the initial wealth is lower (e.g., from €0 to €1000) than if it is higher (e.g., from €10 000 to €11 000). This theory is called the expected utility theory in the literature. It is a theory about decision-making under conditions of risk in which each alternative results in a set of possible outcomes and in which the probability of each outcome is known. Individuals attempt to maximize the expected utility in their risky choices. They calculate the utility of each outcome using probabilities and choose the alternative with the highest sum of weighted outcomes. The attitude of an individual to risk is defined by the form of a utility function. For a risk-averse individual, the function is concave; for a risk-neutral individual, the function is linear; and for a risk-seeking individual, the function is convex. Given a choice between a certain outcome with a utility y and a gamble with the same utility y, a risk-averse individual will choose a certain outcome, a risk-seeking individual the gamble, and a risk-neutral individual will treat the two outcomes equally [3–5, 7–9].

4.1.3   Prospect Theory

The expected utility theory has been considered for a long time the most applicable method but, despite its strengths, it has encountered several problems as a descriptive model for decision-making under conditions of risk and uncertainty. After all, it requires that the decision-maker has access to all the information describing the whole of the situation. Clearly, this information requirement is far too rigorous for most practical applications, in which the probabilities cannot be calculated exactly, or the decision-maker does not yet know what he or she wants, or the series of alternatives is not fully determined, and so on. In addition, there is plenty of experimental evidence to suggest that individuals often fail to behave in the manner defined in the expected utility theory. In fact, the choices made by individuals consistently deviate from those predicted by the model, as is made clear in the experiments of Allais and Ellsberg.[1] The so-called Allais paradox shows that individuals are sensitive to differences in probabilities. Experiments show that an individual's willingness to choose an uncertain outcome depends not only on the level of the uncertainty, but also on the source. People show dependence on a source when they are prepared to choose a proposition from source A but not the same proposition from source B. Regarding the so-called Ellsberg paradox, there are different versions of Ellsberg's experiment. In one version there are two urns, one containing 50 black and 50 red balls, and the other containing 100 balls about which it is only known that they are either red or black. People must choose an urn and a color, then draw a ball from the chosen urn. If the ball matches the color chosen, the subject wins a prize. Ellsberg noted that people display a strong preference for the urn with an equal number of red and black balls, despite the fact that economic analysis suggests that there is no reason to prefer one urn or the other. Ellsberg's classic study therefore shows that people attach a higher value to bets with known probabilities (risk) than to bets with unknown probabilities (uncertainty). This preference is called "ambiguity aversion." The Ellsberg experiment shows that it is possible that decision-makers will not follow the usual rules of probability if they cannot act based on subjective probability. As a result, in such situations the theory of expected utility does not accurately predict behavior. For a more positive appreciation of expected utility theory see Chapter 7. An alternative theory was therefore required to describe the observed empirical deviations [3–5, 7–20].

Kahneman and Tversky's [4] prospect theory is considered an alternative to the expected utility theory and one of the principal theories for decision-making under conditions of risk. According to this theory, individuals are more likely to assess outcomes based on change from a reference point than on an intrinsic value. After all, people tend to think in terms of losses and gains rather than intrinsic values and therefore make choices based on changes relative to a certain reference point. The reference point is often, but not always, the status quo. Assuming that the reference point is zero, gains are represented by positive outcomes and losses by negative outcomes. The concept of a reference point means that people deal with losses differently from gains. Because people code outcomes in terms of a reference point and deal with losses and gains differently, the definition of this reference point or construction of the choice problem is crucial. How people approach the problem – i.e., in terms of losses or gains – affects the preferences independently of the mathematical equality between the two choice problems. Most decision problems take the form of a choice between maintaining the status quo and accepting an alternative, which may be more advantageous in some respects and less advantageous in others. The advantages of an outcome are then assessed as gains and the disadvantages as losses. Individuals usually give more weight to losses than to comparable gains and display more risk-averse behavior in the case of gains and more risk-seeking behavior in the case of losses [2–5, 10, 21–28].

According to prospect theory, each individual attempts to maximize the utility function, defined not in absolute terms but in terms of losses and gains relative to a reference point. This function is usually concave for gains and convex for losses, reflecting the risk-averse behavior for gains and risk-seeking behavior for losses, as shown in Figure 4.1 [2, 3, 8, 10, 21, 24, 27–30].

---

[1] Remark that the following interpretation of the Allais and Ellsberg experiments is the viewpoint expounded by behavioral economists. In Sections 7.3 and 7.4 an alternative interpretation and appreciation of, respectively, the Allais and Ellsberg experiments is offered up.

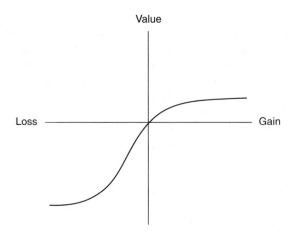

Figure 4.1   Utility function of prospect theory. (Source: Wikipedia, 2016 [31].)

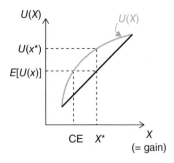

Figure 4.2   Utility function for gains (risk-averse behavior).

Risk-averse behavior for gains is seen in the fact that people choose certain gains rather than uncertain, but larger, gains. In Figure 4.2, a certain gain is shown as CE and the uncertain, larger, gain as x^*. Due to risk-averse behavior, the expected value of the uncertain gain $E[U(x)]$ is lower than the actual utility for this gain $U(x^*)$. The utility function of a risk-averse individual is therefore concave. Risk-seeking behavior in the case of losses, on the other hand, is expressed by the fact that people choose uncertain, higher losses rather than certain, but lower losses. Finally, in the case of risk-neutral behavior, the expected value, $E[U(x)]$, will be equal to the actual utility $U(x^*)$, and the utility function will be linear. These findings can be readily applied to operational safety by inserting the safety utility function S_u in the place of the utility function (see also Figure 3.6).

These findings show that the preference for prospects relating to losses only are the mirror image of preference for prospects relating to gains only. Kahneman and Tversky [4] call this the "reflection effect" around the reference point. This means that the sensitivity to change decreases with increasing distance from the reference point [2, 3, 8, 10, 21, 24, 27–30].

An aversion to risk is reflected in the observed behavior, i.e., people are more sensitive to losses than to gains. The utility function for losses is therefore steeper than that for gains, as

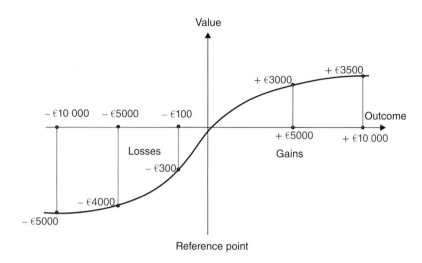

Figure 4.3   The utility function curve. (Source: Reproduced with permission of LessWrong [32].)

shown in Figure 4.3. A loss of €5000 results in a value of −€4000, whereas a gain of €5000 only results in a value of €3000. In absolute terms, therefore, people attach a higher value to a loss than to an equivalent gain. This means that a reduction in loss has a much higher value than an increase in gain of the same amount, so that people express more risk-seeking behavior in the case of losses and more risk-averse behavior in the case of gains. The convex curve for losses and concave curve for gains therefore give a typical utility function an S-form [2, 3, 8, 10, 21, 24, 27–30].

People treat gains differently from losses based on two aspects. The first of these is the reflection effect, as discussed earlier, and the second is "loss aversion." Loss aversion explains why losses weigh more heavily than gains. It implies that people prefer the status quo, or other reference point, to a 50/50 chance of positive and negative alternatives with the same absolute value. The negative utility of giving up a good is therefore experienced as greater than the utility of acquiring a good, so that people often choose the status quo rather than a change. This phenomenon reflects the observed greater sensitivity to losses than to gains. The result is a utility function that is steeper for losses than for gains. One implication of loss aversion is that individuals have a strong tendency to remain in the status quo, as the disadvantages of leaving the status quo weigh more heavily than the advantages. Loss aversion also implies that people place a higher value on a good that they own than on a similar good that they do not own. This overvaluing of property is called the "endowment effect." The process of acquiring a good increases its value. This even applies to trivial goods such as sweets or coffee mugs. One consequence of this phenomenon is that people often ask for more to give up a good than they are prepared to pay for it. This means that the selling price is always higher than the buying price, as the highest price a buyer is prepared to pay to acquire a good will be lower than the minimum compensation required to give up the same good [2, 4, 5, 10, 21, 24, 26–28, 33–36]. Translated into safety language, this means that safety management is not inclined to change existing company safety measures and policies, even if the new safety measures and policies would prove to be better at the same price. Therefore, a minimum safety increase premium (see Section 4.2.5) may be needed to convince safety management to change an existing safety situation.

In addition to the reflection effect, loss aversion, and the endowment effect, the prospect theory also makes use of other effects to describe and explain behavior during decision-making processes. First of all, it is possible to deduce from a number of empirical observations that individuals weigh certain outcomes more heavily than possible outcomes. This is called the "certainty effect." People also weigh small probabilities more heavily, and attach too little weight to average and high probabilities. As a result, highly probable but uncertain outcomes are often treated as if they were certain. This effect is called the "pseudo-certainty effect." Consequently, changes in probability that approach 0 or 1 have a larger impact on preferences than comparable changes in the middle of the range of probability. This effect can be clarified using the following example: during a hypothetical game of Russian roulette, people are prepared to pay more to reduce the number of bullets in the revolver from one to zero than from four to three. There is also the "immediacy effect," which refers to a preference for outcomes that are experienced immediately. Finally, prospect theory also describes a "magnitude effect." This effect refers to the following pattern: people are risk-seeking in the case of small gains and become strongly risk-averse as the size of the gain increases, and people are risk-averse in the case of very small losses and become more risk-seeking as the size of the losses increases [5, 8, 10, 21, 24, 25, 27].

4.1.4 Bayesian Decision Theory

The Bayesian decision theory can be seen as the newest theory, further advancing insights beyond prospect theory, and aims to resolve the consistency problems with, on the one hand, predictive risk analysis and expected utility theory and, on the other, common-sense considerations of risk acceptability in some well-constructed counter-examples in experiments like the ones by Allais and Ellsberg which are typically brought forth by the behavioral economists. This neo-Bernoullian decision theory is a direct reaction to Kahneman and Tversky's prospect theory [27], which was felt to put too high a premium on the sterile empiricism of hypothetical betting experiments, while at the same time failing to appeal to the need for compelling and rational first principles. The Bayesian decision theory differs from the expected utility theory in that an alternative position measure which captures the worst-case, most likely, and best-case scenarios is maximized, rather than the traditional criterion of choice where the expected values are maximized.

It can be demonstrated that this simple adjustment of the criterion of choice accommodates (non-trivially) the Allais paradox and the common-sense observation that type II events are more risky than type I events, as well as (very trivially, as the Ellsberg paradox is arguably not that paradoxical) the Ellsberg paradox (see Sections 7.3 and 7.4, respectively). Moreover, it can be shown that the proposed alternative criterion of choice relative to the traditional criterion of choice leads to a more realistic estimation of hypothetical benefits of type II risk barriers (see Section 8.12).

4.1.5 Risk and Uncertainty

The world that we live in is a world full of uncertainties in which, in most cases, we are unable to predict future events. Nevertheless, people are forced to make predictions given that suitable management responses are required when facing uncertainties, particularly in the field of operational safety. To explain how people deal with uncertainty when making a choice between different alternatives, use has traditionally been made of either formal models or rational analyses. Two types of models are used. The first of these is used mainly in economics and in research into decision-making: a numerical value is assigned to each alternative and the choice characterized by maximization of this value. In doing this, people mainly make use of normative models, such as the expected utility theory, or descriptive models, such as the prospect theory. The second type of model is usually used in law and politics. This approach identifies different reasons and arguments for why a certain decision should be taken. When making a decision, people weigh up the reasons for and the reasons against a certain alternative. A decision therefore depends in part on the value attached to the pros and cons of a particular alternative [1, 37].

A distinction should be made between "uncertainty" and "(negative) risk." The difference between risk and uncertainty is that, in the case of uncertainty, the probability is not fully known. Risk is characterized by an objective probability distribution, while uncertainty bears no relation to statistical analysis [5, 38, 39].

In the modern world, (negative) risk takes two fundamental forms. Risk as a feeling refers to the quick, instinctive, and intuitive response of an individual to danger. Risk as analysis results in logic, reasoning, and scientific deliberation, as applied in risk management. People generally make decisions involving risk by constructing qualitative arguments that support their choice. During a decision-making process, a distinction is made between risky and risk-free choices. Risky choices are made with no prior knowledge of the consequences. As the consequences of such actions depend on uncertainties, this choice can also be regarded as a kind of gamble that results in certain outcomes with varying probabilities. However, the same choice can be described in different ways. For example, the possible outcomes of a gamble can be described as gains and losses compared with the status quo or the initial position. An example of decision-making involving risk is the acceptability of a gamble that results in a certain monetary outcome with a specific chance. A typical risk-free decision concerns the acceptability of a transaction in which a good or service is exchanged for money or labor. Most people will choose certainty in preference to a gamble, even though a gamble has, mathematically speaking, a higher expected value. This preference for certainty is an example of risk-averse behavior. As indicated before, a general preference for certainty rather than a gamble with a higher expected value is called risk aversion, and rejection of certainty for a gamble with the same or a lower expected value is called risk-seeking behavior. Furthermore, as also previously mentioned, people usually display risk-averse behavior when faced with improbable gains and risk-seeking behavior when faced with improbable losses [4, 38–41].

In addition to decision-making under conditions of risk, decision-making under conditions of uncertainty is also possible. Uncertainty describes a situation in which the decision-maker does not have access to any required statistical information. Various definitions of uncertainty are given in the literature. Whichever definition is used, uncertainty is not static, due to the dependence on knowledge. For example, if the amount of available information increases, the uncertainty will be reduced (e.g., going from domain B toward domain D in Figure 2.5). The term uncertainty is mainly used to describe the uncertainty resulting from the consequences of actions that are unknown because of their dependence on future events. This is often called "external uncertainty," as it concerns the uncertainty of ambient conditions beyond the control of the decision-maker. Internal uncertainty refers to uncertainties in the preferences of the decision-maker, the definition of the problem, and the vagueness of the information [38, 42–50].

There are in fact two types of uncertainty – objective and subjective. Objective uncertainty relates to the information about the probabilities, whereas subjective uncertainty relates to the attitude of the decision-maker. There are two aspects with respect to this attitude of decision-makers: their attitude to the gain and their attitude to the negative risk. The attitude of a person to risk is traditionally – in accordance with expected utility theory – defined in terms of marginal utility or the form of the utility function. If a decision-maker is not risk-neutral, then the subjective and objective probabilities will not be equal [5, 49].

Decision-making under conditions of uncertainty takes place either if the prior information is incomplete or if the outcomes of the decisions are unclear. The best strategy for dealing with uncertainty in decision-making is to reduce or completely remove this uncertainty. Uncertainty can be reduced by collecting more information before making a decision or by delaying a decision. If no further information is available, uncertainty can be reduced by extrapolating the available information. Statistical techniques can be used to predict the future based on information relating to current or past events. Another technique is to form assumptions. Assumptions enable experienced decision-makers to take quick and efficient action within their domain of expertise, even if there is a lack of information. If it is considered to be unfeasible or overly expensive to reduce the uncertainty, this uncertainty can be acknowledged in two ways: by taking it into account when selecting an alternative, or by preparing to avoid or face the possible risk involved [42, 46, 48, 50].

If a decision is to be made under conditions of uncertainty, use can be made of various decision strategies, including MaxMin strategy, MaxMax strategy, and the Hurwicz criterion strategy. The choice of model depends on the attitude of the decision-maker. Three different attitudes are described in the literature: pessimistic, optimistic, and neutral. In the case of a pessimistic attitude, the decision-maker only considers the worst case for each alternative. He therefore makes use of the MaxMin strategy, and the alternative with the highest gain in the worst case will be selected. In the case of an optimistic attitude, decision-makers only consider the best case for each alternative. They therefore select the alternative with the highest gain in the best case, using the so-called MaxMax strategy (not to be confused with the so-called "Maxmax hypothetical benefits" – see later in this chapter). The third type is the neutral attitude. In this case, the decision-maker considers each case in the same way and the evaluation value of each alternative is determined by the general conditions. A combination of an optimistic and pessimistic attitude is called the Hurwicz method [45, 47, 49].

It should be noted that the uncertainties involved in making decisions increase in size and impact if decisions need to be made concerning the risks related to type II risks. After all, high-impact, low-probability (HILP) events are characterized by low probabilities and large, extensive, and possibly irreversible consequences. Type II risks are not always well understood, with the result that people do not always know how to deal with them. It is not possible to increase knowledge through experimentation, given that the risks are effectively irreversible in a normal time frame. The classic theories, based on expected utility, therefore do not always work and results should be interpreted with great caution, as the models tend to underestimate events with low probabilities [46, 51], as is demonstrated in Section 7.5.

4.1.6   Making a Choice Out of a Set of Options

When making a decision, a choice usually needs to be made from a set of possible options. This choice is not made randomly. An attempt is made to make a choice that provides the best answer to a number of criteria: the optimum choice. Each choice results in a gain of a certain value, so that the choice is selected with the largest gain. However, in most cases, the gains resulting from a choice are affected by variables of unknown value. It is traditional to assume that, if a rational choice needs to be made between two alternatives that involve uncertainties, the uncertainty is described using probability and the different alternatives are ordered based on the expected utility of the consequences of these alternatives. Selection of one alternative is therefore based on the probability and potential value of a possible outcome [47, 52, 53].

There are two phases involved in making a choice. The first phase involves various mental operations that simplify the choice problem by transforming the representation of outcomes and probabilities. Coding involves the identification of the reference point and the framing of outcomes as deviations from this reference point, and this affects orientation toward risk. After all, a difference between two alternatives will always have a larger impact if it is seen as a difference between two disadvantages rather than a difference between two advantages. The decision can therefore be made to allow the disadvantages of a deviation to weigh more heavily than the advantages. In addition, common elements to the various alternatives or irrelevant alternatives are also often eliminated. This "isolation effect" can result in different preferences, as there are different ways of separating prospects into shared and distinctive elements. This can result in a reversal in preferences and inconsistency. The second phase is the evaluation phase [5, 36].

The traditional utility function assumes that an individual has a well-organized and stable system of preferences and good calculation skills. However, such an assumption is unrealistic in many contexts. In response to this problem, Simon [54] developed the theory of bounded rationality. This implies that, due to the expense and impracticalities involved in choosing between all possible alternatives to achieve the optimum choice, people usually look for the first satisfactory alternative that meets a set of predetermined objectives. The implication is therefore that human behavior should be modeled as satisfying rather than optimizing. Such an approach has several appealing qualities, as working with objectives is, in most cases, very natural [47, 54, 55].

Evaluation of the outcomes is susceptible to framing effects due to the non-linear utility function and people's tendency to compare outcomes with a reference point. The decision frame used to analyze a decision refers to the perception of the decision-maker regarding a possible outcome of a risky decision in terms of gains and losses. Whether or not a certain outcome is seen as a loss or a gain depends on the reference point used to evaluate possible outcomes. The framing effect is seen whenever the same outcome can be considered a loss or a gain, depending on the reference point used. For example, if a company is faced with an accident cost of €5000, which was expected to be €10 000, is this experienced as a loss or a gain? The answer depends, for example, on whether the previous accident cost or the expected value is used as a reference point. Whether or not a decision is considered in terms of a loss or a gain is very important, because people display risk-seeking behavior when faced with losses and more risk-averse behavior when faced with gains [4, 8, 56].

It is therefore clear that emotions play an important role in social and economic decision-making. Individuals, including people who need to make decisions on safety budgets – such as safety managers and middle and top managers – evaluate alternatives subjectively and emotions influence these evaluations. People generally try to control these emotions and to anticipate the emotional impact on future decisions, as emotions can affect such decisions. The impact of emotions is of particular importance when assigning a value to different perspectives [8, 57].

4.1.7   Impact of Affect and Emotion in the Process of Making a Choice between Alternatives

Traditional economic theory assumes that most decision-making involves the rational maximization of the expected utility. Such an approach assumes that people have access to unlimited knowledge, time, and information-processing skills. In the 1970s and 1980s, researchers identified several phenomena that systematically violated the normative principles of economic theory. In the 1990s, emotions were shown to play a significant role in various forms of decision-making, in particular concerning decisions that involved a high level of risk and uncertainty [40, 58–60].

According to Epstein and Pacini [61], people understand reality through two interactive, parallel processing systems. The rational system is a deliberative, analytical system that functions using established rules of logic and evidence (such as the prospect theory). This system makes it possible for individuals to consciously seek knowledge, develop ideas, and analyze. On the other hand, the experiential, emotional system codes reality in images, metaphors, and narratives to which affective feelings are attached. This system enables individuals to learn from experience without consciously paying attention. Individuals differ in the extent to which the rational and experiential systems influence their decision-making. The balance between these two processes is influenced by various factors, such as age and cognitive load, both of which result in a greater dependence on affect. By affect, the "faint whisper of emotion" that influences decisions directly is meant, and is not simply an answer to a prior analytical evaluation [41, 60, 62].

People therefore base their judgment on a certain activity, not just on what they think, but also on what they feel. If they have a good feeling about a certain activity, they assess the risk as low and the benefit as high. However, if they do not have a good feeling, they will judge the risk as high and the benefit as low. This is called "affect heuristic" in the literature. If the consequences are highly affective, as in the case of the lottery or cancer, there is little variation. For example, research by Loewenstein and Prelec [8] showed that people expressed the same feelings regarding winning the lottery whether the chance was 1 in 10 million or 1 in 10 000 [41, 59, 60].

In any case, it is clear that emotions, which can be interpreted and expressed as moral principles, should be much more considered in risk indexing and risk calculations than is the case today. Section 4.10.4 therefore suggests an approach to calculate risks, where both rational and moral factors are taken into account.

4.1.8   Influence of Regret and Disappointment on Decision-making

People always wish that they had made a different decision if the decision they made turns out to have negative consequences. This feeling is even stronger if it turns out that the alternative would have resulted in a more favorable outcome. In such situations, people experience a feeling of regret. Regret is a psychological response to the making of a wrong decision. A "wrong" decision is defined based on the actual outcome and not on the information available at the moment the decision was made. Regret can result from comparison of an outcome with an outcome that would have resulted from making a different decision. Regret therefore has much in common with opportunity cost (see Chapter 5).

Regret can affect people in two ways. First of all, it can lead people to try to undo the consequences of their regretted choice after the decision has been made. Second, if they anticipate the regret they would later feel, it can influence the choice they make before the decision is taken. A new decision-making model was developed based on this anticipatory aspect of regret – the "regret theory." Whereas classical theories assume that the expected utility of a choice depends only on the possible pain or pleasure associated with the outcome of that choice, in regret theory utility also depends on the feelings caused by the outcome of the rejected choices. Regret theory is based on two fundamental assumptions. The first is that people compare the real outcome with the outcome that would have resulted from making a different choice. As result of this, people experience certain feelings: regret, if the choice that they did not make would have resulted in a more favorable outcome; and pleasure, if the choice made has produced the best possible outcome. The second assumption of the regret theory is that the emotional consequences of the decision are anticipated and taken into account when making that decision. Different experiments have shown that such behavior results in a more risk-averse attitude. However, this is not always the case. Although most people do their best to avoid regret, this does not only result in more risk-averse behavior, but can also lead people to display more risk-seeking behavior. An aversion to regret can therefore result in choosing an outcome with the lowest (negative) risk as well as in choosing an outcome with the highest (negative) risk. Experiencing regret and disappointment also has another consequence, which is that avoiding feedback so as not to experience regret and disappointment means that people do not learn from their experiences [3, 13, 59, 63].

As well as regret and pleasure, people can also experience disappointment when making a choice between alternatives. Disappointment is a psychological response to an outcome that does not correspond to expectations. When people make a decision, they can experience these feelings when comparing uncertain alternatives. However, it is not always possible to compare the outcome of a chosen alternative and the outcome of an alternative that was not chosen. Imagine that a choice needs to be made between a certain alternative and an uncertain one. If an alternative is chosen with outcome characteristics with some probability, it is always clear what the result of the other (certain) alternative would be. However, if the certain alternative is chosen, it is not always possible to determine whether this choice resulted in a more favorable outcome or not, because one does not know whether the uncertain alternative would have been realized or not [3, 13, 59, 63].

It is possible to translate the principle of regret theory into industrial practice by using expected hypothetical benefits (see also Section 4.2.4) to aid decision-making. To this end, later in this book, expected hypothetical benefits are explained in greater depth and they are used in cost-benefit analysis (see Chapter 5) and cost-effectiveness analysis (see Chapter 6) approaches for decision-making purposes.

4.1.9   Impact of Intuition on Decision-making

The accuracy of decision-making is often inversely proportional to the speed with which decisions need to be made. To explain the fact that people are able to make high-quality decisions relatively quickly, most researchers focus on intuition. Individuals mainly make use of intuition when they are faced with severe time constraints. After all, intuition is based on our innate ability to synthesize information rapidly and effectively and could therefore be essential in successfully solving complex tasks in a short space of time. However, intuitively making decisions does not always result in accurate decisions, as applying intuition only results in speed, sometimes at the expense of accuracy [62, 64].

4.1.10   Other Influences while Making Decisions

In addition to emotions, there are also other factors that influence decision-making. The knowledge available within the decision context also determines an individual's behavior when choosing between alternatives. Research shows that individuals who are not experts are typically overconfident. They overestimate the quality of their abilities and knowledge and perceive extreme probabilities as not so extreme. Experts and more qualified people, on the other hand, are more sensitive to the risk associated with a hypothetical gamble and display more risk-averse behavior, as shown in the research of Donkers *et al.* [65]. Second, age also has an impact on the decision-making process. The older an individual, the more experience they have and the more capable they are of curtailing overconfidence and therefore responding like an expert. Older people have a more accurate idea of their knowledge and the limits of that knowledge [66, 67].

Finally, gender also plays a large role in decision-making. Many studies have highlighted the fact that men are more risk-seeking than women. The reasons for this difference in approach to risk have already been investigated in various research areas. They can be caused by nature or nurture, or a combination of the two. For example, (young) boys are expected to take risks when participating in competitive sports, whereas (young) girls are often encouraged to be careful. The high-risk choices that men make may therefore be the result of the way in which they were raised by their parents. The same applies to women's risk aversion. Another possible explanation for gender differences is that the way in which risk is experienced relates to the cognitive and non-cognitive differences between men and women. It should be pointed out that women who acquire more competences, become more confident or obtain more knowledge show increased risk-seeking behavior when making decisions. However, such factors have an opposite effect on men, who display more risk-averse behavior with increasing expertise and confidence. Finally, the framing of a decision problem strongly influences the way in which a man makes a decision, but not a woman [14, 66, 68, 69].

4.2   Making Decisions to Deal with Operational Safety

4.2.1   Introduction

In the previous section, the way in which people make decisions and choices, and all the accompanying influencing psychological principles, was discussed. It is obvious that decision theory and making choices under uncertainty are not at all easy to model. The very nature of operational safety management, safety being defined as "a dynamic non-event," makes it very hard to develop a decision-supporting model. Nonetheless, safety managers need to make such decisions, based on their emotions, their intuition, and the available information and its accompanying uncertainty and/or variability. Making decisions to deal with operational safety should be as accurate as feasible, rational, cost-effective, simple yet not over-simplified, no-nonsense and pragmatic, and, most importantly, as effective as possible (so that there are as many non-events as possible). To this end, a number of possible decision-making approaches may be worked out and will be elaborated upon and explained in the remainder of this book. The decision process should lead to adequate and optimized (according to the company's safety management characteristics) risk treatment. The concept of "risk treatment" should be explained at this point. There are several possible ways to treat risks:

1. *Risk treatment option 1: risk reduction*
   (a) Risk avoidance
   (b) Risk control.
2. *Risk treatment option 2: risk acceptance*
   (a) Risk retention
   (b) Risk transfer.

Organizations basically have two main options to deal with risks: they can either accept the risks, or they can reduce the risks. Risk reduction options involve technological, organizational/procedural, and human factor prevention and mitigation measures in order to decrease the risk level. Risk acceptance options, on the other hand, encompass measures that reduce the financial impact of the risks on the organization. There are two sub-options available within each of the two main options [70]. For risk reduction these are risk avoidance and risk control, whereas for risk acceptance, these are risk retention and risk transfer.

4.2.2   Risk Treatment Option 1: Risk Reduction

Risk avoidance and risk control can be achieved in an optimal way by inherent safety or design-based safety. There are five principles of design-based safety. The possible implementation of these five principles is explained for the case of the chemical industry, illustrated in Figure 4.4.

As explained by Meyer and Reniers [73], techniques of risk reduction aim to minimize the likelihood of occurrence of an unwanted event and/or the severity of potential losses as a result of the event, or aim to make the likelihood or outcome more predictable. They are applied to existing processes and situations and can be listed as:

- *Substitution* – by replacing substances and procedures by less hazardous ones, by improving construction work, and so on. Care should be taken that there is not simply a replacement of risk.
- *Elimination of risk exposure* – this consists in not creating or completely eliminating the condition that could give rise to the exposure to risk.
- *Prevention* – combines techniques to reduce the likelihood/frequency of potential losses.
- *Protection/mitigation* – these are techniques whose goal is to reduce the severity of accidental losses before/after an accident occurs.
- *Segregation* – summarizes the techniques used to minimize the overlapping of losses from a single event. This possibility may imply very high costs:
      Segregation by separation of high-risk units
      Segregation by duplication of high-risk units.

Using principles such as elimination, substitution, moderation, and attenuation in the design phase is always better than to use them in an existing situation/process. In the former case, the

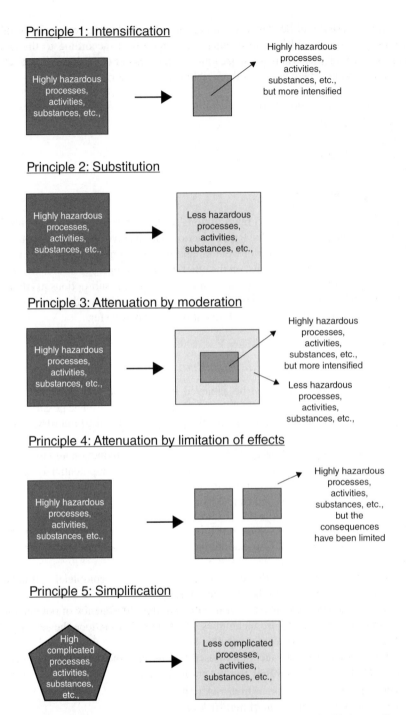

Figure 4.4 Five principles of design-based safety implemented in the chemical industry. (Source: Kletz [71, 72]. Reproduced with permission from Taylor & Francis.)

approach is called design-based safety (as already explained), and in the latter case it is called "add-on safety." Design-based safety is known to be much more cost-effective than add-on safety, certainly in the long term. Hence, from an economic viewpoint, design-based safety should always be preferred.

4.2.3   Risk Treatment Option 2: Risk Acceptance

Risk retention and risk transfer, both sub-options of risk acceptance, only differ in the financial consequences of a risk turning into an accident. Risk retention concerns intentionally or unintentionally retaining the responsibility for a specified risk. Intentional retention is referred to as "risk retention with knowledge" and can be seen as a manner of self-insurance or auto-financing, finance planning of potential losses by an organization's own resources. Unintentional retention is referred to as "risk retention without knowledge," which is caused by inadequate hazard identification and insufficient risk assessment and management.

Usually, self-insurance will only be an option for large organizations able to set up and administrate an insurance fund. It is important to understand that only those losses that are predictable through statistical calculations, thus mainly type I risks, should be retained within an organization. Companies should avoid those risks (of mainly type II), possibly resulting in a major catastrophe or bankruptcy. The advantage of self-insurance is that insurance premiums should be lower, as there are no intermediaries expecting payments for costs, profits, or commissions. In such cases, the insured party also receives the benefit of investment income from the insurance fund that has been established, and keeps any profit accrued from it. Hence, there is an incentive for an organization to practice good risk management principles. However, there is also a downside to auto-financing: a self-insured organization will usually establish its insurance fund based on a statistical normal distribution of losses, being calculated over an extended time period. Hence, as company losses for some insurable type II risks may occur within 20 years but also the next day, there may be insufficient funds accrued to cope with certain losses, especially during the early years of the insurance fund.

Risk transfer involves shifting the responsibility for a specified risk to a third party. Insurance represents an important option for an organization to deal with risks. Risks can – in total or in part – be transferred to external agents, and hence the organization may carry out its business with the reassurance that if an accident occurs that is insured, the insurance company will provide an indemnity.

As risk transfer provides little incentive for a company to significantly reduce risk levels, there used to be a time when "risk management" was considered by company management as "insurance management." This period is rightfully a long time ago. Companies nowadays realize that not all risks are insurable, and that for those risks that are insured, there are often large uninsured consequences, besides those that are insured and that will be compensated. Risks that are typically covered by insurance include employer's liability, public liability, product liability, fire, and transport. Uninsured risks typically include equipment repairs, employee sick pay, loss of production, staff replacement, damage to corporate image, and accident investigations. A more elaborated list of incident and accident costs is given in Section 4.3 and also further in Chapter 5 on cost-benefit analysis.

4.2.4 Risk Treatment

Risk treatment is the selection and implementation of appropriate options for dealing with risk. It includes choosing among risk avoidance, risk reduction, risk transfer, and/or risk retention, or a combination thereof. Often, there will be residual risk which cannot be removed totally as it is not cost-effective to do so. Risk treatment thus involves identifying the range of options for treating risks, assessing these options financially and preparing and implementing treatment plans. The risk management treatment measures may be summarized as:

- *avoid the risk* – decide not to proceed with the activity likely to generate risk;
- *reduce the likelihood* of harmful consequences occurring;
- *reduce the consequences* occurring;
- *transfer the risk* – cause another party to share or bear the risk;
- *retain the risk* – accept the risk and be prepared if an accident occurs.

Risk treatment should be looked at from an economic viewpoint. Financial aspects can play a pivotal role in advocacy and decision-making on risk treatment by demonstrating the financial and economic value of the different treatment options.

Some new concepts need to be defined to be able to develop an approach to decide between different risk treatment options, i.e., the concepts of variability, normalized variability, and hypothetical benefit.

Variability can be defined as the maximum variation between the possible costs of every risk treatment option. Variability can thus be calculated as the difference between the maximum cost and the minimum cost of risk treatment, and doing this for every risk treatment option. As the possible costs per risk treatment option really are different depending on the scenario, variability does not depend on the measurement process used to estimate the costs.

The following formula can be used to determine the variability:

Variability = (maximum cost of treatment option) − (minimum cost of treatment option)

The normalized variability can then be determined by:

Normalized variability = (variability)/(maximum cost of treatment option)

The hypothetical benefit of the risk treatment option can be defined in two ways:

Definition (i) – as the difference between the highest possible costs of an accident in the current situation and those of an accident after applying the treatment measure; hence:

Maxmax hypothetical benefit = (maximum possible accident cost without any treatment)

− (maximum possible accident cost after the risk treatment)

Definition (ii) – as the difference between the costs of retention when doing nothing (taking no action) and those of the possible accident after applying the treatment measure; hence:

Expected hypothetical benefit = (expected cost of retention) − (expected cost

of accident after the risk treatment)

Based on the risk attitude of the decision-maker, one of both hypothetical benefit definitions may be preferred and chosen.

Notice that uncertainty is not the same as variability (as previously explained in this book). Uncertainty is the result of not having sufficient information. It can be considered the result of not being able to measure with sufficient precision, and hence, simply put, it is the result of imprecise methods of measurement [74]. As all measurement methods are imprecise, at least to some degree, all information is accompanied by a high or a low amount of uncertainty. This is the basis of the notion of "risk" and leads to the knowledge that every aspect of decision-making in life is accompanied by risk (see also Chapter 1). It also leads to the distinction between the two types of risk: risks with a lot of information available and hence a low amount of uncertainty (so-called type I risks), and risks with a paucity of information and thus a high amount of uncertainty (so-called type II risks). Risk treatment calculations such as those suggested and illustrated in this section should only be carried out if adequate financial information is available. This relates to type I risks and the more common type II risks.

Let us explain the risk treatment approach by using an illustrative example and exercise for type I risks.

Imagine you are a newly appointed safety manager for a company and you are asked to advise on a risk treatment decision. You are told that the total safety budget for the treatment should not exceed €12 000. A choice has to be made for treating a possible accident happening to one of the storage tanks. Two scenarios have been identified: scenario A and scenario B. In scenario A, the cost to the company could be €250 000, and in scenario B this cost could be €20 000. The probability of the A outcome is 1 in 1000, while the probability of the B scenario is estimated at 1 in 100. The two scenarios can occur separately, but they could also happen at the same time. The possible preventive and protective measures for a budget of €10 000 would lead to a 10-fold outcome reduction, as well as a 10-fold likelihood reduction of both scenarios. Insuring against the accident scenarios, on the other hand, would cost €12 000 with a franchise of €3000. It is also possible to first use the preventive and protective measures at a cost of €6000, leading to a five-fold outcome reduction and a five-fold likelihood reduction of both scenarios, and afterwards to insure against the remaining risk which would cost €7500 with a franchise of €1000.

What risk treatment measure would you recommend, if you know that the company sets the maximum normalized variability level at 75%?

To solve this problem, we first summarize the accident scenarios:

$$\text{Scenario A} - \text{consequences}_A = €250\,000; \text{probability}_A = 0.001$$
$$\text{Scenario B} - \text{consequences}_B = €20\,000; \text{probability}_B = 0.01.$$

Next, we calculate all the costs, the variability and the normalized variability, and the hypothetical benefits for every treatment option:

1. *Risk treatment 1 = risk retention (= do nothing):*
   (a) Total expected accident cost $= 250\,000 \times 0.001 + 20\,000 \times 0.01 = €450$
   (b) Variability $= \{€250\,000\,[\text{accident A (= max cost) happens)}]\} - \{€0\,[\text{nothing happens (= min cost)}]\} = €250\,000$

(c) Normalized variability = €250 000/€250 000 = 1 = 100%

(d) Maxmax hypothetical benefit of risk treatment measure according to definition (i) = €250 000 − €250 000 = €0

(e) Expected hypothetical benefit of risk treatment measure according to definition (ii) = €450 − €450 = €0.

2. *Risk treatment 2 = prevention and protection*:

(a) Expected possible cost of accident after prevention = 25 000 × 0.0001 + 2000 × 0.001 = €4.50

(b) Cost of risk treatment measure "prevention and protection" = €10 000

(c) Total expected cost = €10 004.50

(d) Variability = [€25 000 + €10 000 (worst-case accident happens, despite the prevention measures taken)] − [€10 000 (nothing happens)] = €25 000

(e) Normalized variability = €25 000/€35 000 = 0.714 = 71.4%

(f) Maxmax hypothetical benefit of risk treatment measure according to definition (i) = €250 000 − €25 000 = €225 000

(g) Expected hypothetical benefit of risk treatment measure according to definition (ii) = €450 − €4.5 = €445.50.

3. *Risk treatment 3 = risk transfer (= insurance)*:

(a) Expected possible cost of accident after insurance = 3000 × 0.001 + 3000 × 0.01 = €33

(b) Cost of risk treatment measure "insurance" = €12 000

(c) Total expected cost = €12 033

(d) Variability = [€12 000 + €3000 (worst-case accident happens, but the company is insured against it)] − [€12 000 (insurance premium)] = €3000

(e) Normalized variability = €3000/€15 000 = 0.2 = 20%

(f) Maxmax hypothetical benefit of risk treatment measure according to definition (i) = €250 000 − €3000 = €247 000

(g) Expected hypothetical benefit of risk treatment measure according to definition (ii) = €450 − €33 = €417.

4. *Risk treatment 4 = prevention and protection + insurance*:

(a) Expected possible cost of accident after prevention + insurance = 1000 × 0.0002 + 1000 × 0.002 = €2.2

(b) Cost of risk treatment measure "prevention" = €6000

(c) Cost of risk treatment measure "insurance after prevention" = €7500

(d) Total expected cost = €13 502.20

(e) Variability = [€13 500 + €1000 (worst-case accident happens, despite the prevention measures taken, but the company is insured against it)] − [€13 500 (nothing happens)] = €1000

(f) Normalized variability = €1000/€14 500 = 0.0689 = 6.89%

(g) Maxmax hypothetical benefit of risk treatment measure according to definition (i) = €250 000 − €1000 = €249 000

(h) Expected hypothetical benefit of risk treatment measure according to definition (ii) = €450 − €2.2 = €447.8.

A figure for better visualization in case definition (i) for the Maxmax hypothetical benefits is used can then be drafted based on these calculations, and a risk treatment option recommendation can be provided to company management. Figure 4.5 illustrates the results.

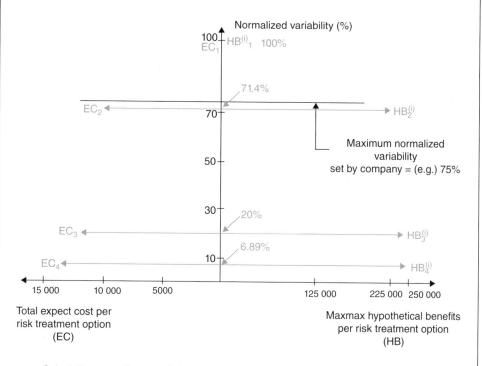

Calculation according to definition (i):

Risk treatment option 1: Maxmax $HB^{(i)}_1 - EC_1 = 0 - 450 = -450$
Risk treatment option 2: Maxmax $HB^{(i)}_2 - EC_2 = 225\ 000 - 10\ 004.5 = 214\ 995.5$
Risk treatment option 3: Maxmax $HB^{(i)}_3 - EC_3 = 247\ 000 - 12\ 033 = 234\ 967$
Risk treatment option 4: Maxmax $HB^{(i)}_4 - EC_4 = 249\ 000 - 13\ 502.2 = 235\ 497.8$

Figure 4.5 Risk treatment options visualized while employing definition (i) for the Maxmax hypothetical benefits.

Figure 4.5 indicates that only options 2–4 are characterized by a normalized variability under the 75% limit which was specified by company management. However, the budget needed for applying risk treatment option 4 (€13 500) surpasses the available budget (€12 000), and thus option 4 is not affordable. Further, risk treatment option 3 has the highest difference between the Maxmax hypothetical benefit and expected costs (€234 967). Hence, based on the information given and definition (i) used for calculating the Maxmax hypothetical benefits, risk treatment option 3, insurance, is recommended.

If definition (ii) is used for expected hypothetical benefits, it is not useful to balance the hypothetical benefits against the total expected costs, and therefore in this case, taking into consideration the conditions eliminating options 1 and 4 as described in the previous paragraph, risk treatment option 2 has the highest expected hypothetical benefits (€445.5) according to definition (ii), and thus option 2, prevention and protection, would be recommended.

In the case of definition (i), calculation of the Maxmax hypothetical benefits, emphasis is placed on the consequences of an event, irrespective of the probabilities. In the case of definition (ii), calculation of the expected hypothetical benefits, probabilities are considered. Therefore, definition (i) may be interesting for decision-makers putting more emphasis on consequences (or in case there is a lack of probability information, such as with some type II risks), while definition (ii) would probably be more interesting for decision-makers taking consequences as well as probabilities into consideration (and disposing over the probability information).

It should be noted by the reader that, besides the choice of the definition to determine the hypothetical benefits, which can be rather subjective, the risk treatment calculations approach explained in this section is highly rational and only financial considerations are used to make the calculations and recommendations. No moral aspects, such as equity, justification of the risk, and other ethical aspects, are taken into account in this approach. Therefore, if used, it should be applied with this knowledge in mind. Ethical aspects will be discussed later in this chapter.

4.2.5 The "Human Aspect" of Making a Choice between Risk Treatment Alternatives

Different factors play a role when taking decisions and making choices between various alternatives. The decision-making process becomes even more complicated in an organizational context. Managers need to make decisions concerning issues that affect not only their personal lives, but also the company they work for, its employees, its surroundings, and more. The majority of management problems of any enterprise, including safety investment portfolio management, is characterized by uncertainties and lack of necessary information. The difficulty in predicting future parameter values and future events has an impact on the evaluation of safety investment projects. Uncertainty, or the lack of information, thus obviously hampers managerial decisions. To decrease the level of uncertainty, managers may use historic data, trends, statistical information, qualitative expert judgments, past achievements, and all other information available to the company [37, 75].

Managerial behavior and decision-making sometimes have little affinity with real skills and available resources. The way people think and people behave is much more linked to situational perception and the ability to control processes and potential results. Managers may make different decisions in a similar situation as a result of a different interpretation and evaluation of

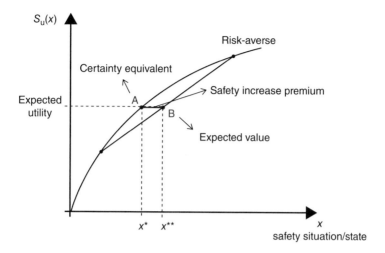

Figure 4.6 Safety increase premium determination.

the situation at hand. Decisions can be divided into two groups: opportunities and threats. Man-agers usually view controllable situations as opportunities and uncontrollable ones as threats. However, what is interpreted as "controllable" and as "uncontrollable" depends entirely on the person, and varies strongly from manager to manager. Therefore, it is possible that one man-ager sees a situation characterized by a certain level of uncertainty as an opportunity, while another manager sees the same situation as a threat [75–77].

4.2.5.1 Safety Increase Premium

Preferences toward safety can be described (as already indicated in Chapter 3), by a safety utility function and provide, for example, an idea of the safety state difference one is will-ing to suffer to leave the person indifferent between an uncertain (but higher) safety state (point B in Figure 4.6) and a certain (but lower) safety state (point A in Figure 4.6). "Uncer-tain" in this regard indicates that management feels less certain about the effectiveness of the higher safety situation than that of the lower safety situation, and asks, "Is it worth the trouble or money?"

The safety increase premium can then be calculated as the extra budget that is required to cover the difference between safety state x^{**} and x^*, because a certain safety situation x^* delivers the same expected utility as the uncertain safety situation x^{**} and its corresponding expected value. Hence, the safety premium can be regarded as the amount of money that safety management needs to invest to get into a higher safety situation (but one perceived as more uncertain with respect to its effectiveness). It is the minimum amount that risk-averse man-agement (cf. the shape of the safety utility function) would need to spend on safety to leave it indifferent between the uncertain and the certain safety situations.

4.3   Safety Investment Decision-making – a Question of Costs and Benefits

4.3.1   Costs and Hypothetical Benefits

Accidents do bear a cost – and often not a small one. An accident can be linked to a variety of direct and indirect costs. Table 4.1 gives a non-exhaustive list of potential costs that might accompany accidents.

Hence, by implementing a sound safety policy and by adequately applying operational risk management, substantial costs can be avoided, namely all costs related to accidents that have never occurred (i.e., the non-events), the so-called "hypothetical benefits." However, in current practice, companies place little or no importance on the "hypothetical benefit" concept due to its complexity. In Section 4.2.4 an attempt has been made to make the concept more tangible. In Section 8.12 the concept is explored even more in depth, as hypothetical benefits are equated and modeled as a maximum investment willingness.

Non-quantifiable costs are highly dependent on non-generic data such as individuals' characteristics, a company's culture and/or the company as a whole and even socio-economic circumstances. Rather, the costs assert themselves when the actual costs supersede the quantifiable costs. In economics (e.g., in environment-related predicaments) monetary evaluation techniques are often used to specify non-quantifiable costs, among them the contingent valuation method and the conjoint analysis or hedonic methods. In the case of non-quantifiable accident costs, various studies have demonstrated that these form a multiple of the quantifiable costs [73].

The quantifiable socioeconomic accident costs (see Table 4.1) can be divided into direct and indirect costs. Direct costs are visible and obvious, while indirect costs are hidden and not immediately evident. In a situation where no accidents occur, the direct costs result in direct hypothetical benefits, whilst the indirect costs result in indirect hypothetical benefits. The resulting indirect hypothetical benefits comprise, for example, not having sick leave or absence from work, not having staff reductions, not experiencing labor inefficiency and not experiencing change in the working environment. Figure 4.7 illustrates this reasoning.

Although hypothetical benefits seem to be rather theoretical and conceptual, they are nonetheless important if you are to fully understand safety economics. Hypothetical benefits resulting from non-occurring accidents or "non-events" can be divided into different categories at the organizational level, depending on the preferences and characteristics of the organization and its safety management. Meyer and Reniers [73], for example, mention five classes of hypothetical benefits. The first category concerns the non-loss of work time and includes the non-payment of the employee who at the time of the accident adds no further value to the company (if payment were to continue, this would be a pure cost). The non-loss of short-term assets forms the second category and can include, for example, the non-loss of raw materials. The third category involves long-term assets, such as the non-loss of machines. Various short-term benefits, such as non-transportation costs and non-fines, constitute the fourth category. The fifth category consists of non-loss of income, non-signature of contracts or non-price reductions. In Chapter 5, hypothetical benefits are elaborated in greater depth. Independent of the number of classes or categories used, the visible or direct hypothetical benefits generated by the avoidance of costs resulting from non-occurring accidents only make up a small portion of the factors responsible for the total hypothetical benefits resulting from non-occurring accidents.

Next to tangible benefits, intangible benefits might exist. Intangible benefits may manifest themselves in different ways, e.g., non-deterioration of image, avoidance of lawsuits, and the

Table 4.1   Non-exhaustive list of quantifiable and non-quantifiable socioeconomic consequences of accidents

Interested parties	Non-quantifiable consequences of accidents	Quantifiable consequences of accidents
Victim(s)	Pain and suffering Moral and psychic suffering Loss of physical functioning Loss of quality of life Health and domestic problems Reduced desire to work Anxiety Stress	Loss of salary and bonuses Limitation of professional skills Time loss (medical treatment) Financial loss Extra costs
Colleagues	Bad feeling Anxiety or panic attacks Reduced desire to work Anxiety Stress	Time loss Potential loss of bonuses Heavier workload Training and guidance of temporary employees
Organization	Deterioration of social climate Poor image, bad reputation	Internal investigation Transport costs Medical costs Lost time (informing authorities, insurance company, etc.) Damage to property and material Reduction in productivity Reduction in quality Personnel replacement New training for staff Technical interference Organizational costs Higher production costs Higher insurance premiums Administrative costs Sanctions imposed by parent company Sanctions imposed by the government Modernization costs (ventilation, lighting, etc.) after inspection New accident indirectly caused by accident (due to personnel being tired, inattentive, etc.) Loss of certification Loss of customers or suppliers as a direct consequence of the accident Variety of administrative costs Loss of bonuses Loss of interest on lost cash/profits Loss of shareholder value

Meyer and Reniers [73]. Reproduced with permission from De Gruyter.

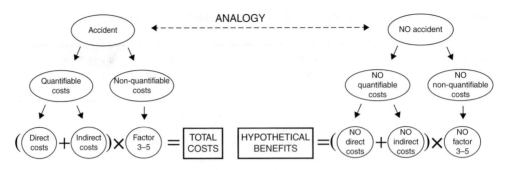

Figure 4.7   Analogy between total accident costs and hypothetical benefits.   (Source: Meyer and Reniers [73]. Reproduced with permission from De Gruyter.)

fact that an employee, thanks to the safety policy and the non-occurrence of accidents, does not leave the organization. If an employee does not leave the organization, the most significant benefit arises from work hours that other employees do not have to make up. Management time is consequently not given over to interviews and routines surrounding the termination of a contract. The costs of recruiting new employees can, for example, prove considerable if the company has to replace employees with experience and company-specific skills. Research into the reasons that form the basis of recruitment of new staff reveals a difference between, on the one hand, the recruitment of new employees due to the expansion of a company and, on the other hand, the replacement of staff who have resigned due to poor safety policies.

4.3.2   Prevention Benefits

An accident that does not occur in an enterprise, thanks to the existence of an efficient safety and prevention policy within the company, was referred to as a non-occurring accident or a non-event. Non-occurring accidents (in other words, the prevention of accidents) result in the avoidance of a number of costs and thus create hypothetical benefits, as explained previously. An estimation of the amount of input and output of the implementation of a safety policy is thus clearly anything but simple. It is impossible to specify the costs and benefits of one exceptional measure. The effectiveness (and the costs and benefits) of a safety policy must be regarded as a whole.

Hence, if an organization is interested in knowing the efficiency and the effectiveness of its safety policy and its prevention investments, in addition to identifying all kinds of prevention costs, it is also worth calculating the hypothetical benefits that result from non-occurring accidents. By taking all prevention costs and all hypothetical benefits into account, the true prevention benefits can be determined.

4.3.3   Prevention Costs

In order to obtain an overview of the various kinds of prevention costs, it is appropriate to distinguish between fixed and variable prevention costs, on the one hand, and direct and indirect prevention costs on the other.

Fixed prevention costs remain constant irrespective of changes in a company's activities. One example of a fixed cost is the purchase of a fireproof door. This is a one-off purchase and the related costs are not subject to variation in accordance with production. Variable costs, in contrast to fixed costs, vary proportionally in accordance with a company's activities. The purchase of safety gloves can be regarded as a variable cost due to the fact that the gloves have to be replaced sooner when used more frequently or more intensively due to increased productivity levels.

Variable prevention costs have a direct link with production levels. Fixed prevention costs, by contrast, are not directly linked to production levels. A safety report, for example, will state where hazardous materials must be stored, but this will not have a direct effect on production. The development of company safety policy includes not only direct prevention costs, such as the application and implementation of safety material, but also indirect prevention costs such as development and training of employees and maintenance of the company safety management system.

A non-exhaustive list of prevention costs is as follows:

- staffing costs of company HSE department;
- staffing costs for the rest of the personnel (time needed to implement safety measures, time required to read working procedures and safety procedures, etc.);
- procurement and maintenance costs of safety equipment (e.g., fire hoses, fire extinguishers, emergency lighting, cardiac defibrillators, pharmacy equipment);
- costs related to training and education with respect to working safely;
- costs related to preventive audits and inspections;
- costs related to exercises, drills, simulations with respect to safety (e.g., evacuation exercises, etc.);
- a variety of administrative costs;
- prevention-related costs for early replacements of installation parts, and so on;
- maintenance of machine park, tools, and so on;
- good housekeeping;
- investigation of near-misses and incidents.

In contrast to quantifying hypothetical benefits, companies are usually very experienced in calculating direct and indirect (non-hypothetical) costs of preventive measures.

4.4   The Degree of Safety and the Minimum Overall Cost Point

As Fuller and Vassie [70] indicate, the overall (or total) cost of a company safety policy will be directly related to an organization's standards for health and safety, and thus its degree of safety. The higher the company's health and safety standards and its degree of safety, the greater the prevention costs, and the lower the accident costs within the company. Prevention costs will rise exponentially as the degree of safety increases, because of the law of diminishing returns, which describes the difficulties of trying to achieve the last small improvements in performance. If a company has already made huge prevention investments, and the company is thus performing very well on health and safety, it becomes ever more difficult to further improve its health and safety performance with one extra "unit" (in economics also called "marginal" improvement). On the contrary, the accident costs will decrease exponentially as

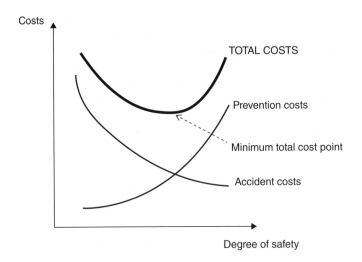

Figure 4.8 Prevention costs and accident costs as a function of the degree of safety (qualitative figure).

the degree of safety improves because, as the accident rate is reduced, and hence accident costs are decreased, there is ever less potential for further (accident reduction) improvements (see Figure 4.8).

From a purely microeconomic viewpoint, a cost-effective company will choose to establish a degree of safety that allows it to operate at the minimum total cost point. Figure 4.8 illustrates this reasoning. At this point, the prevention costs are balanced by the accident costs, and at first sight the "optimal safety policy" (from a microeconomic point of view) is realized. It is important that for this exercise to deliver results that are as accurate as possible, the calculation of both prevention costs and accident costs should be as complete as possible, and all direct and indirect, visible and invisible, costs should be taken into account. However, Figure 4.8 indicates that, in an optimal cost-effective situation, one should not strive for a zero-accident situation. But is this really always the case? As will be explained further, this is possibly not always the case.

As it happens, an observation is that companies who have aspired to zero accidents globally over very long periods are still in business, and indeed even highly profitable. Their safety and profitability records are usually better than those with fatalistic beliefs that some accidents are inevitable and that they should/can be considered part of the cost-effectiveness policy of the company. Figure 4.8 was actually first produced in the quality management context and originally looked like Figure 4.9. Despite the reasoning that an optimum overall cost point is not at zero accident level, many manufacturers that did aim for zero quality defects (no matter the theory) largely achieved their aims, despite what is shown in Figure 4.9.

Panopoulos and Booth [78] sought to challenge the conclusions drawn from Figure 4.9, namely that zero accidents could "never" be the optimum outcome. To this end, they assumed that the starting point of prevention costs for a safety minimum overall cost graph needs to be zero accidents, achieved at finite cost, not infinite cost (as this must also be true in real industrial practice). From that initiating point, the trends in accident and prevention costs as accident numbers increase may be explored. It should be noted that the novel presentation method is

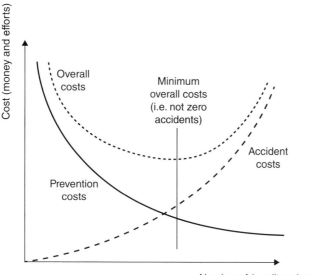

Figure 4.9 Conventional cost-benefit analysis for determining the tolerable accident level in quality management.

not profound. Its key purpose is to challenge the existing ideas regarding the minimum overall cost curve (cf. Booth [79]).

Figures 4.10 and 4.11 present graphical minimum overall cost curves, assuming finite values of prevention costs for zero accidents.

Figure 4.11 shows the case where zero accidents offer the cheapest option. Everything turns on the relationship between accident and preventive costs as the number of accidents goes up, and where the minimum overall cost lies.

What, then, are the factors that determine whether Figure 4.10 or 4.11 applies? Figure 4.10 is identical to Figure 4.9 but drawn in a different way, as in the latter the cost of each accident is relatively modest, and prevention costs rise dramatically as zero is approached. This might typically be where safety expenditure is not cost-effective. In contrast, Figure 4.11 shows zero accidents as the optimum case. Here each accident is more expensive than shown in Figure 4.10, and the costs of moving from a small number of accidents to zero are relatively modest. A question that arises is: what are the practical circumstances where zero accidents is the minimum cost option?

As Booth [79] explains, the arguments for zero quality defects provides the clue, but also the challenge. Frequent, albeit minor, quality defects threaten sales (and product safety, where the effects can be catastrophic). Moreover, quality management was revolutionized when the change was made from rectifying defects "at the end of the production line" to the more cost-effective approach of quality checks being incorporated into every task – so achieving zero defects at the end of the line via a dramatically better preventive approach. Returning to accidents, zero accidents are more likely to offer the best business case where the costs of an accident are high, as with type II events. Hence, Figure 4.11 provides a good argument for zero type II events or zero major accidents, while the pursuit of zero type I events or zero

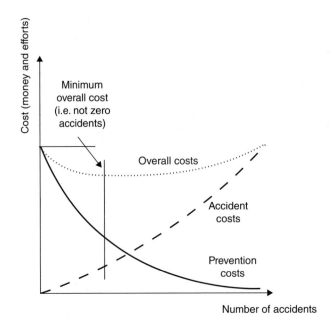

Figure 4.10 Conventional cost-benefit analysis (Figure 4.9 redrawn) showing zero accidents as not the optimum case.

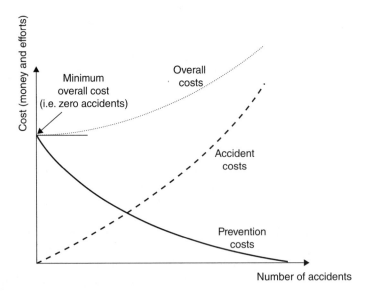

Figure 4.11 Conventional cost-benefit analysis showing zero accidents as the optimum case.

occupational accidents should be treated like Figure 4.10. In summary, using a cost-benefit analysis to determine a minimum overall cost is only suitable for one type of risk (type I risks) and not for other risks (type II). But the cost-benefit exercise results will still be largely improvable, and the next paragraph and section explain why and how.

Earlier in this section, the phrase "from a microeconomic point of view" was mentioned, due to the fact that victim costs and societal costs should, in principle, also be taken into account in the overall/total costs calculations. If only microeconomic factors are important for an organization, only accident costs related to the organization are considered by the company, and not the victim and societal costs. Hence, even if all direct and indirect accident costs are determined by an organization, there will be an underestimation of the full economic costs, due to individual and macroeconomic costs. True accident costs are composed of the organizational costs, as well as the costs to the victim and to society.

Organizations should thus make a distinction between the different types of risk, and between real accident costs and hypothetical benefits. Costs of accidents that happened ("accident costs") and costs of accidents that were avoided and never happened ("hypothetical benefits" due to non-events) are completely different in nature: many more non-events happen than accidents. Nonetheless, their analogy is clear and therefore they are easily confused when making safety cost-benefit considerations. Curves such as those in Figure 4.8 are only valid for type I risks, as the curve of "accident costs" (displaying the costs related to a certain degree of safety) can only be empirically determined if a large number of accidents happened, and thus if sufficient data are available. It is impossible to determine empirically the costs of disasters linked to a certain degree of safety, simply because not that many catastrophes happen, which would be necessary to obtain sufficient information to draw such a curve. Hence, the curves displayed in Figure 4.8 cannot be drawn – and should not be used – for type II risks. At best, a thought experiment can be conducted, with a resulting Figure 4.11 being applicable to type II events.

4.5   The Type I and Type II Accident Pyramids

Heinrich [80], Bird [81], and James and Fullman [82], among other researchers, mentioned the existence of a ratio relationship between the numbers of incidents with no visible injury or damage, and those with property damage, those with minor injuries, and those with major injuries. This accident ratio relationship is known as "the accident pyramid" or "the safety triangle." Figure 4.12 illustrates such an accident pyramid (or safety triangle). Accident pyramids unambiguously indicate that accidents with higher consequences are "announced" by accidents with lesser consequences. Hence the importance of, among other things, awareness and incident analyses for safety managers.

Studies [80–82] found different ratios (varying from 1 : 300 to 1 : 600), depending on the industrial sector, the area of research, cultural aspects, and so on. Although the statistical relationship between the different levels of the pyramid has never been proven, the existence of the accident pyramid obviously has merits from a qualitative point of view. From a qualitative interpretation of the pyramid, it could be possible to prevent serious accidents by taking preventive measures aimed at near misses, minor accidents, and so on. Hence, these "classic" accident pyramids clearly provide an insight into type I accidents, where a lot of data are

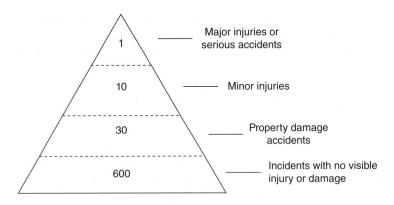

Figure 4.12   The Bird accident pyramid ("Egyptian").

at hand. In brief, the assumptions of the "old" safety paradigm emanating from the accident pyramid (see Figure 4.12) hold that:

1. As injuries increase in severity, their number decreases in frequency.
2. All injuries of low severity have the same potential for serious injury.
3. Injuries of differing severity have the same underlying causes.
4. One injury reduction strategy will reach all kinds of injuries equally (i.e., reducing minor injuries by 20% will also reduce major injuries by 20%).

Using injury statistics, Krause [83] indicated that while minor injuries may decline in companies, serious injuries can remain the same, hence casting credible doubts on the validity of the safety pyramid concept. In fact, research indicated that only some 20% (21% to be precise) of the type I incidents have the potential to lead to a serious type I accident. This finding implies that if one focuses only on "the other 80%" of the type I incidents (i.e., the 80% of the incidents that are unlikely to lead to serious injury), the causative factors that create potential serious accidents will continue to exist and so will the serious accidents themselves.

Therefore, Krause [83] proposed a "new" safety paradigm with the following assumptions:

1. All minor injuries are not the same in their potential for serious injury or fatality. A subset of low severity injuries come from exposures that act as a precursor to serious accidents.
2. Injuries of differing severity have differing underlying causes.
3. Reducing serious injuries requires a different strategy than reducing less serious injuries.
4. The strategy for reducing serious injuries should use precursor data derived from accidents, injuries, near misses, and exposure.

Based on research by Krause [83] on the different types of uncertainties/risks available, and on several disasters (such as the BP Texas City Refinery disaster of 2005), the classic pyramid shape thus needed to be refined and improved. Instead of an "Egyptian" pyramid shape, such as the classic accident pyramid studies suggest, the shape should rather be a "Mayan" one, as illustrated in Figure 4.13.

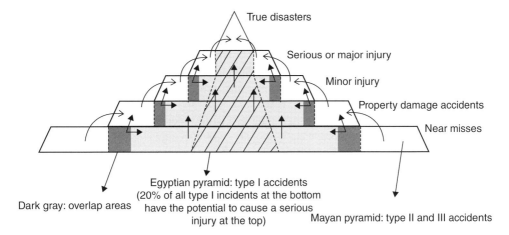

Figure 4.13 Mayan accident pyramid shape. (Source: Meyer and Reniers [73]. Reproduced with permission from De Gruyter.)

First, only 20% of the near misses at the bottom of the pyramid have the potential to become serious accidents. This is indicated by the hatched part of the light gray zone in Figure 4.13. Second, the shape of the pyramid should be Mayan instead of Egyptian. The Mayan pyramid shape shows that there is a difference between type I and type II risks; in other words, "regular accidents" or occupational accidents (and the incidents going hand-in-hand with them) should not be confused with "major accidents" or catastrophes. Not all near misses have the potential to lead to disaster, but only a minority of unwanted events may actually end up in a catastrophe. Obviously, to prevent disasters and catastrophes, risk management should be aimed at both types of risk, and certainly not only at the large majority of "regular" risks. Hopkins [84] illustrates this by using a bi-pyramid model, consisting of two pyramids partially overlapping. One pyramid represents type I risks, leading at most to a serious accident (e.g., a lethality), but not to a true catastrophe, and the other pyramid represents type II risks, with the potential to lead to a disaster. The overlap is present (and also in the Mayan pyramid of Figure 4.13 represented by the dark gray areas) because in some cases, unwanted events may be considered as warnings or incidents for both pyramids. Hopkins [84] indicates that the airline industry was the pioneer for this kind of type I and type II risk thinking. In that industry, taking all necessary precautionary measures to ensure flight safety is regarded as fundamentally different from taking preventative measures to guarantee employee safety and health. Two databases are maintained by airline companies: one is used to keep data of near-miss incidents affecting flight safety and the other stores information regarding workforce health and safety. Hence, in this particular industry, it is clear that workforce injury statistics say nothing about the risk of an aircraft crash. This line of type I and type II risk thinking can (and should) be implemented in every industrial sector.

4.6 Quick Calculation of Type I Accident Costs

The Bird pyramid can be used to make a rough estimate of the total yearly costs of occupational accidents within an organization, based on the number of "serious" accidents, where serious accidents are defined as accidents where the victim has at least 1 day off work [so-called lost time injury (LTI) accidents]. If, as the Bird pyramid suggests, different sorts of type I accidents are discerned, different metrics for different kind of accidents are used.

4.6.1 Accident Metrics

For accidents defined as occurrences that result in a fatality, permanent disability or time lost from work of 1 day/shift or more, usually the LTI frequency rate (LTIFR) is used and these accidents are referred to as "LTI accidents." The LTIFR is the number of lost time injuries per million hours worked, calculated using the following equation:

LTFIR = Number of LTIs over the accounting period × 1 000 000/total number

of hours worked in accounting period

Hence, the LTIFR is how many LTIs occurred over a specified period per 1 000 000 hours (or some other number; 100 000 is also often used) worked in that period. Mostly, the accounting period is chosen to be 1 year. By counting the number of hours worked, rather than the number of employees, for example, discrepancies in the incidence rate calculation as a result of part-time workers and overtime are avoided. However, this metric using the employees instead of the number of hours worked [called the lost time injury incidence rate (LTIIR)] is also used in many organizations. To calculate the LTIIR, which is the number of LTIs per 100 employees, the following equation is used:

LTIIR = Number of LTIs over the accounting period × 100/average number

of employees in accounting period

Next, the severity rate, which takes into account the severity per accident, is also often used. Depending on how this is expressed, at least the information from above and the number of work days lost over the year are needed. Often the severity rate is expressed as an average by simply dividing the number of days lost by the number of LTIs. Another way of calculating the LTI severity rate (LTISR) is to use the following equation (the figure 1 000 000 may be replaced by any other figure; it just tells us that the LTISR in this case is expressed per million hours worked):

LTISR = Number of work days lost over the accounting period × 1 000 000/total number

of hours worked in accounting period

Also, the medical treatment injury (MTI) frequency rate is often measured. This frequency rate measures how often MTIs are occurring. It is expressed as the number of MTIs per million hours worked:

MTIFR = Number of MTIs over the accounting period

× 1 000 000/total number of hours worked in accounting period

Finally, the total recordable injury frequency rate (TRIFR) measures the frequency of recordable injuries, i.e. the total number of fatalities, LTIs, MTIs, and restricted work injuries

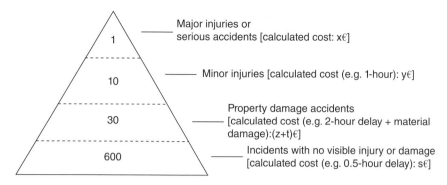

Figure 4.14 The Bird accident pyramid with costs per type of accident. (Source: Meyer and Reniers [73]. Reproduced with permission from De Gruyter.)

occurring. It is expressed as the total number of fatalities, LTIs, MTIs, and restricted work injuries per million hours worked:

TRIFR = Number of recordable injuries (fatalities + LTIs + MTIs

+ restricted work injuries) over the accounting period

× 1 000 000/total number of hours worked in accounting period

4.6.2   A Quick Cost-estimation Approach for Type I Risks

Companies usually have a good understanding of the cost per sort of type I accident. For example, the average cost of an LTI accident is calculated on a yearly basis by organizations, and equals €x. Similarly, the average costs of the different sorts of accidents can be determined. Figure 4.14 shows some theoretical ways of calculating the different costs.

Based on these average costs, a rough estimate of the total yearly cost due to type I accidents can be made. Every company usually knows very well the number of serious accidents that happen annually, but the numbers of other type I accidents with lesser consequences are often unknown. Nonetheless, a rough estimation of all accident costs can be made using the type I (e.g., Bird) pyramid principle. In principle, an organization should draft its own pyramid, and use these ratios. Assume that the number of serious accidents is N, and hence this rough calculation can be done for any company. Table 4.2 displays how to calculate the total yearly accident costs.

Table 4.2 shows that, based on the ratio between "serious" type I accidents and other sorts of accidents, a rough estimate can be made of the total average accident costs for a given year if the number of incidents and accidents in that year is known, and if the average cost per sort of accident is known. Table 4.2 doesn't take into consideration the fact that only around one-fifth of all accidents lead to serious accidents (see Section 4.5), as this knowledge has no impact on the overall costs of all incidents happening in the organization. It does, however, have an impact on the amount of costs avoided purely related to serious accidents, and this should be taken into account when calculating those avoided costs.

This method can be used by a company to estimate, for example, the accident cost line displayed in Figure 4.8, when calculating the costs for different levels of safety. Clearly, the definition of a "level of safety" would have to be agreed upon within the company, and estimations may then be carried out based on this definition.

To illustrate the quick calculation method for type I events, consider the following example.

Assume that an organization has determined that the following ratios are valid for the company: $1 : 5 : 60 : 450$. Further assume that a serious accident with at least 1 day off work costs, on average, €40 000. Furthermore, a minor injury costs the company, on average, €2000 (due to work delay etc.), whereas property damage has an average cost of €1000 (due to lost time and delay) plus €7000 (pure material losses), and incidents represent an average delay cost of €50. What is the total yearly accident cost if there are five serious accidents per year. If the organization were able to bring the serious accident figure down from five to two in the following year, what would be the amount of money saved annually due to non-events of type I in this company?

To solve this problem, the formula given in Table 4.2 is used, but adapted for the ratios of the company, to determine the overall cost for N serious accidents: $40\,000 + 5 \times 2000 + 60 \times (1000 + 7000) + 450 \times 50 = €552\,500$. Thus the total yearly accident cost in the organization amounts to €552 500 × 5 = €2 762 500. Furthermore, avoiding three serious accidents (going from five to two) the next year would benefit the company by some $3 \times (40\,000 + 5/5 \times 2000 + 60/5 \times 8000 + 450/5 \times 50) = €427\,500$. More importantly, this calculation shows how hypothetical benefits may represent considerable amounts of money for a company. If a company only has a small number of serious accidents, this can be seen as making high (hypothetical) profits as many non-events happen, leading to high non-costs.

4.7   Quick Calculation of Type II Accident Costs

4.7.1   Introduction to a Study on Type II Event Decision-making

It is very difficult, if not impossible, to elaborate on a "quick calculation" tool for estimating type II accident costs. In fact, it could be very dangerous as well. It is possible that company

Table 4.2   Quick calculation of the total yearly costs based on the number of "serious" type I accidents

Sort of incident/accident	Bird pyramid	Number of incidents/accidents	Cost per sort of incident/accident	Cost
Serious	1	N	x	Nx
Minor injury	10	$10N$	y	$10Ny$
Property damage	30	$30N$	$z + t$	$30N(z + t)$
Incident	600	$600N$	s	$600Ns$
			Total cost $= N[x + 10y + 30(z + t) + 600s]$	

managers might reduce the decision-making about type II accidents to a simple mathematical formula, and such a practice would be irresponsible, unacceptable, and wrong. If there is one thing that history teaches, it is that one should be extremely careful with safety investment decision-making where extremely low-frequency events are involved. Hence, instead of working out a method for the quick calculation of type II accidents, we present the results of a study by Van Nunen *et al.* [85] on decision-making in the case of type II events. It is more of a "psychological" problem than a rational one: the hypothetical benefits achieved in the case of averting a true type II accident are almost always higher than the safety investment costs (SICs) of successful accident avoidance.

The subject of the study concerned the investigation of the influence of probability change in averting a certain loss due to a major (type II) accident. The question that the respondents needed to answer was basically whether they preferred an investment in safety (prevention measures – representing an uncertain gain) or an investment in production (representing a certain gain), for varying levels of potential losses accompanied by varying probabilities of avoiding these losses. The research thus aimed to discover which parameters of consequence and probability would persuade people that investing in production was worth the major accident risk, and which parameters would make the risk or uncertainty unacceptable, leading them to invest in major accident prevention. Knowledge as to how these decisions are made by people in general is obviously important for the understanding and management of activities involving potential type II events [86].

Concerning the decision-making under risk, as already mentioned, accidents with major consequences are included, as well as a variety of probabilities of occurrence. If both the probability of occurrence and the disaster potential are perceived as high, one would normally reject such risks [87]. The acceptance of accidents with high disaster potential and a low probability of occurrence, so-called HILP accidents (i.e., type II accidents), is far less straightforward [46, 51]. It could be argued that the occurrence of large-scale accidents is unacceptable regardless of their probability. On the other hand, it can be argued that safety has a cost that makes it impossible for organizations to spend an unlimited budget reducing or eliminating all (major) accident scenarios. Therefore, a certain level of risk and uncertainty has to be accepted [88].

In addition to the decision-making under risk (the outcomes as well as the probabilities of each outcome are known), decision-making under uncertainty can also be considered. In an uncertain situation, the probability of an accident occurring is completely unknown and/or the amount of the loss as a result of the accident is also unknown. When these uncertainties remain, subjective judgments are inevitable [87], which makes it difficult to predict the decision-making.

It should be noted that the general results of the choices between the production investment and the prevention investment were compared with the cumulative expected values of these investments. As mentioned at the beginning of this chapter, the theory of expected value can provide a reference for discussing how the decisions are taken [86].

The problem setting is as follows. If a respondent has chosen a production investment, the company will make, during the next 5 years, an extra yearly production profit of €500 000. So, after a 5-year period, the production investment will have resulted in an extra profit of €2 500 000 (interests aside).

The cumulative expected value of the prevention investment depends on the probability that an accident will occur and on the associated amount of losses. The calculation of the cumulative expected value is given for the following example: assume a prevention investment that avoids a major accident with probability of 1 in 100 000 and losses worth €20 000 000 per year during the next 5 years. A probability of 1 in 100 000 corresponds to a probability of 0.001%. This brings the cumulative expected value to $(0.001\% \times 5) \times €20 000 000$. So, after 5 years, the cumulative expected value will be €1000. The same calculation was carried out for all other combinations of probabilities and losses due to major accidents considered in the questionnaire.

In Table 4.3, the investment (production or prevention) entailing the highest cumulative expected value is given. Based on the answers of the respondents, for each given probability and possible loss, it was calculated whether the largest group of respondents opted for the production or the prevention investment. The distribution of the matrix from the answers of the respondents was then compared with the matrix from Table 4.3.

As previously stated, several factors can have an influence on the decision-making process. Some factors were thus included in the analyses to determine whether people would decide in a more risk-averse or risk-seeking way. The results were compared according to gender (male or female). Based on the Short Rational-Experiential Inventory, respondents with a high rational thinking style were compared with those with a low rational thinking style, and respondents with a high intuitive thinking style were compared with those with a low intuitive thinking style. Based on the Brief Sensation-Seeking Scale, the choices of the respondents with high sensation-seeking behavior were compared with those with low sensation-seeking behavior.

4.7.2 Results of the Study on Type II Event Decision-making

In Table 4.4, the proportion of respondents opting for the production or prevention investment is given. Table 4.5 shows whether the largest group of respondents opted for the production investment or for the prevention investment. In this table, there is also a comparison between the choices of the respondents and the cumulative expected values of these investments: the distribution of the matrix from the answers of the respondents is compared with the matrix in Table 4.3. For the unknown probabilities of occurrence and the unknown losses of the accident, the cumulative expected value cannot be calculated, which makes this comparison impossible for these investments (in italics in Table 4.5).

According to calculated cumulative expected values, the higher the accident probability and the higher the possible loss, the more advantageous it is to choose a prevention investment.

Table 4.3 Investment with the highest cumulative expected value

		Probability that an accident will occur			
		1 in 100 000	1 in 10 000	1 in 100	1 in 10
Loss of the accident	€20 000 000	Production	Production	Production	Prevention
	€50 000 000	Production	Production	Production = prevention	Prevention
	€100 000 000	Production	Production	Prevention	Prevention
	€500 000 000	Production	Production	Prevention	Prevention
	€1 000 000 000	Production	Production	Prevention	Prevention
	€10 000 000 000	Production	Prevention	Prevention	Prevention

If the accident probability is equal to 1 in 100 000, then, according to the theory of expected values, it is always better to opt for the production investment, regardless of the possible loss of the accident (considering the numbers that were used in the survey). The survey results indicate that the respondents make decisions in a more risk-averse way than can be anticipated by the theory of expected values. The majority of the respondents opt for the production investment only when the possible loss of a major accident is equal to €100 000 000 or lower. Regarding

Table 4.4 Proportion of respondents who opt for the production or the prevention investment

			Probability that an accident will occur				
			1 in 100 000 (%)	1 in 10 000 (%)	1 in 100 (%)	1 in 10 (%)	Unknown (%)
Loss of the accident	€20 000 000	Production	89.4	82.2	46.7	20.0	75.9
		Prevention	10.6	17.8	53.3	80.0	24.1
	€50 000 000	Production	80.2	68.7	25.4	7.7	56.9
		Prevention	19.8	31.3	74.6	92.3	43.1
	€100 000 000	Production	63.2	40.2	11.1	5.4	34.0
		Prevention	36.8	59.8	88.9	94.6	66.0
	€500 000 000	Production	30.9	19.0	4.7	2.2	16.3
		Prevention	69.1	81.0	95.3	97.8	83.7
	€1 000 000 000	Production	15.1	9.7	2.0	2.0	9.2
		Prevention	84.9	90.3	98.0	98.0	90.8
	€10 000 000 000	Production	9.7	6.2	1.7	1.7	4.7
		Prevention	90.3	93.8	98.3	98.3	95.3
	Unknown	Production	43.2	34.2	12.0	7.2	5.0
		Prevention	56.8	65.8	88.0	92.8	95.0

Table 4.5 Largest group of the respondents who opt for the production or the prevention investment

		Probability that an accident will occur				
		1 in 100 000	1 in 10 000	1 in 100	1 in 10	*Unknown*
Loss of the accident	€20 000 000	Production	Production	Prevention*	Prevention	*Production*
	€50 000 000	Production	Production	Prevention*	Prevention	*Production*
	€100 000 000	Production	Prevention*	Prevention	Prevention	*Prevention*
	€500 000 000	Prevention*	Prevention*	Prevention	Prevention	*Prevention*
	€1 000 000 000	Prevention*	Prevention*	Prevention	Prevention	*Prevention*
	€10 000 000 000	Prevention*	Prevention	Prevention	Prevention	*Prevention*
	Unknown	*Prevention*	*Prevention*	*Prevention*	*Prevention*	*Prevention*

The distribution of the investment of the respondents that does not match the distribution of the investment with the highest cumulative expected value is indicated by an asterisk. In cases where a comparison between the distribution of the investment of the respondents and the distribution of the investment with the highest cumulative expected value is not possible, italics are used.

the higher probabilities of occurrence, i.e., 1 in 10 000 and 1 in 100, the respondents behave in a more risk-averse way than the cumulative expected values predict. This means that these respondents, being laypeople, also expect company safety managers to be cautious and choose prevention instead of production in such cases. It thus becomes obvious that moral principles should be seen as an important part of decision-making regarding type II risks, in order for them to be acceptable by laypeople. Klinke and Renn [87] explain this risk-averse behavior through the moral obligation that people have to prevent harm to human beings and the environment. They argue that risk refers to the experience of something that people fear or regard as negative. Another explanation can be found in the framework of Random Regret Minimization, which postulates that, when choosing, people anticipate and aim to minimize regret, and therefore behave in a more risk-averse way [89] (see also in the introductory section of this chapter).

As already mentioned, it is extremely difficult to predict decision-making in an uncertain situation where there is no knowledge whatsoever about the probability and/or the losses. It turns out that when there is uncertainty concerning the possible loss of an accident, the majority of respondents opt for the prevention investment, regardless of the probability of the accident occurring. As the accident's probability increases, the number of respondents that choose the production investment decreases. The number of respondents choosing the production investment decreases where the possible losses increase. When there is complete uncertainty, i.e., where both the accident probability and the possible loss are unknown, the majority (95%) opts for the prevention investment. With complete uncertainty, the risk is mostly categorized as intolerable, as the consequences might be catastrophic. Therefore, the risk-averse attitude is perceived as appropriate [87]. Notice that this finding also provides an argument for the precautionary principle.

4.7.3   Results by Gender

Table 4.6 shows the proportion of respondents opting for the production or the prevention investment by gender. When there is a significant difference ($P < 0.05$) between investments of the male respondents and the female respondents, this is indicated with an asterisk.

There are significant differences between male and female respondents for the lower probabilities of occurrence (1 in 100 000 and 1 in 10 000) in combination with higher possible losses (€500 000 000, €1 000 000 000, and €10 000 000 000). For these combinations, men behave in a more risk-seeking way, i.e., they are more likely to opt for the production investment. When the possible loss increases, this difference between men and women decreases. The results are consistent with existing studies indicating that men behave in a more risk-seeking way than women (e.g., [14, 90, 91]). When it comes to making decisions regarding type II events, men display more risk-seeking decision-making behavior than women. This study also shows that men – regardless of whether they have to take decisions – have higher levels of rational thinking style and sensation-seeking than women, and that women – regardless of whether they have to take decisions – have higher levels of intuitive thinking style (see also the following section).

4.7.4   Rational and Intuitive Thinking Styles

The study revealed that for the combinations of a probability of occurrence of 1 in 10 with possible losses of €20 000 000, €50 000 000, and €100 000 000, the respondents with a high rational thinking style behave in a more risk-averse way than those with a low rational thinking style. For the combination of a probability of occurrence of 1 in 100 000 with an unknown possible loss, respondents with a low intuitive thinking style follow the overall results, i.e., a prevention investment. However, the majority of the respondents with a high intuitive thinking style opted for a production investment for the combination of a probability of occurrence

Table 4.6 Proportion of respondents who opt for the production or the prevention investment per gender

				Probability that an accident will occur				
				1 in 100 000 (%)	1 in 10 000 (%)	1 in 100 (%)	1 in 10 (%)	Unknown (%)
Loss of the accident	€20 000 000	Production	Male	88.6	82.3	48.7	17.7	80.1
			Female	89.9	82.2	45.3	21.5	73.3
		Prevention	Male	11.4	17.7	51.3	82.3	19.9
			Female	10.1	17.8	54.7	78.5	26.7
	€50 000 000	Production	Male	84.2	68.4	25.9	5.7	52.2
			Female	77.7	69.0	25.1	8.9	59.9
		Prevention	Male	15.8	31.6	74.1	94.3	47.8
			Female	22.3	31.0	74.9	91.1	40.1
	€100 000 000	Production	Male	65.2	44.9	11.4	4.4	33.8
			Female	61.9	37.2	10.9	6.1	34.1
		Prevention	Male	34.8	55.1	88.6	95.6	66.2
			Female	38.1	62.8	89.1	93.9	65.9
	€500 000 000	Production	Male	37.6[a]	25.3[a]	3.8	1.9	19.6
			Female	26.7[a]	15.0[a]	5.3	2.4	14.2
		Prevention	Male	62.4[a]	74.7[a]	96.2	98.1	80.4
			Female	73.3[a]	85.0[a]	94.7	97.6	85.8
	€1 000 000 000	Production	Male	20.4[a]	13.4[a]	1.9	1.9	12.1
			Female	11.7[a]	7.3[a]	2.0	2.0	7.3
		Prevention	Male	79.6[a]	86.6[a]	98.1	98.1	87.9
			Female	88.3[a]	92.7[a]	98.0	98.0	92.7
	€10 000 000 000	Production	Male	14.0[a]	9.6[a]	1.3	1.9	5.7
			Female	6.9[a]	4.0[a]	2.0	1.6	4.0
		Prevention	Male	86.0[a]	90.4[a]	98.7	98.1	94.3
			Female	93.1[a]	96.0[a]	98.0	98.4	96.0
	Unknown	Production	Male	45.9	37.4	11.5	5.8	5.7
			Female	41.6	32.1	12.3	8.2	4.5
		Prevention	Male	54.1	62.6	88.5	94.2	94.3
			Female	58.4	67.9	87.7	91.8	95.5

[a]There is a significant difference ($P < 0.05$) between the investments of the male and female respondents.

of 1 in 100 000 with an unknown possible loss. These differences are statistically significant. Another significant difference between the high and low intuitive thinking styles is found for the combination of a probability of 1 in 10 000 with an unknown possible loss: respondents with a low intuitive thinking style are more likely to behave in a risk-averse way than respondents with a high intuitive thinking style.

Butler *et al.* [91] also have done research on how intuition and reasoning or rationality affect the decision-making under risk and uncertainty. They obtained similar results: people with a high intuitive thinking style are more risk-seeking and people with a high rational thinking style are less risk-seeking. Based on the terminology of Stanovich and West [92], individuals with a high rational thinking style rely on effortful, deliberative reasoning and systematic processing of information. On the other hand, when decisions are based on intuition, there is no systematic comparison of alternatives: a decision is taken at glance, by rapidly evaluating the main features of the problem [91]. An explanation for the finding that people with a high intuitive thinking style behave in a more risk-seeking way could be the fact that the high speed of intuitive thinking puts intuitive thinkers at a comparative advantage in situations involving high risk and uncertainty, making them less averse. Intuition can handle severe uncertainty so that individuals who are better at using intuition may feel more comfortable dealing with uncertainty and risk and thus develop higher tolerance for both [91, 93–96]. The downside is that they are only human, and, like anyone, they may become complacent or over-optimistic, which could make them blind for disaster. Hence, the high reliability organization (HRO) principles, as discussed in Chapter 2, should be applied to overcome such a possibility.

For the combination of a probability of 1 in 10 000 and an unknown possible loss, respondents with a high sensation-seeking style make decisions in a more risk-seeking way than respondents with a low sensation-seeking style. As already shown in a number of studies, higher levels of self-reported sensation-seeking are indeed associated with greater risk taking on various domains (e.g., [97–100]).

4.7.5 Conclusions of the Study on Type II Event Decision-making

Decision-making under risk and uncertainty is obviously not straightforward, and the acceptance of risk and uncertainty is influenced by a number of factors. Knowledge of these decision-making processes is important for the understanding and management of activities involving potential major accident events. This study illustrates that people behave in a more risk-averse way as compared with what would be expected based on the theory of expected values, in the case of decision-making involving type II risks. Only for low accident probabilities combined with low potential losses do the respondents take more risk-seeking decisions and opt for production investments. Concerning decision-making under uncertainty, the respondents also make risk-averse decisions: only for known and relatively low possible losses do they opt for the production investment; otherwise they always choose prevention. Under complete uncertainty, almost all respondents behave in a risk-averse manner. The study also showed that men are more likely to behave in a risk-seeking way than women, that people with a high intuitive thinking style are less risk-averse than those with a low intuitive thinking style, that people with a high rational thinking style are more risk-averse than those with a low rational thinking style, and that respondents with a high sensation-seeking style behave in a more risk-seeking way than those with a low sensation-seeking style. In other words, women and people with a low intuitive thinking style, a high rational thinking style, or a low sensation-seeking style will take more cautious decisions regarding investments when coming to a choice between production and prevention in the case of type II risks.

4.8   Costs and Benefits and the Different Types of Risk

The optimum degree of safety required to prevent losses is open to question, both from a financial and economic point of view and from a policy point of view. As explained previously, developing and executing a sound prevention policy involve prevention costs, but remember that the avoidance of accidents and damage leads to hypothetical benefits. Consequently, in dealing with safety, an organization should try to establish an optimum between prevention costs and hypothetical benefits. This can be done by determining the minimum overall cost point (by calculating the prevention and the accident cost curves), but only if sufficient data are available, i.e. in the case of type I risks and accidents.

It is possible to further expand upon costs and benefits of accidents in general in terms of the degree of safety. The theoretical degree of safety can vary between $(0 + \varepsilon)\%$ and $(100 - \varepsilon)\%$, wherein ε assumes a (small) value suggesting that "absolute risk" or "absolute safety/zero risk" in a company is, in reality, not possible. The economic break-even safety point, namely the point at which the prevention costs are equal to the resulting hypothetical benefits, can be represented in graph form (see Figure 4.15). The graph distinguishes between two cases: type I and type II risks. In the case of type I risks, the hypothetical benefits resulting from non-occurring accidents or non-events are considerably lower than in the case of type II risks, especially if the definition of Maxmax hypothetical benefits is used. This means essentially that the company reaps greater potential financial benefits from investing in the prevention of type II accidents than in the prevention of type I accidents. When calculated, prevention costs related to type II risks are, in general, also considerably higher than those related to type I risks. Type II accidents are mostly prevented by means of expensive technical studies and the furnishing of (expensive) technical state-of-the-art equipment and maintenance. Type I accidents are therefore more associated with the protection of the individual, first aid, the daily management of safety codes, and so on.

As the company invests more in safety, it can be assumed that the degree of safety within the company will increase. Moreover, the higher the degree of safety, the more difficult it becomes to improve upon this (i.e., to increase it) and the curve depicting investments in safety thereafter displays asymptotic characteristics. This was also explained before. Moreover, as more financial resources are invested in safety from the point of $(0 + \varepsilon)\%$, higher hypothetical benefits are obtained as a result of non-occurring accidents or non-events. These curves will display a much more level trajectory due to the fact that marginal prevention investments do not produce large additional benefits in non-events. The hypothetical benefits curve and the prevention costs curve dissect at a break-even safety point. If there are greater prevention costs following this point, hypothetical benefits will no longer balance prevention costs. A different position for the costs and benefits curves can be expected for the different types of accident.

It should be noticed that in order to calculate the hypothetical benefits of type I risks, the focus can be placed either on the consequences, or on the consequences as well as the probabilities (if those probabilities are available; cf. definitions (i) and (ii), respectively, in Section 4.2.4). They will clearly produce different curves. In the case of calculating the hypothetical benefits of type II risks, this can also be done, but as probabilities are much less reliable for such risks, and as people perceive the consequences of such risks as much less acceptable, it

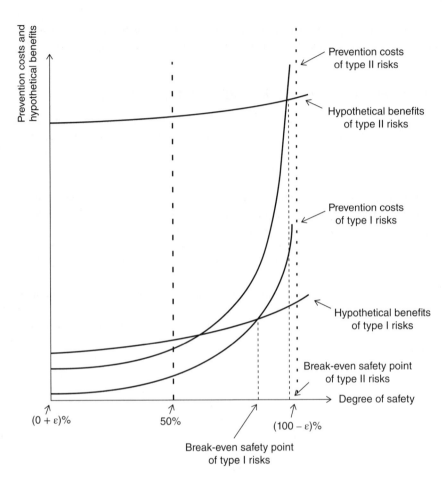

Figure 4.15 Economic break-even safety points for the different types of risk (qualitative figure). (Source: Meyer and Reniers [73] Reproduced with permission from De Gruyter.)

may be advisable to place more focus on the consequences. The calculation of hypothetical benefits has already been discussed in this chapter, but the concrete avoided costs leading to the hypothetical benefits will be explained more in depth in Chapter 5.

Figure 4.15 illustrates the qualitative benefits curves for the different types of risk. It is clear that hypothetical benefits and prevention costs relating to type I risks are considerably lower than those relating to type II risks. The break-even safety point for type I risks is likewise lower than that for type II risks. Figure 4.15 also shows that in the case of type II risks where a company might be subject to extremely high financial damage, the hypothetical benefits are even higher and the break-even safety point is likely to be located near the $(100 - \varepsilon)\%$ degree of safety limit. This supports the argument that in the case of such risks, almost all necessary prevention costs can be justified and zero type II risk should be strived for: the hypothetical benefits are nearly always higher than the prevention costs of such type II events.

Thus for type II risks, it is possible to state that a very high degree of safety should be guaranteed in spite of the cost of corresponding prevention measures. Of course, the uncertainties associated with these types of accident are very high, and therefore, economic analyses of such events can nevertheless be justified. Managers are not always convinced that such a major accident might happen in their organization, which is the main reason why the necessary precautions are not always taken, and why organizations are not always prepared. Therefore, economic analyses can be carried out for such events, but they should be different from those carried out for type I events and their results should be treated with a lot of caution and reservation.

In the case of type I risks, regular economic analyses such as cost-benefit and cost-effectiveness analyses may be carried out, and the availability of sufficient data should lead to reliable results, based on which "optimal" (and rational) safety investment and precaution decisions can then be taken.

4.9   Marginal Safety Utility and Decision-making

As explained earlier, safety utility can be regarded as the satisfaction that safety management receives from a certain safety situation or state thanks to operational safety measures. When the utility concept was explained in Chapter 3, the overall utility of a safety state, and its accompanying operational safety measures, was assumed. Nonetheless, the increase in (overall) safety utility as a result of the increase in one extra unit of operational safety state, a concept called the "marginal safety utility," should be central to safety decision-making. Usually, the overall safety utility does not give any indication as to what optimized decisions to make regarding, for instance, extra safety measures. The marginal safety utility providing information about the additional satisfaction obtained from applying an additional safety measure does so.

Similar to any other goods and services, also in safety science, the marginal safety utility of a certain type of safety measure decreases when applying more of this type of safety measure (assuming *ceteris paribus*; i.e., all the other types of safety measures stay the same). Sometimes it is possible that the marginal utility initially increases, and only afterwards decreases. Usually, most goods are characterized by a very prompt stage of decreasing marginal utility.

Looking at the engineering factors of safety from The Egg Aggregated Model (TEAM) of Chapter 2, three types of observable operational safety measures exist to improve safety: technological safety measures, organizational/procedural safety measures, and personal safety measures related to human factors ("TOP"). Sometimes, strategic safety measures are explicitly mentioned, as they are applied at the design stage to achieve inherent safety. With strategic measures included, the acronym becomes "STOP" (a term originating from the DuPont chemical company). To explain the relationship between marginal safety utility and decision-making, we assume that only an optimal allocation of the TOP measures, for the purposes of add-on safety, is strived for.

If, for example, technological safety measures are focused on, the principle is illustrated in Table 4.7. Assume that one technological measure costs, on average, €200.

In case of the illustrative example given in Table 4.7, the marginal utility associated with an increase from 0 to 5 units of technological measures is 200; from 5 to 10 units it is 160; from 10 to 15 units it is 140; and from 15 to 20 units it is 80. These numbers are consistent with

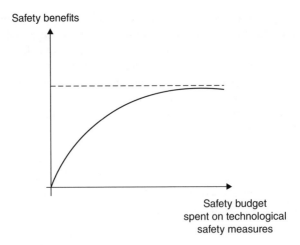

Figure 4.16 Law of diminishing marginal rate of return on investment for technological safety measures.

the principle of diminishing marginal utility. As more and more technological safety measures are applied, additional amounts of such measures will yield smaller and smaller additions to safety utility, or smaller and smaller returns on investment. To make the illustrative example more concrete for the reader, it is possible to look upon the number of technological safety measures as the amount of the safety budget spent on technological safety measures, and the safety utility can then be compared with safety benefits obtained. Using these interpretations, it is possible to draft Figure 4.16.

Fuller and Vassie [70] indicate that safety measures do indeed show a diminishing marginal rate of return on investment: further increases in the number of a type of safety measure become ever less cost-effective, and the improvements in safety benefits per extra safety measure exhibit a decreasing marginal development. In other words, the first safety measure of the "technology" type provides most safety benefit; the second safety measure provides less safety benefit than the first, and so on. Hence, to be most efficient there should be safety measures chosen from different types (technology, organization/procedures, and people/human factor). Of course, there may be differences in the safety benefit curves for the different types of safety measures. Figure 4.17 shows the increased safety benefits from choosing a variety of safety measures.

Table 4.7 Utility and marginal utility: an illustrative example

Number of technological safety measures	Total safety utility of technological safety measures	Marginal utility of technological safety measures	Marginal utility of technological safety measures per euro
0	0	—	—
5	1000	200	1
10	1800	160	0.8
15	2500	140	0.7
20	2900	80	0.4

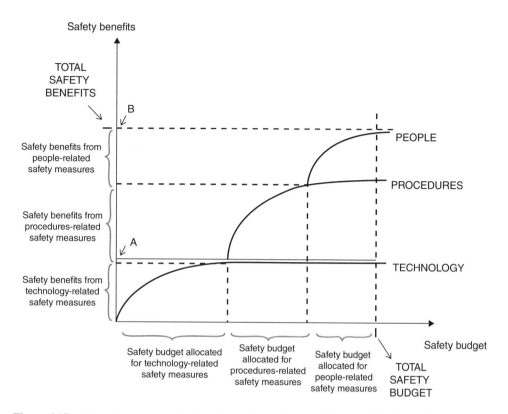

Figure 4.17   Allocation strategy for the safety budget. (Source: Meyer and Reniers [73]. Reproduced with permission from De Gruyter.)

Figure 4.17 shows that, if the total budget available is spread over a range of technology-, organizational/procedures-, and people/human factors-related safety measures, the overall safety benefit can be raised from point A (only investment in technology-related safety measures) in Figure 4.17 to point B (investment in TOP-related safety measures) in Figure 4.17. Hence, spreading the safety budget over different types of safety measures is always more efficient and effective than only focusing on one type of safety measure. This "equal marginal principle" is an important concept in microeconomics. Utility maximization is achieved when the budget is allocated so that the marginal utility per euro of expenditure is the same for each type of safety measure.

The usefulness of this theory with respect to decision-making and safety budget allocation can be illustrated by a numerical example. First, the curve of diminishing marginal rate of return on investment needs to be determined for each type of safety measure. The form of such a curve is given by

$$y = \frac{ax}{x+b}, \qquad \text{for } a, b > 0.$$

The parameters a and b determine the exact shape of the curves. The parameter a indicates the maximum value that the curve is approximating. If, for example, the maximum safety benefit is normalized and set to be 100% or 1, then $0 < a \le 1$. For instance, if $a = 0.5$, then the maximum safety benefit displayed by this curve is 50%. It would be possible for safety experts to assess that technological measures can maximally lead to 50% of safety benefits, organizational measures to 20% of safety benefits, and human factor measures to 30% of safety benefits. Any other combination is also possible and depends on a number of factors such as, for example, the industrial sector and the safety measures already in place. The parameter b indicates how fast the curve is approximating the maximum value. The higher b is, the faster the curve approximates its maximum value. Obviously, b is a measure displaying the level of efficiency of the measures.

Assume the following functions for the curves of diminishing marginal rate of return on investment for technology, organization, and people, respectively:

$$y_T = \frac{0.5x_T}{x_T + 5000}, \qquad y_O = \frac{0.2x_O}{x_O + 200}, \qquad y_P = \frac{0.3x_P}{x_P + 2000}$$

Furthermore, the safety budget is, for example, set to €20 000. Hence, since x_i represents the safety budget for a safety measure of type i, the condition $x_T + x_O + x_P = 20\ 000$ can be drafted. The following maximization problem thus arises:

$$\underset{x_T, x_O, x_P}{Max} \left[\frac{0.5x_T}{x_T + 5000} + \frac{0.2x_O}{x_O + 200} + \frac{0.3x_P}{x_P + 2000} \right]$$

subject to

$$x_T + x_O + x_P \leq 20\ 000$$

To solve the problem, the Lagrange function is formulated (for using the Langrange method to solve maximization problems, see calculus and analysis books; e.g., [101] or [102]):

$$L = \frac{0.5x_T}{x_T + 5000} + \frac{0.2x_O}{x_O + 200} + \frac{0.3x_P}{x_P + 2000} + \lambda(20\ 000 - x_T - x_O - x_P)$$

The four first order conditions look as follows:

$$\frac{\partial L}{\partial x_T} = \frac{2500}{(x_T + 5000)^2} - \lambda = 0$$

$$\frac{\partial L}{\partial x_O} = \frac{40}{(x_O + 200)^2} - \lambda = 0$$

$$\frac{\partial L}{\partial x_P} = \frac{600}{(x_P + 2000)^2} - \lambda = 0$$

$$\frac{\partial L}{\partial \lambda} = 20\ 000 - x_T - x_O - x_P = 0$$

Solving this system of equations gives $x_T \approx 11,700, x_O \approx 1920, x_P \approx 6350$. Hence, under the assumptions, the safety budget of €20 000 should be allocated to technological measures for some €11 700, to organizational measures for some €1920, and to people-related measures for some €6350. Therefore, using a budget of €20 000 allows a safety benefit of some $(0.35 + 0.18 + 0.23)\% = 76\%$ to be achieved.

4.10   Risk Acceptability, Risk Criteria, and Risk Comparison – Moral Aspects and Value of (Un)safety and Value of Human Life

Before explaining cost-benefit analyses in the next chapter, there are some important concepts that need to be clarified. Difficult questions that will arise when carrying out economic analyses include, for instance, what is an "acceptable risk"; is it possible to put a value on everything, even on human life; what are possible risk criteria; and is it possible to compare completely different risks with one another? These issues are discussed in this section.

4.10.1   Risk Acceptability

Risk acceptability is an extremely difficult and complex issue. The problem starts with the questions, acceptable for whom, and from whose perspective? In other words, who suffers the consequences when something goes awfully wrong and disaster strikes, on the one hand, and who gains the benefits when all goes well and profits are made, on the other? Are the risks equally spread? Are the risks justified/defensible and, even more importantly, are they just? Moral aspects always start to emerge when discussing the acceptability of risk, in other words, when asking the question "how safe is safe enough?"

Independent of the personalities of people, the discussion and debate about what risks and levels of risk are acceptable continues. Some people argue that one fatality is one too many, and others interpret the prevailing accident rate in areas of voluntary risk-taking as a measure of the level of risk that society as a whole finds acceptable [103]. Obviously, both viewpoints have their merits. No one will argue against the fact that one death is one too many. But it also seems reasonable to assume that voluntary risk levels can be used as some sort of guideline for involuntary risk levels. However, both perspectives are far from usable in the real industrial world. On the one hand, lethal accidents do happen and should not be economically treated as if they were not possible. So being part of industrial practice and real-life circumstances, fatalities should be taken into account in economic analyses. On the other hand, research indicates that voluntary risk cannot be compared with involuntary risk. People are much less willing to suffer involuntary than voluntary risks, and the risk perception that people have about involuntary risks is much higher than that about voluntary risks [104]. Hence, these two sorts of risk (i.e., voluntary and involuntary risks) should not be used interchangeably in economic analyses, and risk criteria should not be based on them.

In this book, *risks* and *dealing with risks* are viewed from a microeconomic decision-making perspective. Hence, the focus is on the optimal operational safety decision, based on the information available and thereby taking economic issues into account. Risks are relative, and making the best resource allocation decision for operational safety in a company means avoiding as much possible loss as possible within a certain safety budget. However, due to the "acceptability" aspect of risks this seems easier than it is in reality. One of the best known and most used principles in this regard is the proactive concept of "as low as reasonably practicable" (ALARP). The term implies that the risk must be insignificant in relation to the sacrifice in terms of money, time, or trouble required to avert it. Hence, ALARP means that risks should be averted unless there is a gross disproportion between the costs and the benefits of doing so. The difference between ALARP and other terms such as "best available technology" (BAT), "best available control technology" (BACT), and "as low as reasonably achievable" (ALARA) can best be described as the observation that to follow the ALARP approach, not only should the technology be available, but also the costs of prevention should be reasonable. BAT and ALARA demand to do work, research, engineering, and so on, to make the prevention work, irrespective of the costs. The term "so far as is reasonably practicable" (SFAIRP) is used in health and safety regulations in the UK, and should not be confused with ALARP, which has no legislative connotation. The SFAIRP term has been defined in UK courts. The formal definition given by Redgrave [105] is:

> 'Reasonably practicable' is a narrower term than 'physically possible', and implies that a computation must be made in which the quantum of [operational] risk is placed in one scale and the sacrifice involved in the measures necessary for averting the risk (whether in money, time or trouble) is placed in the other, and that, if it be shown that there is a gross disproportion between them – the risk being insignificant in relation to the sacrifice – the defendants [persons on whom the duty is laid] discharge the onus upon them [of proving that compliance was not reasonably practicable]. Moreover, this computation falls to be made by the owner at a point of time anterior to the accident.

An acronym similar to ALARP (but with slightly different meaning as regards the word "reasonable") is "best available technology not entailing excessive costs" (BATNEEC). Other acronyms focusing on the legislative (and reactive) part are "cheapest available technology not invoking prosecution" (CATNIP) and "as large as regulators allow" (ALARA – second meaning) .

It is clear that there is a diversity of acceptability approaches out there, but the most well known and the most used is undoubtedly ALARP. ALARP is also the most useful on an organizational level, as the risk level of being ALARP is influenced by international guidelines and directives, the progress of societal knowledge of risk, the continuous evolution of people's perception of what "risk" constitutes, political preferences and other forms of regulation in different industries. Hence, the ALARP principle allows us to have a different risk judgment depending on industrial sectors, risk groups, risk-prone areas, and even organizations.

The relationship between the ALARP principle and the terms "intolerable," "tolerable," "acceptable," and "negligible risk" is illustrated in Figure 4.18.

Figure 4.18 shows that it is possible to distinguish between "tolerable" and "acceptable." The subtle distinction between both terms is used to indicate that a level of risk is never accepted, but it is tolerated in return for other benefits which are obtained through activities, processes, and so on, generating the risk [106]. Figure 4.18 illustrates that the activities characterized by high levels of risk are considered intolerable under any normal circumstances. At the lower end, there is a level below which risk is so low that it can be ignored – i.e., it is not worth the cost of actively managing it. There is a region between these two limits, called the ALARP region, in which it is necessary to make a trade-off between the risk and the cost of further risk reduction.

The practical use of ALARP within one single organization can be demonstrated by using the risk assessment decision matrix (see also [73]). The goal of all reduction methods is summarized in Figure 4.19, where one can observe that risks that are considered unacceptable (with the visual help of the risk matrix) have to be brought down to an acceptable level or at least to one that is tolerable.

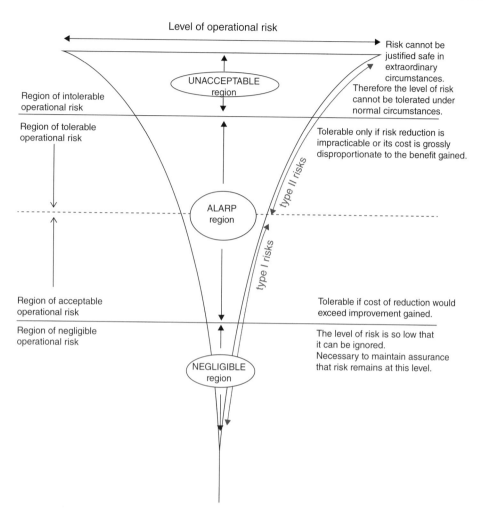

Figure 4.18 The "as low as reasonably practicable" (ALARP) principle and its relationship with the terms "intolerable," "tolerable," "acceptable," and "negligible."

In the middle zone of the matrix, two intermediate regions, "2" and "3", between the unacceptable zone "1" (needing immediate action) and the negligible zone "4" can be noted in diagonal. These zones are the ones where the ALARP principle occurs. Using the risk matrix, it is thus easily possible for an organization to make a distinction between the tolerable region "2" and the acceptable region "3" as noted in Figure 4.19, and to link certain actions to it, specifically designed by and for the company.

In any case, the ALARP principle needs a decision rule (or criteria) specifying what is "reasonable" for the person, group of persons, or organization using it. The decision will ultimately depend on a combination of physical limits, economic constraints, and moral aspects. The advantage is that the principle works in practice, but the downside is that it will sometimes

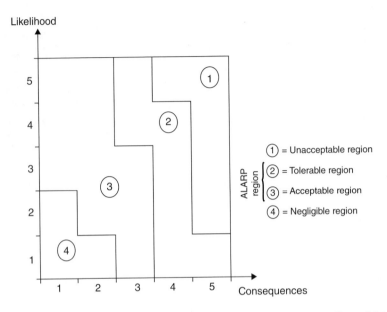

Figure 4.19 The "as low as reasonably practicable" (ALARP) principle and the effect of risk reduction represented in the risk matrix.

lead to inconsistencies in effective valuation of various risks, and have inefficiencies in the sense that it may be possible in some cases to save more lives at the same cost, or the same number of lives at a lower cost [74]. The next section discusses some criteria that are widely used to make decisions using ALARP.

There are a number of disadvantages or issues concerning the using of economic (e.g., cost-benefit) analyses for ALARP. For example, there is a difference between the willingness to pay (WTP) and willingness to accept (WTA) evaluations (see also Section 4.10.4.3), both approaches that are often used to put values on (un)safety. There are also problems of understanding "gross disproportionality" and of putting a value on human life. Furthermore, sometimes risks are only replaced by (exported to) another facility not falling under the same (or under any) ALARP calculation.

4.10.2   Risk Criteria and Risk Comparison

Risk criteria are related to the definition of risk and the target exposed to the risk. Within a single organization, there can be a number of possible targets: employees, groups of employees, material assets of the company, non-tangible assets of the company (such as reputation, image), intellectual property of the company, and also surrounding communities and the society as a whole.

As this book is only concerned with operational risks, and not, for example, with voluntary risks, natural-technological risks, or terrorist risks, no policy factor varying with the degree to which participation in the activity is voluntary and with the perceived benefit needs to be introduced into the risk calculation. All operational risks are in essence involuntary, for citizens living nearby the industrial activity but also for the employees working at the organization. However, there is a difference between a company's employees and its surrounding communities: i.e., the benefits arising from the industrial activity. The perceived risk will thus be different between both groups of people. This is the problem of equity or distribution of risk. Those people who take the risk are often not the same as those who bear the risk. As Ale *et al.* [107] indicate, when the rewards of risk-taking go to the risk creators and the costs are carried by the risk bearers, a strong incentive is created for a small group to take large risks at the expense of others. Moreover, the increase in productivity, turnover, and/or profits as a total may be maximized, but some employees may have more profit than others while bearing less risk. Therefore, this moral aspect should be considered when carrying out an economic analysis and establishing risk criteria. Section 4.10.4 treats this question in more depth.

If one considers only human losses (injuries and fatalities), there are different possible approaches to calculating risk (using a strictly rational point of view), to subsequently base the necessary risk criteria on human-related risks. Two very well-known approaches that are widely used are the calculation of the "location-based (individual) risk" and the "societal risk." An individual risk provides an idea of the hazardousness of an industrial activity. A societal risk takes the exposure of the population in the vicinity of the hazardous activity into account in the calculation.

4.10.2.1   Location-based (Individual) Risk

An individual risk can be defined in general as "the frequency with which a person may expect to sustain a specified level of harm as a result of an adverse event involving a specific hazard." Hence, the individual risk can be used to express the general level of risk to an individual in the general population, or to an individual in a specified section of the community. Legislators often use levels of individual risk with "specified level of harm" equal to "fatality" (such risks should actually be called "individual fatality risks") as a regulatory approach to setting risk criteria. In the process industries, for example, the individual (fatality) risk is defined as "the risk that an unprotected individual would face from a facility if he/she remained fixed at one spot 24 hours a day 365 days per year." Therefore, the risk is also sometimes called location-based (individual) risk. A location-based risk describes the geographic distribution of individual risk for an organization. It is shown using so-called iso-risk curves, and is not dependent on whether people or residences are present (see Figure 4.20).

Location-based risk is used to assess whether individuals are exposed to a greater than acceptable risk in the locations where they may spend time (e.g., where they live or work). It does not directly provide information on the potential loss of life, nor does it distinguish between exposure affecting employees and the general (surrounding) population. In many countries, an additional fatality risk from industrial activities of 10^{-6} (1 per million)/year to a person exposed to this fatality risk is considered to be a very low level compared with risks that are accepted every day, and hence used as a risk criterion. The reasoning for this risk criterion is based on what is known about risk perception. On the one hand, the individual risk of being killed while driving a car is estimated (very generally) as 10^{-4}/year. If the level of (individual) risk is higher than driving a car, it is perceived as unacceptable. On the other hand, the individual risk of being struck by lightning is estimated as 10^{-7}/year, and this risk level is perceived (in general) to be so low that it is acceptable.

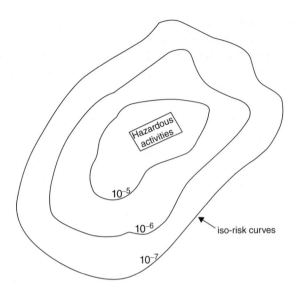

Figure 4.20 Example of iso-risk curves showing the distribution of location-based (individual) risk surrounding an enterprise.

Risks of the same type can thus be compared with one another to agree upon a risk criterion. There are actually a number of reasons for comparing risks with each other. The most important reason might be that by comparing risks, it is possible to improve the understanding and the perception of risks, both by the general public and by experts. It is difficult to grasp the meaning of "one in a million," although people do tend to have a natural feeling and intuition for categorizing risks into low–medium–high, for example. By comparing lesser known risks with more familiar risks, the familiarity and understanding of known or older risks can be transferred to new and unknown risks. Table 4.8 helps to place different risks in perspective, by quantifying the one-in-a-million (10^{-6}) risk of dying from various activities.

Table 4.8 Some "one-in-a-million" risks of dying from various activities

1. Smoking two cigarettes (risk of heart disease included)
2. Eating 100 g servings of shrimp
3. Eating 35 slices of fresh bread
4. Eating 350 slices of stale bread
5. Eating one-half basil leaf (weighing 1 g)
6. Drinking 70 pints of beer in 1 year
7. One-quarter of a typical diagnostic chest X-ray
8. A non-smoker living in a home with a smoker for 2 weeks

All of these risks are risks relevant to adults consuming this amount. Children consuming half this amount would be at comparable risk.
Wilson and Crouch [74]

Individual fatality risks in an industrial context are calculated by multiplying the consequences and the frequency of undesired events. For example, if the severity of an industrial accident is such that there is a probability p of killing a person at a specified location [the probability merely takes into account the level of lethality due to the accident (e.g., due to heat radiation or due to a pressure wave), but does not consider population figures], and the accident has a annual frequency of f, then the individual fatality risk at this particular location due to this event is $p \times f$ per year. Where there is a range of incidents that expose the person at that point to risk, the total individual fatality risk is determined by adding the risks of the separate incidents/events.

Contours of identical risk level (such as those in Figure 4.20) can then be plotted around an industrial activity, and can be used to present the risk levels surrounding the activity. Different levels of individual fatality risk can then be used by companies to discern between different company sites. A distinction can be made, for example, among production areas, administration areas, commercial areas, parks and fields, fire department, and care units .

4.10.2.2 Societal Risk or Group Risk

Calculating the individual fatality risk around a specific industrial activity does not make a distinction between, for example, the activity taking place somewhere in the desert or in a major city center. The calculated individual risk will be the same, regardless of the number of people exposed to the activity. However, it is evident that in reality the level of risk will not be identical in both situations: the level of risk will be higher if the industrial activity takes place in the city, rather than in the desert. This is a result of people living in the neighborhood of the activity, and thus being exposed to the danger. To take the exposed population figures into account, a so-called societal risk is calculated. The societal risk, also called group risk, is defined as "the probability that a group of a certain size will be harmed – usually killed – simultaneously by the same event or accident" [108]. It is presented in the form of a so-called "FN curve." Each point on the line or curve represents the probability that the extent of the consequence is equal to or larger than the point value. These curves are found by sorting the accidents in descending order of severity, and then determining the cumulative frequency. As both the consequence and the cumulative frequency may span several orders of magnitude, the FN curve is usually plotted on double logarithmic scales. A log–log graph as depicted in Figure 4.21 is obtained. For a concrete example on how to calculate and plot an FN curve, see Section 4.11.2.3.

The different steps in Figure 4.21 show the probabilities (cumulative frequencies) of different scenarios. The societal risk is designed to show how risks vary with changing levels of severity. FN curves are therefore often used for comparison with a "boundary line" representing a norm following legislation. This boundary line is also called the societal risk criterion. As an example, a boundary line fixed by Flemish law is given in Figure 4.21. This type of boundary was originally proposed by Farmer in 1967, as explained in Duffey and Saull [109], in a paper that launched the concept of a boundary line for use in the probabilistic safety assessment of nuclear power plants. The boundary line actually represents the transition region between the tolerable region and the unacceptable region. For example, in theory (not taking moral aspects into consideration at this time), a hazard may have an acceptable level of risk for 20 fatalities, but may be at an unacceptable level for 100 fatalities. Usually, such a boundary line, if it exists, is determined by the authorities, but it can also be fixed by individual companies for internal use.

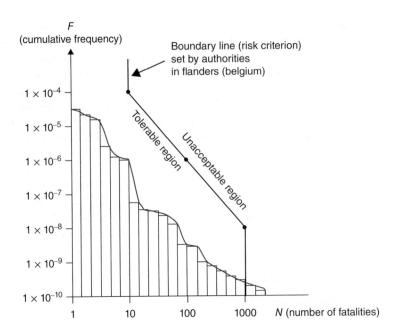

Figure 4.21 Illustrative example of an FN curve.

In some jurisdictions, there are two defined boundaries (societal risk criteria) on the FN graph, i.e., separating the zones of high risk (unacceptable region), intermediate high risk (tolerable region), and intermediate low risk to low risk (acceptable region and negligible region) (see also Meyer and Reniers [73]). It is common, where an industrial activity is calculated to generate risks in the tolerable region, to require the risks to be reduced to an ALARP level (see also the previous section), provided that it can be proved in some way that the benefits of the activity that produces the risks are seen to outweigh the generated risks.

Mathematically, the equation for an FN risk criterion may be represented as [110]:

$$FN^a = k$$

or:

$$F = kN^{-a}$$

where F is the cumulative frequency of N or more fatalities; N is the number of fatalities; a is an aversion factor (usually between 1 and 2); and k is constant. From the equation, it is clear that the slope of the societal risk criterion (when plotted on a log–log basis) is a and represents the degree of aversion to multi-fatality events embodied in the criterion:

$$\begin{cases} a < 1 : \text{risk-seeking} \\ a = 1 : \text{risk-neutral} \\ a > 1 : \text{risk-averse} \end{cases}$$

For instance, if $a = 1$, the frequency of an event that results in 100 fatalities or more should be 10 times lower than the frequency of an event that results in 10 or more fatalities. If $a = 2$, the frequency of an event that results in 100 or more fatalities should be 100 times lower than the frequency of an event that results in 10 fatalities or more.

In enterprises encountering numerous hazards with severe potential consequences, computer programs are used to calculate the risk levels on a topological grid, after which they are used to plot contours of risk on the grid. These contours are used to display the frequency of exceeding excessive levels of hazardous exposure. For example, Reniers *et al.* [111] indicate the availability of software tools that will prepare contours, for example, for the frequency of exposure to nominated levels of heat radiation, explosion overpressure and toxic gas concentration. So-called quantitative risk assessment (QRA) software is used to plot FN curves.

One should keep in mind that all these software tools provide an insight into the possible scale of a disaster and the possibility of its occurrence, but they do not offer adequate information for the optimal prevention of catastrophic accidents. Moreover, if the number of events observed in the past is not sufficient to estimate significant frequency values, as in the case of type II events, a simple histogram plotting the absolute number of past events versus a certain type of consequence is often used instead of a risk curve. However, type II event predictions are extremely difficult to make, simply because of the lack of sufficient data. Although it is therefore not possible to take highly specific precautions based on statistic predictive information in such cases, engineering risk management leading to a better understanding of relative risk levels and to an insight into possible accident and disaster scenarios is essential to prevent such disastrous accidents and should therefore be fully incorporated into industrial activities worldwide.

From the previous paragraph, it can thus be argued [109] that the uncertainty about the events should be given much more emphasis in risk criteria and trade-offs. Actual outcome data from major catastrophes in modern society indicate that the desired risk levels are often not achieved, and that the actual outcomes have greater uncertainty and hence higher risk. Concrete examples are: the Concorde aircraft crash with over 100 fatalities over 27 years of operation; the Columbia and Challenger shuttle losses with 14 fatalities over an interval of 17 years between the flights; the Texas city refinery explosion with 15 fatalities over about 30 years of operation; the Chernobyl reactor explosion with 30 fatalities directly, and many more indirectly, over 30 years of plant life.

In any case, Sunstein [112] mentions that as the severity of an event increases, people become more risk-averse. In particular, once the death threshold is passed, it appears that the community has a much greater aversion to multiple fatality accidents. This is the reason why the aversion factor is often chosen to be higher than 1. The public in general is risk-averse, or, more accurately, consequence-averse, meaning that more severe consequences (with the same frequency) weigh heavier in the decision-making process than more frequent events (with the same total consequence).

In fact, there has been an ongoing debate about giving equal weight to the frequencies and consequences of accidents. By not distinguishing between one accident causing 50 fatalities and 50 accidents each causing one fatality over the same period of time of operation, one fails to reflect the importance society attaches to major accidents. Evans and Verlander [113], for example, claimed that the economic rule of constant utility (iso-utility) was violated by choosing an aversion factor higher than 1, and that therefore decisions made on any other criterion than the expectation value (EV) of the number of victims is wrong. However, recently Bedford [114] showed that an FN curve with an aversion factor higher than 1 is not risk-averse but consequence-averse, and thus no economic laws are violated. Nonetheless, as Ale *et al.*

[107] mention, multiple fatalities can create greater disutility than single fatalities. Obviously, 30 people killed in a major explosion at a chemical plant operating for 20 years will lead to social debate and parliamentary enquiries, while the same number killed on the road in 20 days does not cause any societal ripple.

Linking the multiple fatality aversion with an economic analysis approach is rather difficult. This was demonstrated by the European Commission cost-benefit analysis results, indicating that the live-saving benefits of safety case regimes disappeared from view because the monetized value of life (VoL) (see Section 4.10.4) lost in major accidents was insignificant in comparison to other costs. Nonetheless, disasters can (and often will) result in multiple fatalities, and an economic analysis ignoring this fact is clearly problematic. Sunstein [112] therefore argues that preventing major accidents cannot be based on conventional cost-benefit analyses of lives lost. In fact, a rule of thumb is that 10 fatalities occurring at the same time are $100\,(10^2)$ times worse than one death, and 100 fatalities at the same time are $10\,000\,(100^2)$ times worse than one fatality [115]. This can be translated into monetary values in the following way. Assuming that it is possible to put a value on human life (see Section 4.10.4 for more details), and that the value is €10 million, then the value of 10 lives lost at the same time equals €1 billion, and the value of 100 lives lost together equals €100 billion, and so on. By treating single fatalities differently from multiple fatalities for assigning monetary values, such a rule of thumb could obviously make a big difference in cost-benefit calculations. However, organizations should decide for themselves whether or not to use such a rule of thumb (or a similar one). In any case, it is a means to take into consideration the public's disaster aversion, and to recognize as a company that multiple deaths would affect society – and also the company itself – far more than the multiple would suggest.

Moreover, it is worthwhile mentioning a special case, i.e., the observation that several industrial activities from nearby companies in the same industrial area may each generate a low level of societal risk, each falling in the acceptable region, whereas their combined societal risk might fall within the high-risk zone (unacceptable region) of the chart in Figure 4.19, if these industrial activities were all to be grouped for the purposes of the calculation. As society places a special value on the prevention of large-scale loss of life, such multiple plant disasters should also be considered when carrying out economic analyses. However, this falls outside the scope of this book, which is concerned with operational safety economics within single organizations.

4.10.3   Economic Optimization

The relationship between risk criteria and an economically optimal situation can also be formulated as an economic decision problem, analogous to the method proposed by van Danzig [116] for flood defenses and by Jonkman *et al.* [117] for tunnel safety. In view of economic optimization, the total (hypothetical) safety benefits of a system (HB_{tot}) are determined by calculating the expected hypothetical benefits due to safety investment costs (SIC), that is, $E(HB_{SIC})$, and then subtracting these SICs (or expenditure for a safer system) from these expected hypothetical benefits. In the optimal economic situation, the total hypothetical benefits in the system are maximized:

$$\underset{P_{loss}}{Max}(HB_{tot}) = \underset{P_{loss}}{Max}[E(HB_{SIC}) - SIC]$$

subject to

$$P_{loss} \leq P_{risk\ criterion}.$$

Assuming that the SICs and the expected hypothetical benefits are both a function of a probability of loss, P_{loss} (e.g., the probability of dying combined with the probability of failure), it is possible to determine the optimal probability of loss of a system, subject to the condition of a predefined risk criterion. Moral aspects such as equity or justification of risk can then be taken into account in defining the risk criterion by the organization (see also Section 4.10.4).

For a thorough discussion of the SICs and the hypothetical benefits, see Chapter 5. When the probabilities of different accident scenarios are known, the expected hypothetical benefits can thus be assessed, and the optimal safety investments can be derived within the bounds of a predefined risk criterion. Obviously, although it is possible to take only material damage into the valuation problem, it is sometimes necessary, and therefore it should also be possible, to take the value of human life into the economic optimization problem. However, the valuation of safety and the valuation of human life both raise numerous ethical and moral questions, and they have been a subject of the safety literature for decades. The next section treats these sensitive subjects in more detail.

4.10.4   Moral Aspects and Calculation of (Un)safety, Monetizing Risk and Value of Human Life

4.10.4.1   Moral Aspects and Calculation of (Un)safety

This section on moral aspects and the calculation of safety, together with the next section on the monetizing of risks and the valuation of human life can be seen as the most difficult sections of the entire book. They are not difficult because of the mathematical formulae involved, but rather because the sections are highly complex due to the very important moral aspects interwoven with the questions of calculation of safety and valuation of human life. This ethical dimension may quickly turn into a philosophical discussion with very diverse viewpoints and opinions. Needless to say, the topic is highly debated, controversial, and sensitive to many people. That being said, a book on the economics of operational safety cannot simply omit or gloss over the moral aspects, or the valuation of human life, as something trivial in the calculation of safety. A thorough discussion and a viewpoint are thus needed. This section delivers just that.

As noted by van de Poel and Fahlquist [118], the psychological literature on risk perception has established that laypeople include contextual elements in how they perceive and understand risks (e.g., Slovic [104]). Such contextual elements include voluntariness, familiarity, dread, exposure, catastrophic potential, controllability, perceived benefits, and time delay (future generations). The fact that there is a difference in risk perception between laypeople and experts is often interpreted as the result of the irrationality of laypeople. However, such an interpretation indirectly assumes that the expert opinion, or the technical conceptualization of risk, is the correct one and that laypeople should simply be educated to the same viewpoint. However, Slovic [104] and Roeser [119, 120] explain that the contextual elements considered by laypeople are actually relevant for a better perception of the reality of risk, and are thus important for the acceptability of risks and for adequately managing risks.

Sunstein [121] notes that when asked to assess the risks and benefits associated with certain items, people tend to think that hazardous activities contain low benefits, whereas beneficial activities contain low hazards. Rarely do people consider an activity to be both highly beneficial and hazardous, or to be both benefit-free and hazard-free. Slovic [122] even suggests a so-called "affect heuristic," by which people have primarily an emotional reaction to certain situations, processes, activities, products, and so on. It follows that emotions and feelings are essential elements to consider when dealing with operational risks and appreciating operational safety. Besides objective consequences and probabilities of potential scenarios, moral aspects (possibly fed by emotions) should thus play an important role in the calculation of safety. The subsequent activities of hazard identification, risk analysis (including estimation/calculation of consequences and likelihood), risk evaluation, and risk prioritization, the process known as risk assessment, only serve provide a rational idea of all the prioritized (rational) risks (for a certain scope of the safety study), in order to make decisions about what risks to deal with first, second, and so on (hence, safety budget allocations). However, to improve the safety budget allocation decision process, more than just rational risk prioritizations should be carried out; safety appreciation prioritizations should also be executed, where moral aspects are taken into account as well as the rational aspects.

If moral aspects were considered in the safety calculation process, economic analyses would perhaps be much less contested or debated, and more accepted as a valid approach to ensure that company policy is driven not by pure ratio or pure emotion, but by a combination of both, and thus a full appreciation of all aspects of safety, rational as well as emotional, is strived for. The risk thermostat, originally developed by Wilde [123] and adapted by Adams [103, 124, 125], can be applied to an organization, and the importance of balanced safety appreciations becomes immediately obvious in this regard. This is explained in the following paragraphs.

Everyone, including decision-makers in companies, has a propensity to take risks within appropriate limits. This propensity varies from one company to another and is influenced by the potential level of the rewards and losses arising from the organization's risk-taking decisions. Companies' so-called "perceptions of risk" are determined by the safety appreciations carried out in the company. Furthermore, companies' risk policies represent a balancing act in which the perceptions of risk are balanced against the propensity to take risks. Benefits and losses arise from an organization taking risks, and the more risks (in general, thus also the more operational risks) an organization takes, the greater the potential level of costs and benefits. Figure 4.22 shows this risk thermostat model applied to organizations.

The conceptual model illustrated in Figure 4.22 explains the impact of the safety appreciation process on the position of the company's risk policy balance. If moral aspects are not taken into consideration in the safety calculation, and, as a result, the costs from the company's risk policy are determined to be quite low (compared with the possible costs for the company if moral aspects were taken into account), the perception of a certain risk (e.g., a HILP risk) will be low, and the balance will tend toward risk-seeking behavior. In the case of type II risks, in particular, the risk thermostat conceptual model shows the importance of safety calculations taking moral aspects into account.

The literature ([126–130]) indicates the moral concerns that are often mentioned: voluntariness, the balance and distribution of benefits and risks (over different groups and over generations), and the availability of alternatives. These ethical dimensions can be used to enlarge the technical conception of operational risk, given in Chapter 2, in an extended formula of a "qualitative and quantitative" risk index, a so-called Q&Q risk index. The following formula for the Q&Q risk index (R^*) is proposed to this end:

$$R_i^* = \frac{(L_i C_i^{\,a})}{\beta_i (E_i F_i^{\,b})} = \frac{R_i}{\beta_i A_i} \tag{4.1}$$

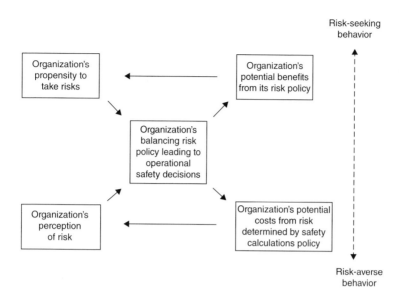

Figure 4.22   The risk thermostat model applied to organizations. (Source: Adams [103]. Reproduced with permission from John Wiley & Sons.)

where

$R_i^* = $ Q&Q Risk index of event/scenario i

$\beta_i = $ the policy factor that varies according to the degree which participation

in the risk due to event/scenario i is voluntary

$E_i = $ Acceptability of the principle used to apportion liabilities for undesired

consequences for event/scenario i (Equity principle)

$F_i = $ Acceptability of the procedure by which collective consent is obtained

to those who must bear the consequences of event/scenario i (Fairness principle)

$b = $ factor expressing the availability of alternatives in combination

with the anti-recklessness of management

$L_i = $ Likelihood of occurrence of event/scenario i

$C_i = $ Magnitude of the consequences of event/scenario i

$a = $ aversion factor towards consequences

$R_i = $ risk of event/scenario i, calculated using a rational approach

(with consequence and likelihood estimation)

$A_i = $ Acceptability of event/scenario i following an emotional approach

It is important to realize that the risk aversion factor a should be seen as a general factor for people with respect to their aversion to the consequences of risks, whatever the nature and background of the risks. The factors determining the acceptability, A, of a risk, however (i.e., F, E, and b), are risk-specific and depend on the nature and background of the risk. The factors E and F are also mentioned and defined by Rayner [131] in an early attempt to extend the well-known rational risk formula $R = L \times C$. In the formula for an operational risk index that is proposed above, rationally calculated risk is combined with emotional aspects to determine a proxy for operational risk within an organization. E is a parameter representing the moral principle of "equity," taking the emotions concerning the balance and distribution of benefits and risks into consideration. F is a parameter providing an idea of the moral principle of "fairness" or justification related to the operational risk of an event or scenario. Using this formula, the Q&Q risk index, R^*, of an undesired event (or the risk of an event happening) is lower the more acceptable this event is, the higher the voluntariness of suffering the risk accompanying the event, and the lower the rationally calculated risk. As β and A are both dimensionless numbers, this R^* index is expressed in the same units as rational risk. Hence, if rational risk is expressed in monetary units per time units (e.g., €/year), R^* is also expressed in these units. The R^* index could thus be used as an approach to calculate risk within an organization, using both the well-known rational data such as probabilities, consequences, and exposure, and the less frequently used moral principles such as fairness, equity, and anti-recklessness of management.

If company management wanted to use this approach, it would have to quantify the factors, E, F, a, and b. Notice that the parameters β, E, and F should be expressed as indices, and numbers for each of these parameters should be sufficiently different from each other to be relevant in the formula. A protocol within the company or within society should be designed to unambiguously determine the approach to quantify every parameter. Notice that if it were possible that a parameter β, E, or F could be chosen to be zero, then the R^* level for the event would be infinite, and thus the event could never be made safe enough, whatever the safety investments to influence (lower) the likelihood and/or consequences. Therefore, this possibility is best excluded.

As an example, the following qualitative indices could be suggested for the parameters β, E, and F:

- Very low – 0.1
- Low – 0.5
- Medium – 1.0
- High – 1.5
- Very high – 2.0.

The parameters a and b are also suggested to be situated in the interval [0, 2], to avoid stressing the aversion factor and the anti-recklessness factor too much in the risk formula.

The R^*_i index can then be used to fix a (maximum) company risk criterion for the probability of loss (see also the previous section) by determining the maximum probability for a certain (company-specific) Q&Q risk level (as given/fixed by the maximum R^*_i number).

An illustrative example can be given to further explain the risk calculation formula. Assume that a comparison is needed for two accident scenarios, i.e., a major industrial accident with one fatality within a single chemical plant and a large-scale domino effect disaster within a cluster of chemical plants, affecting at least three companies and entailing 10 fatalities. Assume we want to know the difference in risk between both scenarios and taking moral principles into consideration. Table 4.9 provides the different (illustrative) quantitative (L and C) and qualitative (a, b, E, and F) parameter values, for example, determined by software (quantitative parameters) or by expert solicitation (qualitative parameters) for both scenarios. As there are many people killed in the second (domino effect) scenario, the illustrative example follows the rule of thumb explained in Section 4.10.2.2, and sets the aversion factor a equal to 2 in the disaster scenario. In the major accident scenario, people are assumed to be consequence-averse, but only slightly.

Table 4.9 Parameter values for scenarios "major industrial accident" and "domino effect disaster."

Major industrial accident	Domino effect disaster
$L = 10^{-3}$/year	$L = 10^{-9}$/year
$C = €8\,000\,000$	$C = €40\,000\,000$
$a = 1.2$	$a = 2$
$\beta = 0.5$	$\beta = 0.1$
$E = 1.0$	$E = 0.5$
$F = 0.5$	$F = 0.1$
$b = 0.7$	$b = 0.2$

From Table 4.9, the so-called "expected losses" (only considering consequences and likelihood) for both scenarios are (Eq. 2.1): €8000/year for the major industrial accident scenario, and €0.04/year for the domino effect disaster scenario. Thus, obviously in this case the extremely low probability of the domino effect scenario plays a very important role to turn the decision in favor of safety investments for preventing major industrial accidents instead of domino effect accidents.

However, for the major industrial accident scenario using the formula (Eq. 4.1) for R^* and using the parameters from Table 4.9, the "rational" risk (taking only the consequences, the likelihood, and the aversion factor into consideration) is calculated to be, (Eq. 2.2), €192 180/year, and the R^* index equals, (Eq. 4.1), €624 366/year. In the case of the domino effect disaster, the rational risk equals €1 600 000/year, while $R^* = €50\,793\,651$/year. These numbers are already much higher, and in this case, they are in favor of safety investment for domino effect prevention. Some further observations can be made. Clearly, using only quantitative data (being the consequences, the likelihood, and the risk aversion factor for an accident scenario), the disaster scenario entails the highest risk, but the difference between both scenarios is rather limited. Especially if the probabilities are mainly considered by managers (due to psychological principles discussed in the introduction of this chapter) in the decision-making process, prevention of the major industrial accident may be strongly preferred over prevention of the domino effect disaster ($10^{-3} >>> 10^{-9}$). However, if moral principles are fully considered and quantified as displayed in Table 4.9, this illustrative example shows that the Q&Q risk index for the domino effect disaster scenario might become much higher than that of the major industrial accident scenario. Moreover, whereas the multiplication factor between rational risk and R^* in the case of the major accident scenario is only 3, the multiplication factor between rational risk and R^* in the case of the domino effect disaster scenario sums to 30, in turn demonstrating the importance of taking into account principles such as fairness and equity when calculating risk. Such a picture may have an impact on a managerial safety investment decision, or it may not, but at least it offers more information for the decision-making process, leading to a more informed decision.

Clearly, this example should only be seen as illustrative, showing how moral principles might be integrated into risk calculation and hence influence safety decision-making with respect to the allocation of safety budgets. The rational risk calculations will always be very important to obtain a thorough idea of the real situation with respect to risk prioritization, but moral principles can help to give new and thorough insights into of the perceived situation with respect to risk prioritization.

4.10.4.2   Micromorts – the "Units of Death"

The so-called "micromort" can be seen as a "unit of death," as it is defined as a one-in-a million probability of death. Death can be sudden, or it can be chronic. As operational risks in this book are looked upon as being acute and not chronic (this is dealt with by health management), the micromort can be used in terms of acute death. A one-in-a-million probability of death can also be explained as throwing 20 coins simultaneously into the air and them all coming up heads in one go, which corresponds to the probability of a micromort (1 in 10^6 is roughly 1 in 2^{20}).

The micromort allows completely different risks to be compared with one another, as the units provide an idea of the riskiness of various day-to-day activities. For instance, if the risk of dying while carrying out a certain operation/task/handling in an organization is calculated to be 1 in 100 000, one can expect 10 fatalities in every million operations/tasks/handling carried out in the organization. This can be described as 10 micromorts per operation/task/handling. This number can then be compared with other average risks (which are perhaps more familiar to laypeople) such as traveling 160 km by bicycle or traveling 3700 km by car, which also corresponds to 10 micromorts. Micromorts can thus obviously be employed to compare and communicate small risks.

As an example, as the number of people dying in the Netherlands per year from non-natural causes is approximately 6000 and the number of inhabitants in the Netherlands is roughly 17 000 000 people, the average risk of dying in the Netherlands due to unnatural causes is $6000/(17 \times 365) \approx 1$ micromorts/day.

So a micromort can be seen as the average "ration" of lethal risk that people spend every day, and which they do not worry about unduly.

Another application of micromorts is the measurement of values that humans place on risk. Consider, for example, the amount of money one would have to pay a person to get him or her to accept a one-in-a-million chance of death, the so-called willingness to accept (WTA), or, conversely, the amount that someone might be willing to pay to avoid a one-in-a-million chance of death, the so-called willingness to pay (WTP; see also the following section). However, utility functions are often not linear, and the more that people have spent on their safety, the less they will be willing to spend to further increase their safety, and hence one should be very careful in upscaling numbers. A €50 valuation of a micromort does not necessarily mean that a human life (corresponding to 1 million micromorts) should be valued at €50 000 000. Conversely, if human life is valued at, say, €5 000 000 (see also the following section), this does not necessarily mean that a micromort value corresponds to €5. The concept becomes even more difficult to use if accidents are considered with a number of fatalities at once.

4.10.4.3   Monetizing Risk and Value of Human Life

The risk thermostat model applied to organizations, explained in the previous section shows how operational safety decisions come about. An organization's balancing risk policy is assumed to seek an optimal trade-off between the potential benefits and the potential costs induced by the risk policy. There is one problem though: costs and benefits are numerous, various, multifaceted, and often considered incommensurable. There is no established or accepted theory about the balancing act of an organization's operational risk policy. There is, however, a large body of literature in economics on cost-benefit analysis that insists that decisions can be rational only if they are the result of mathematical calculation in which all relevant considerations have been rendered measurable; the common unit for measurements preferred by economists is money.

The next chapter treats cost-benefit analyses in depth, but it is useful to indicate in this section that there may be some problems with monetizing risk, as is sometimes required for determining the costs and the benefits associated with operational safety decisions. As Adams [103] explains, the purpose of cost-benefit analyses is to produce a Pareto improvement, if possible. A Pareto improvement is a change that will make at least one stakeholder better off and no one worse off. As any operational safety project that will produce this result is rare to nonexistent, economists commonly modify the objective toward producing a "potential" Pareto improvement. Hence, the purpose is then that the "winners" of the safety project compensate the "losers" out of their winnings and in such a way that they would still have something left over. In order to answer this question, there is a need to compare, in some way, the winnings and the losses, leading to the concept of risk monetization.

However, not everything can be translated directly into monetary terms, e.g., body parts or human life. For those matters that cannot be monetized directly, it is assumed that their valuation into money can be inferred from the values of those matters that can be monetized directly. However, there is a lot of controversy and debate surrounding this topic. Kelman [132], for example, argues that the mere act of attaching a price to a quality or condition changes it fundamentally. He reasons that unpriced values (e.g., a hand that is lost due to an industrial accident) provoke solidarity, as they are directly rooted in the feelings and experiences shared among social beings; by introducing prices (e.g., €50 000 as the price of a hand lost) these things are reduced to the level of commodities about which people are socialized to respond to individualistically and even competitively. This is on a societal level. In an organizational setting, however – which is the focus of this book – and with the aim to provide inputs into company decision processes about taking various safety actions to reduce risks, this argument can be challenged. In organizations, employees have the goal of making profits. Hence, every working day they trade off all kinds of operational risks against the profits that are made through the company's activities. It would therefore be irrational for employees to respond differently to a risk or opportunity that they know well and that they deal with day by day (thus systematically), just because a number is attached to it. Employees within an organization have chosen to work within this particular organization, to carry out specific tasks, and to implicitly accept the risks of the activities in return for a paycheck. Another argument is that people make decisions, but they also have decisions made for them. This may be true, but in an organization, if safety is calculated as explained before (using moral aspects) and people are treated from the point of view of respect and self-leadership, the monetization of risks is possible/defensible/allowed .

Another way of looking at this is that monetizing risks in an organizational setting is all about measuring preferences, not about measuring intrinsic values of operational safety. Economic values and intrinsic values are different. The use of a monetary standard in an organization is convenient and should be looked upon from a pragmatic perspective, i.e., it is employed to measure relative values (as opposed to actual values, because they are incommensurable). The question that organizations really want to answer by using risk monetization techniques, is as follows: Are the benefits (accidents avoided) following the taking of safety measures big enough to outweigh the operational safety investment decisions (costs of safety actions), or not? The answer may then be used as one (but, of course, only one) input to a decision process. As Wilson and Crouch [74] explain, in many cases, the comparison between risks is easy. The risk may be so large that once it is perceived, it needs action; this is the case if the risks are situated in the unacceptable region. In another case, the risk may be so small that it is not worth the trouble thinking about what actions to take; this is the case if risks are situated in the negligible region. Such cases are not worth highlighting as there is no controversy. Still other risks may belong to one of these two regions, but they may not be perceived in this way, due to there being no calculations or incorrect calculations/perceptions. The challenge for company decision-making, however, lies in the risks belonging to the ALARP region.

The coefficient for the equivalent of a monetary cost of a risk inevitably causes controversy, argument, and debate. When used by a company, it should, however, be regarded as an approximation to the organization's safety management's idealized safety utility function. By using such a coefficient, the goal is to obtain a measure of the utility of a risk by assigning a numerical value per unit of risk, usually expressed in monetary terms. Let's refer to this coefficient as gamma (γ). The coefficient does not need to be a constant, independent of the size and nature of the risk. This is explained in the following.

Consider a risk R to a company. The value of γ is determined by the level of the risk, which in turn depends on the circumstances. Thus, γ is function of R, implying it should be written as $\gamma(R)$. Besides being consistent with theories accompanying utility models and models of perception, it can easily be understood. Let's assume that a company has a fixed safety budget to spend, and all factors in the risk calculation formula are constant except for the probability p. Consider a company faced with a major accident scenario probability of, for example, one in a million (10^{-6})/year. The company would be prepared to invest a certain amount of the available (fixed) safety budget in order to tackle the risk. In case of a larger risk, e.g. of a probability of 10^{-4}, the company may be willing to invest a proportionally larger amount of the available safety budget (in the example, 100-fold larger), in exchange for dealing with the risk. However, if the risk were to become high enough, the limit of the available fixed budget that can be employed for safety investments might be reached, and the amount per unit of risk should logically decrease thereafter. Figure 4.23 displays this $\gamma(R)$ curve.

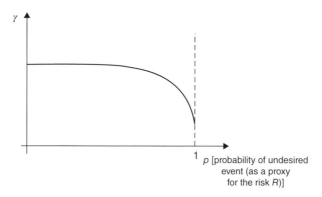

Figure 4.23 Safety investment (to deal with the risk) per unit of risk, in the case of a fixed safety budget.

All by all, every risk can be monetized in some way with finite resources (although they may sometimes be very high) and thus a fixed budget can be used – but there is one exception, human life, which is considered priceless. However, the figure changes when the budget is not fixed to a certain maximum amount. Replacing a "company" with an "individual," and the "probability of a major accident scenario" with the "probability of dying," the coefficient assigning a numerical value per unit of risk can be determined by asking people. The VoL or the value of statistical life (VoSL) is an extremely sensitive and much-debated subject. As the VoSL raises emotive feelings among many people, the concept is also often framed in terms of "the value of a life saved," "the cost of a fatality averted," or "the value of preventing a statistical fatality." Nonetheless, as will become clear, efforts to quantify human life are always accompanied by inconsistencies and ethical concerns.

The value of a human life can be discussed in different ways. Obviously, a life is valued much more than the total worth of the various chemical elements composing a body. Dorman [133] indicates that from the time of Hammurabi (eighteenth century BC), attempts have been made to establish the "value" of lives of different classes of people, primarily for the purposes of punishment. A prince would be worth so many peasants in the harsh calculation of Hammurabian justice. In recent history, the most important reason for setting a value on life stems from the problem of establishing awards in wrongful death judgments. The reasoning goes that it is obviously not enough to demand reimbursement for the direct economic costs of dying (e.g., burial and funeral rites), as this does not cover the opportunity cost, i.e., the income that a deceased person would have earned had they lived on. The justice system therefore gravitated toward a procedure that came to be called the "human capital" approach: calculating the present value of the income stream foregone due to premature death. This approach uses the principle that the value of a person's life increases with earning ability and expected longevity, and decreases with age and the interest rate chosen for discounting. Leonard [134] cites cases dating back to 1916 in which expected future earnings were used to compensate wrongful death. As indicated by Dorman [133], this approach can be contested, as from a social viewpoint and the principle that every life should be equal, it is perverse to value citizens only in their capacity to produce. Moreover, there are many reasons for doubting that a person's income necessarily corresponds to their economic contribution to society.

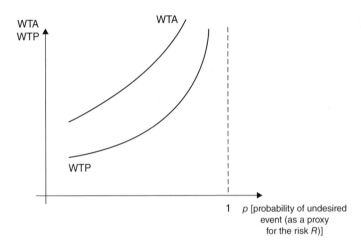

Figure 4.24   Willingness to accept (WTA) and willingness to pay (WTP) (no restrictions to the monetary possibilities imposed).

But how can life then be valued? Economic theorists took up this question in the 1970s and 1980s and devised theoretical models based on the premises of expected utility theory. In particular, these economists were interested in the relationship between an individual's WTP for a reduction of the risk of early death and the extra earnings that reduction would make possible, in order to demonstrate the insufficiency of the human capital approach. The measurements that were used to test the theories were done by a process called "contingent valuation," i.e., asking people. Indeed, the most straightforward approach to the complex problem of valuing changes in fatal (and non-fatal) risk would appear to be that of simply asking a cross-section of the population to state their own values.

In the case of operational safety, the way that people are asked is very important, as the monetary values that individuals place on a certain issue, and especially if that issue concerns human life, depend on how the question is posed. These are referred to as the "willingness to pay" and "willingness to accept" framing of the questions. The WTP refers to how much someone is prepared to pay to prevent a loss occurring, whereas the WTA refers to how much someone is prepared to accept in compensation for the loss of a benefit. In theory, the WTP value should be identical to the WTA value. In practice, it turns out that the WTA values are invariably higher than the WTP values, when individuals are asked to make preference choices. The key reason for this difference is an individual's preference to gain something rather than to lose something, as the negative impact of a loss is considered to be greater than the positive impact of an equivalent gain (cf. the theory on "loss aversion"; see also Section 4.1.3). Figure 4.24 shows that the values of WTP and WTA increase as the probability of the risk increases (when all factors in the risk calculation formula are kept constant, except for the probability p).

Figure 4.24 also illustrates that, if there is no maximum amount fixed, as the probability of the risk becomes very substantial it is plausible that no amount of money would be satisfactory. Jones-Lee developed a theory in 1989, improving on Dehez and Drèze [135] and Jones-Lee and Poncelet [136], based on the effect that changes in expected longevity have on the quality (not quantity) of potential life, which therefore sidesteps the confusion between accumulating years and accumulating utility. Jones-Lee's goal was to be able to quantitatively measure the lowest value of a risk increase consistent with an infinite compensation required to induce one to accept it, and the degree of risk aversion of an individual. In this way, it would be possible to construct a quantitative estimate for the required compensation to induce an individual to accept a certain risk increase. However, besides the differences in WTP and WTA values, there are other unique difficulties associated with the necessary measurements for determining the value of human life by asking individuals, such as, for example, in what narrative context such a demand makes sense, and how responses will change from one context (e.g., chemical industry) to another (e.g., cycling). Furthermore, if the probability of death is very low, people tend to assume that the adverse event will occur to someone else rather than to themselves, and their WTP value reflects this view (see also Fuller & Vassie [70]). Individuals' valuations of other people's lives also are likely to be lower than the value people put on their own lives. Valuations of life and people's WTP for improved control measures to prevent future loss of life also invariably increase after serious accidents (even if the probability of occurrence remains exactly the same). Moreover, when asking people, many respondents refuse to make such trade-offs, offering zero bids for risk reduction or saying that no level of compensation will induce them to accept an increase in risk. How does one interpret these refusals? And, as also indicated by Dorman [133], how much credence should be placed in responses to purely hypothetical questions, when respondents can simply throw out numbers with no apparent consequences for themselves and/or others? Finally, it is important to notice that valuations of life often vary between organizations, industrial sectors, and countries. These variations can be related to cultural and religious criteria, a company's ability to pay, the dread fear associated with some types of death, and the economic strength of different countries.

It becomes clear that there is no widely accepted method of measuring the VoL, and there probably never will be. But as it was settled earlier in this section that risk monetization is allowed and defensible, the question remains: how can averted fatalities be considered in decision-making for safety budget allocations in organizations?

In healthcare economics, the quality-adjusted life-year (QALY) is often used when trying to formally assess the value-for-money of a healthcare intervention. The QALY is a widely used measure of the burden of a disease, which includes both the quality and the quantity of life saved. It is based on the number of years of life that would be added by an intervention. Each year of perfect health is assigned the value of 1, down to the value of 0 for death. If extra years would not be lived in full health, for example, if the person were to lose a hand, or be blind, and so on, then the extra life-years are given a value between 0 and 1 to take this into account. There are some important drawbacks of the model, leading to the observation that it is hard, if not impossible, to employ in an organizational context. QALYs are designed for cost–utility analyses, and do not provide information on the valuation of life, for instance. Furthermore, QALYs require utility-independent, risk-neutral, and constant proportional trade-off behavior, all conditions that are hard to meet in organizational contexts. To use QALYs, organizations should also be able to assess the health states of their individual employees for possible safety investments they are able to choose from, which is most probably impossible. If one further considers the fact that perfect health is very hard to define, and that some health states may be worse than death (at least for a number of people), the meaning and usefulness of QALYs may be contested even more.

How can the averting of fatalities then be used to measure preferences for safety budget allocations? Let's return to the WTP attempts to value human life. Notice that WTP studies are used in the remainder of this section, because WTA studies are much less reliable than WTP studies due to the massive number of refusals to accept compensation in WTA studies. Based on a large number of value-to-life WTP surveys, different countries have put forward different values per statistical life, to be used, for example, in societal cost-benefit analyses. Table 4.10 provides some VoSL values found in a small selection of different countries.

In fact, over recent decades, a large variety of studies have been carried out worldwide to determine the VoSL, and an equally large variety of values have been found. Research by Button [141] indicates that, at the time of the research, the official value of a human life varied between €12 000 and €2.35 million. In the cross-country paper by Miller [142], amounts vary between as low as US $40 000 in Bangladesh to as high as US $8.28 million in Japan (both in 1995 dollars). But Miller also shows minimum and maximum amounts of the VoL for individual countries, diverging substantially. Other studies mention other amounts (see, e.g., Sunstein [143]). Roughly, in current euro values, the amounts from the various studies range between around €50 000 to €25 million, if looked upon from a worldwide perspective. Hence, the variation between lowest and highest VoSLs is characterized by a factor of 500. Although the values are so different, it is possible to find some systematicity in the figures. In fact, Miller [142] shows that across countries, the average VoSL varies almost linearly with income. Hence, the so-called "human capital" approach was refused in part because it simply translated greater worldly success into a higher personal "worth" or value, but now the same results appear in the contingent valuation approach. But this should not be surprising, as any attempt to derive values from consumer demand in the marketplace will attach greater weight to the most affluent consumers. In other words, highly paid workers have a higher perceived VoL.

Nonetheless, the problem of using a VoSL in an organizational context can be solved. To use the VoSL in a company situated in a specific country, it is important to realize that, as already mentioned, the value is a relative one, and not intrinsic. This means that different values may be used by different plants belonging to one and the same organization, but situated in different countries. As the values need to be used in a relative way (to make decisions about budget allocation), all the other monetized risk values also depend on the country where the monetization exercise is carried out. Moreover, depending on the probability of the risk, a

Table 4.10   Some "value of statistical life" (VoSL) numbers used in different countries

Country	VoSL (approximate values) (€ millions)	References
Australia	3.0	Australian Government [137]
The Netherlands	5.8	IENM [138]
United Kingdom	2.4	HSE [139]
United States of America	7.0	DoT [140, 141]

All amounts have been recalculated to 2014 euros, for the purposes of easy comparison.

different VoSL might need to be employed. This is explained by Sunstein [143]. Instead of saying that "a VoSL value is €5 000 000," for example, it would be more accurate to say that "for risks of 1/10 000, the median WTP, in the relevant population, is €500." Or that for risks of 1 in 100 000, the median WTP value is €50. However, it is absolutely not clear whether people's WTP to reduce risks is linear. Suppose that people would be willing to pay €50 to reduce a risk of 10^{-5}. Does it follow from this that people would be willing to pay only €5 to eliminate a risk of 10^{-6}, or that, to eliminate a risk of 10^{-3}, people would be prepared to pay €5000? Not necessarily. As the consequence remains the same (i.e., death), it is necessary to compare probabilities of death with one another to have an idea of how people perceive probabilities; i.e., do people display probability-seeking, -neutral, or -averse behavior? Notice at this point that it was already established that people are consequence-averse. Thus, they are, from a relative point of view, prepared to pay more for higher consequence risks (*ceteris paribus*). This was also discussed in the section on group risk: multiple fatalities count more than single fatalities (see Section 4.10.2.2).

People are very bad at dealing with probabilities, especially if they are extremely low (so-called type II risks). Nonetheless, if people were asked about their WTP based on explicit probability numbers, it may be assumed that they are rationally prepared to pay more for lowering higher-probability than lowering lower-probability type II risks. If this is the case, the VoSL for a 10^{-4} risk would have to be higher than the VoSL for a 10^{-6} risk. In fact, the VoSL figure is mostly used by agencies around the globe for the range of risks between 1 in 10 000 and 1 in 1 000 000. In Europe, for example, if a figure of €6 milion is used (e.g., in the Netherlands), it may be more optimal to use several figures: €8 million for risks of 1 in 10 000; €6 million for risks of 1 in 100 000; and €4 million for risks of 1 in 1 000 000. A minimum and a maximum amount would need to be fixed preferably between the 10^{-4} risk and the 10^{-6} risk, so that for risks of 1 in 10 000 000, for example, the VoSL would be €4 million in the Netherlands. In this way, it is possible for an organization to make budget allocation decisions in a relative way, and while using local economic information and taking both the probabilities and the consequences of risks into account.

Ale *et al.* [107] indicate that in the UK, such an approach, i.e., using a higher VoSL for risks close to the limit of intolerability, has been employed in the past. However, from the more recent amended Railways and Other Guided Transport systems (Safety) Regulations 2006, it seems that this difference is no longer used. Nonetheless, there is no reason – scientific or otherwise – given in favor of or against this approach.

4.11   Safety Investment Decision-making for the Different Types of Risk

4.11.1   Safety Investment Decision-making in the Case of Type I Risks

In the case of type I risks, it is possible to specify a concrete break-even safety point for each individual company. First the degree of safety pertaining to the company should be determined. As this degree of safety is relative to the other companies working in the same industrial sector, a defendable assumption would be that a 50% degree of safety corresponds to the LTIFR in the industrial sector of the country or region where the company is situated and that a maximal degree of safety of $(100 - \varepsilon)\%$ corresponds to a corporate LTIFR of 0. Remember from Section 4.6 that LTI frequency rates are used in the case of type I accidents as lagging indicators, and that the safety measured is thus actually "reactive safety," so it

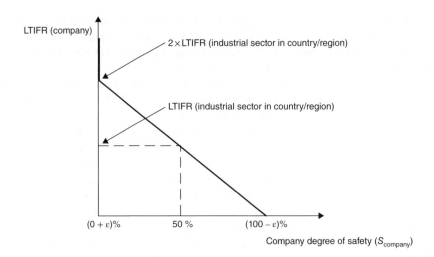

Figure 4.25 Lost time injury frequency rate (LTIFR) of an enterprise in terms of its degree of reactive safety.

might also be possible to see the degree of safety as some measure of "un-safety." Nonetheless, for the sake of simplicity, it is called the "degree of safety." Figure 4.25 illustrates a linear relationship between the degree of safety of a company and its LTIFR. Other linear relationships are possible and safety managers may thus use a company-specific relationship.

The illustrative relation from Figure 4.25 can be set out as follows:

$$S_{company} = 1 - \frac{(LTIFR\ company)}{2 \times (LTIFR\ of\ the\ industrial\ sector\ in\ country/region)},$$

wherein it is assumed that a minimum degree of safety is equal to $(0 + \varepsilon)\%$ (hence $S_{company} \geq 0$) and wherein $\varepsilon \approx 0$.

The model proposed is designed to help with objective investment decision-making for type I risks. In order to provide management decision-making with an objective input, the model should prove that a safety investment for type I risks, corresponding to the degree of safety pertinent to the enterprise, delivers the benefits expected. To achieve this, each company needs to specify in advance the desired break-even safety target ratio (or "BEST ratio"), defined by dividing the calculated expected hypothetical benefits (using definition (ii) of Section 4.2.4) by the SICs. This procedure resembles existing approaches in companies for determining the cost-effectiveness ratio (see also Chapter 6).

The BEST ratio should depend on the degree of safety determined for the company, in the sense that less safe companies (lower safety degree) are more in need of safety investments, and the SICs may thus, in a relative way, be higher than the expected hypothetical benefits for such low-safety companies, certainly when compared with safer companies (higher safety degree). For example, Table 4.11 can be developed and employed by a company safety manager.

By using a table such as Table 4.11, an enterprise can facilitate the attaining of its objective of an envisioned degree of company safety related to type I risks. Notice further that the relationship shown in Figure 4.25 does not need to be linear and other choices can be made by a company. Therefore, the overall working approach needs to be well considered in advance.

In order to obtain an objective comparison from a corporate point of view, the condition of achieving the BEST ratio can then be used to specify whether the safety investment is recommended or should be reworked. The model suggested in this section thus provides input for corporate management decision-making on safety investments for type I risks. Following an eventual reformulation of the safety investments as a result of applying the model (e.g., the model results in a ratio lower than the BEST ratio specified in Table 4.11), the economic exercise should be redone.

To illustrate this approach, assume, for example, a chemical company examining the extent to which investing in safety during the creation of a new chemical installation would prove profitable (or not). The life span of a new chemical installation requiring no major adaptation is estimated at 10–20 years, on average 15 years. If taking the life span of an installation into consideration, a one-off investment/benefit can be assumed to correspond to an investment/benefit of 15 years and is thus written off after this period. The lifetime per investment and per hypothetical benefit should also be calculated. The life of all investments/benefits can then be reduced to a 3-year operating period of the chemical installation. In other words, 3 years is also the duration used to determine the average LTIFR. The discount rate may, for example, be chosen to be 5% annually. Assume further that LTIFR (company, average of 3 years) = 4.7 and LTIFR (industrial sector, average of 3 years) = 11.3. Assume also that the safety investments (calculated for a 3-year period) amount to €80 000 and that the expected hypothetical benefits [calculated using definition (ii) of Section 4.2.4 for a 3-year period] amount to €170 000 (for concretely determining safety investments and avoided accident costs or hypothetical benefits, see Chapter 5). The BEST ratio Table 4.11 is used.

With these illustrative figures, the degree of safety of the company is 0.79 (dimensionless number), which means that a minimum BEST ratio of 1.5 (dimensionless number) should be reached according to Table 4.11. The ratio between the expected hypothetical safety benefits and the total SICs is equal to 2.1 (dimensionless number), thus indeed satisfying the recommendation (as it is > 1.5). Hence, in the case of this illustrative example, no additional economic study would be needed with a view to reformulating safety investments, and the suggested SIC is recommended.

Table 4.11   Illustrative example of break-even safety target (BEST) ratios related to company safety degrees

Company safety degree for type I risks	Minimum BEST ratio
0–0.25	0.2
> 0.25–0.50	0.5
> 0.50–0.70	1.0
> 0.70–0.90	1.5
> 0.90–1.0	2.0

4.11.2   Safety Investment Decision-making for Type II Risks

In the case of type II risks, it is impossible to use LTIFRs to help decision-making, because they provide no information at all about the number of disasters (i.e., type II accidents) that have happened. Therefore, another approach of dealing with such risks needs to be elaborated and employed.

The technique that can be used, is expounded in the Section 5.7. The approach uses a so-called disproportion factor (DF; see Sections 4.11.2.1 and 8.12) and recommends, based on an organization's evaluation of the numerical value of the DF where the net present value (see Chapter 5) becomes zero, whether or not to invest in a prevention decision for type II risks. Hereafter, it is explained how the DF being considered acceptable/usable by the company can be determined, taking the FN curve (explained in this section) as well as societal concerns (Section 4.11.3) into consideration.

4.11.2.1   Disproportion Factor

Before giving more specifications about the DF, it is important to define the so-called proportion factor (PF), which is a term used to indicate the ratio of the costs to the benefits. The PF can then be compared with the numerical value of the DF in order to determine whether the risk reduction measure is "grossly disproportionate" or not.

If

$$PF = \frac{\text{Costs}}{\text{Benefits}} > DF,$$

it means that the safety investment is called "disproportionate" or "grossly disproportionate," and a further risk reduction would be too costly if compared with the extra benefit gained from the safety investment. Hence, in such cases, it would be recommended not to do the investment.

Concerning the range of values that this factor can take, DF is necessarily greater than 1 because otherwise it would mean a bias against safety (or in favor of "un-safety"). According to Goose [144], the value of the DF is rarely over 10 and should never be higher than 30.

Theoretically, the DF is higher when the risk increases and is infinite when the risk level reaches the intolerable region, meaning in fact that the risk must be reduced no matter the cost.

4.11.2.2   Numerical Estimation of the DF

A scheme of calculation has been suggested by Goose [144], using three intermediate factors, called the "how factors," which, when multiplied together, give an estimation of the value of the DF. They are referred to as "how bad," "how risky," and "how variable." They are calculated through the use of three values which can be extracted from an FN curve (see Sections 4.10.2.2 and 8.12). These values are as follow:

1. the sum of the failure rates, expressed in events per year (Σ FR);
2. the "EV" which is sometimes also called potential loss of life and represents the average number of casualties expected per year;
3. the maximum potential fatalities (N_{max}), which is the worst-case scenario concerning the number of fatalities for a single event;
4. a fourth value can be calculated with the ratio of EV to Σ FR, representing the average number of fatalities per event (N_{av}).

It should be noted that this method is based on fatality risk alone and the factors are not fully independent, which introduces non-linear weighting. For example, harm is counted twice in the first two factors.

There are different calculation models developed by Goose [144]. Only the "suggested method" by Goose will be explained further. The approach to calculate the three "how" factors is as follows. The "how bad" factor represents the effect of $N_{av} = \frac{EV}{\Sigma FR}$, the average number of fatalities per event, on the DF:

$$\text{"How bad"} = \log_{10} N_{av}$$

The "how risky" factor represents the effect of EV, on the DF:

$$\text{"How risky"} = \log_{10}(10^5 \times EV)$$

The "how variable" factor represents the effect of the ratio $\frac{N_{max}}{N_{av}}$ on the DF:

$$\text{"How variable"} = \log_{10}\left(\frac{N_{max}}{N_{av}}\right)$$

The DF is then computed by multiplying the "how" factors and adding 3 (dimensionless) to this product:

$$DF = \log_{10}(N_{av}) \times \log_{10}(10^5 \times EV) \times \log_{10}\left(\frac{N_{max}}{N_{av}}\right) + 3 \qquad (4.2)$$

4.11.2.3 Hypothetical and Illustrative Example of the Calculation of an FN Curve-based DF

In this example, the goal is to show how an FN curve is plotted and how in practice it is possible to extract the three values necessary to estimate the DF from the curve. A hypothetical example will be used in order to illustrate the subject. Throughout the example, N will represent the estimated number of fatalities in each event and f will represent the frequency of each event. Before beginning, care should be taken that there are not different values of frequency corresponding to the same N. If this were the case, a simple addition for each frequency of the same N would solve this issue.

As explained in Section 4.10.2.2, an FN plot is used to present societal risks and plots the cumulative frequency F as a function of the number of casualties N. F is the sum of frequencies of all events where the outcome is N or higher. Data are recovered from records of historical incidents or from quantitative risk analysis. Conventionally, a log–log plot is used because of the wide range of values for F and N.

Step 1. Plotting the FN Curve

In order to plot an FN curve, the fatalities have to be sorted in decreasing order and therefore N_{max} becomes the first input in the table. A possible presentation of the data is shown in Table 4.12 and an example of an FN curve once it has been plotted is shown in Figure 4.26.

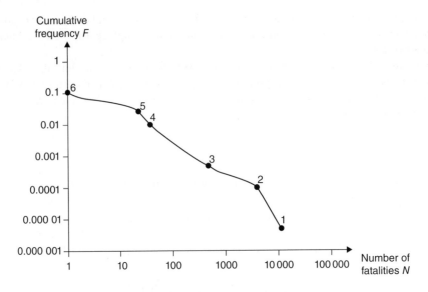

Figure 4.26 Example of a plotted FN curve.

Step 2. Extraction of Useful Data

The goal at this point is to find the value of Σ FR, EV, and N_{max} in order to use them to calculate the DF. These data can be derived from Figure 4.26, as illustrated in Figure 4.27.

In Figure 4.27, it is possible to see how to graphically extract the first 2 values Σ FR and N_{max}. In this case: Σ FR = 0.110605 and N_{max} = 13 000, Table 4.12.

The extraction of the EV is more difficult because theoretically it is the frequency-weighted average of the casualties:

$$EV = \sum_{i=1}^{N_{max}} f_i N_i.$$

with N_i and f_i the number of casualties and frequency of each event.

It is possible to see in Table 4.13 how to correctly calculate the EV from a data table. Unfortunately, an FN curve is rarely given with its corresponding data table. However, the EV can

Table 4.12 Illustrative example of a data table

Events	Event frequency f (per year)	Event outcome N (number of fatalities)	Cumulative frequency Σ FR
1	0.000 005	13 000	0.000 005
2	0.000 1	4 600	0.000 105
3	0.000 5	540	0.000 605
4	0.01	44	0.010 605
5	0.02	27	0.030 605
6	0.08	1	0.110 605

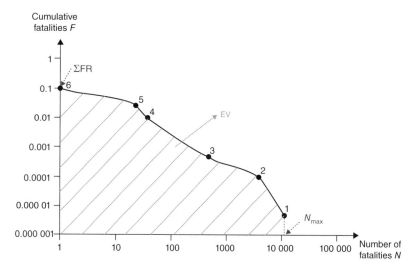

Figure 4.27   Useful data in an FN curve.

also be seen as the area under the curve, which is useful, and the reason stems from the fact that when there is a lot of data available, the following equation is true:

$$EV = \sum_{i=1}^{N_{max}} F_i(N_i).$$

with $F(N_i)$ the cumulative frequency for each event.

Step 3. Calculating the Disproportion Factor
The three options for the DF assessment will be dealt with separately and results will be gathered at the end. Table 4.14 sums up the data which will be used all along the estimation.

Table 4.13   Calculation of expectation value from the data table

Events	Event frequency f (per year)	Event outcome N (number of fatalities)	Expectation value for each event EV
1	0.000 005	13 000	0.065
2	0.000 1	4 600	0.46
3	0.000 5	540	0.27
4	0.01	44	0.44
5	0.02	27	0.54
6	0.08	1	0.08
		Sum of the values gives total expection value	**= 1.855**

Therefore:

$$\text{"How bad"} = \log_{10}(N_{av}) = 1.2,$$

$$\text{"How risky"} = \log_{10}(10^5 \times \text{EV}) = 5.3,$$

$$\text{"How variable"} = \log_{10}\left(\frac{N_{max}}{N_{av}}\right) = 2.9.$$

Hence, (Eq. 4.2):

$$\text{DF} = \text{How bad} \times \text{How risky} \times \text{How variable} + 3 = 22.$$

4.11.3 Calculation of the Disproportion Factor, taking Societal Acceptability of Risks into Account

In 1978, Fischhoff *et al.* [145] described a psychometric study in which it was demonstrated that feelings of dread were the major determinant of public perception and acceptance of risk for a wide range of hazards. A psychometric questionnaire was used in order to correlate nine characteristics of risks, resulting in two main factors. The factor "dread risk" included the following items: perceived lack of control, catasrophic potential, inequitable distribution of risks and benefits, fatal consequences and dread. The "unknown risk" factor consisted of the items observability, experts' and laypeople's knowledge about the risk, delay effect of potential damage (immediacy) and novelty (new-old).

In a study by Huang *et al.* [146], the goal was to find correlations between public perception of the chemical industry and its acceptance in the matter of risks. Based on a survey administered from 1190 participants, four different factors related to social acceptance were found, each one being subdivided into two sub-factors. The first factor, "knowledge," consisted of a newness factor and a knowledge factor. The second factor, "benefit," consisted of the benefit and immediacy together. "Effect" was matched with the third factor and divided between social effect and dread. Finally, "trust" included controllability and trust in governments.

The four factors were then linked with social acceptance through a regression analysis. It is important, however, to specify that knowledge should be understood from the perspective of citizens and that the social effect actually implies "How many citizens are exposed to the risk?" Moreover, the factor "trust in government" also concerns policymakers in a broader sense.

In a study by Gurian [147], the focus was more on how risks are perceived and considered by the general public and society, rather than focusing on the industry's point of view. Based on theoretical explanations, Gurian [147] defined three factors that influence risk perception. The first one, "dread," included gut-level (related to intuition), the emotional reaction due to

Table 4.14 Useful data for estimation

Cumulative frequency	Expectation value	N_{max}	N_{av}	N_{max}/N_{av}
0.110 605	1855	13 000	17	775

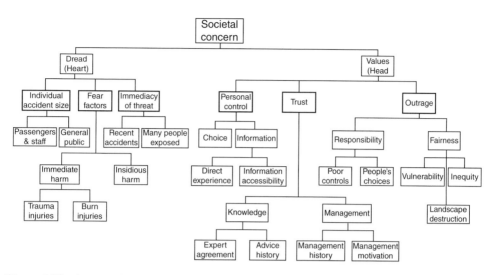

Figure 4.28   Acceptability model developed. Source: Adams [148]. Reproduced with permissions from Springer.

risk, threat to future generations, control over the risk, equitability and catastrophic potential. The second factor, "familiarity," comprised delayed effects, newness, understood by science or not, and encountered often by the public or not. Finally, the last factor was about the "number of people exposed to the risk."

Adams [148] used the risk thermostat model to describe the perception of risks and, again, a new classification of risks is given depending on three main factors: voluntariness, individual control, and profit-motivated. The acceptability model developed by Adams and resulting from the risk thermostat model is shown in Figure 4.28.

This study is very similar to those of Gurian [147], Fischhoff *et al.* [145] and Huang *et al.* [146], as it focused on the perceived levels of risks to laypeople.

A study by Baker and Dunbar [149] takes societal concerns into account from a risk managers' point of view. The Baker model can be seen as an adaptation of a previous study by Mansfield [150]. Mansfield's more complex model describes a variety of societal concerns in the decision-making process. Baker and Dunbar [149] constructed a systematic framework for the Rail Safety and Standards Board (RSSB), which helped to determine the extent of societal concern in specific cases. They even mentioned that they believe a systematic framework is required so that the factors that have been considered in any facilitated evaluation are completely transparent. However, this model is focused on the transport and rail industry and cannot be used directly for industry in general.

Figure 4.29 shows how the model is structured with the six high-level concern factors. As found in several studies, they use spider diagrams to present the results, which make for an interesting visual representation.

In a study by Sandman [151], the goal was to develop software that allows a prediction and management solutions to deal with what they call "outrage," which is closely related to the notion of societal acceptability. Sandman defined risk as the sum of hazard and

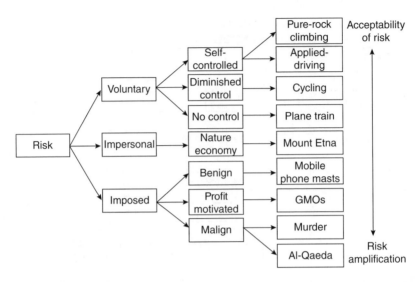

Figure 4.29 Structure of the model developed for the rail industry. Source: Baker and Dunbar [149]. Reproduced with permission of Rail Safety and Standards Board.)

outrage, with hazard being the standard, well-known term used by risk managers (hazard = magnitude × probability) and outrage being related to the public's perception of risk. The idea is that risk assessors see risk only as hazard, e.g., only with the expected annual mortality, whereas the public sees it as a combination of hazard and outrage. Twelve outrage items are then given and can be described in this case using two words at a time. The first one stands for lower chances of outrage and the second for higher chances of outrage: voluntary versus coerced, natural versus industrial, familiar versus exotic, not memorable versus memorable, not dreaded versus dreaded, chronic versus catastrophic, knowable versus unknowable, controlled by us versus controlled by them, fair versus unfair, morally irrelevant versus morally relevant, trusted versus not trusted, and responsive process versus unresponsive process.

4.11.3.1 Societal Acceptability Indicators with Respect to Disaster-related Risks

In this section, indicators provided in the literature influencing societal acceptability with respect to disaster-related risks are described and a brief comment and description are given for the approach being described. In fact, the indicators mentioned in the literature all play a role when it comes to determining the possible acceptability of a level of risk. They are, however, not always relevant when the focus should be on disaster-related risks, and from the viewpoint of company safety management. The framework of this work is therefore constrained to the industrial sector making risk assessments for major accident prevention. That is why only those indicators concerned with societal acceptability related to major accident risks will be presented, and it will also be explained how these indicators can be organized and grouped together, and finally how they can used in the approach that we describe and propose. In fact, the main goal in this part is also to draw parallels between different indicators used in literature which could refer to the same idea or concept and therefore be merged under the same indicator. The 11 indicators are as follows:

Indicator no. 1 – Trust in governments, experts, and company (technical community)

This indicator reflects the confidence in those responsible for understanding and managing the risks. It is sometimes also seen as "credibility," meaning that there is regulation and effective enforcement. In the literature, the indicator is mentioned in Mansfield [150], Baker and Dunbar [149], Sandman [151], and Huang *et al.* [146].

Indicator no. 2 – Allocation of risks and benefits

This indicator is concerned with whether there is a fair distribution of benefits for the bearers of risk. It is mentioned in Huang *et al.* [146] with a discussion about local economic development. The tricky point here is that an industrial activity might always be seen as more beneficial for the company (from the risk bearers' point of view) and therefore might create a bias in the scoring model to be developed in this section. It might also be correlated with profit motivation [148]. In the literature, the indicator is also mentioned in Otway [152], Baker and Dunbar [149], Sandman [151], and Huang *et al.* [146].

Indicator no. 3 – History of bad advice

The indicator concerns the history of the company regarding the given risk and generally if there were recent accidents involving the company itself. The indicator can also be referred to as the history of bad practices. In the literature, it is mentioned in Mansfield [150] and Baker and Dunbar [149].

Indicator no. 4 – Common-dread

An indicator such as "common-dread" is rather subjective and it might be possible to group it with personal experience (or difficulties in conceptualizing the risk exposure). The indicator can also be referred to as fear of harm [149], emotional reaction, or gut-level [147]. In the literature, it is mentioned in Fischhoff *et al.* [145], Mansfield [150], and Sandman [151].

Indicator no. 5 – Man-made versus natural causes

This indicator is related to whether the accident is due solely to natural causes or whether it is a modern technological catastrophe. For example, an accident caused only by hard-to-predict natural causes may be more accepted than one caused by technical failure. In the literature, it is mentioned in Otway [152], Mansfield [150], Baker and Dunbar [149], and Sandman [151].

Indicator no. 6 – Scientific knowledge

This indicator is also referred to as "uncertainties." It is defined as the experts' knowledge and agreement regarding the considered risk. In the literature, it is mentioned in Fischhoff *et al.* [145], Otway [152], Mansfield [150], Baker and Dunbar [149], Gurian [147], and Sandman [151].

Indicator no. 7 – Lack of personal experience

Sometimes grouped with personal knowledge, this indicator is also a sensitive and subjective factor. In the literature, it is mentioned in Otway [152], Mansfield [150], and Baker and Dunbar [149].

Indicator no. 8 – History of the risk itself

This indicator is related to the history of the hazard itself (for risk bearers), e.g., if there was a recent accident involving the same hazard. It can also be referred to as the frequency of the hazard. In the literature, it is mentioned in Baker and Dunbar [149] and Sandman [151], which describes it as comprising accidents that linger in the public's mind. It is also closely related to personal experience as well as news, fiction symbols and even signals such as an odor.

Indicator no. 9 – Public's knowledge

This indicator is also referred to as uncertainties to the risk bearers or accessibility to reliable information (and even difficulties in conceptualizing the risk exposure). It is hard to assess because it is quite subjective. In the literature, it is mentioned in Fischhoff *et al.* [145], Otway [152], Mansfield [150], Baker and Dunbar [149], Sandman [151], and Huang *et al.* [146].

Indicator no. 10 – Environmental/ecological impact

It is difficult to assess how important this indicator is to people, even though environmental issues now raise many more concerns than previously. In the literature, it is mentioned in Mansfield [150] and Baker and Dunbar [149].

Indicator no. 11 – Lack of personal control over the hazard/situation

This indicator can be seen as the possibility of avoidance thanks to personal skills, for example, or the existence of alternatives for the risk bearers. In the literature, it is mentioned in Fischhoff *et al.* [145], Otway [152], Baker and Dunbar [149], Adams [148], Sandman [151], and Huang *et al.* [146].

4.11.3.2   Factor Analysis and Resulting Classifications of Relevant Factors

In order to classify the different indicators, a factor analysis was conducted, the results of which led to the development of the scoring system that follows [153]. The purpose of the study was to examine the factor structure of the societal acceptability of risks (SAR) scale [153]. The first factor consisted of four items (note that the "items" are refereed to as "indicators" in the previous section) and explained 24.73% of variance, the second and third factors each consisted of three items, explaining 11.78% and 10.10% of the variance, respectively, and the fourth factor consisted of only one item that explained 9.61% of the variance. Factor I (four items) consists of items related to the awareness of the company of the risk and fairness of dealing with it; factor II (three items) consists of items related to trust in risk bearers; factor III (three items) consists of items related to the characteristics of the major hazard and its consequences; and factor IV refers to trust in safety management. Figure 4.30 illustrates the results of the confirmatory factor analysis.

The factors presented as relevant have therefore been sorted and included into a scoring system which is explained in the following. The classification of each indicator into the main factors (F_I, F_{II}, F_{III}, and F_{IV}) has been done depending on their order of consistency in the factor analysis, but the same weight in the scoring system has been applied in the end due to small differences between them.

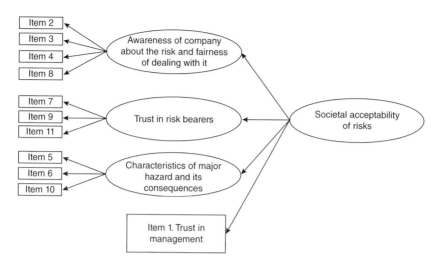

Figure 4.30   Second-order factor structure of the societal acceptability of risks.

4.11.3.3   Scoring System Development

Level of Societal Concern

The model developed in this section is a scoring system that can be filled in directly by safety management, e.g., by board members of a safety committee from the point of view of assessing societal concerns that might arise before or after an accident occurred, and that may thus be important when deciding on safety investments. The idea is therefore to try to picture how workers in a company or outsiders might view the safety management setup in their workplace and what kind of concerns they might have. There are, indeed, cases where the costs due to a strong public reaction were very high. This system therefore identifies, from the perspective of the company managing the hazard, the potential factors influencing this type of situation.

The different factors as listed in the previous section can be grouped into subcategories. Ideally, each subcategory tackles a "key idea" of SAR. As the standard calculation of the DF as proposed by Goose [144] already takes into account the maximum number of casualties (N_{max}) and the average number of fatalities per year (EV) and per event (N_{av}), these factors (which are often described as the severity of consequences) should not be considered again in the model that is being developed. As already explained, the goal of the model is to offer an approach regarding how to include moral aspects in the DF, taking societal acceptance of major hazards into account for the industrial sector in general. It therefore incorporates each relevant factor into a scoring system that gives the global level of concern as an output. Depending on the level of societal concern, the DF is then modified accordingly.

Furthermore, the developed model needs to be designed keeping in mind that it should always possible to use a neutral score for each question. This implies that if one scores between 0 and 4, e.g., 2, this should indicate some kind of neutrality. In this way, it is possible for the user of the scoring system to "skip a question," e.g., if the indicator seems inappropriate or it does not fit the considered case.

F_I *(Four Indicators) – "Awareness of Company about the Risk and Fairness of Its Manage-ment" (16 Points Maximum):*

F_I-1 – History of the risk itself for the risk bearers

The indicator "history of the risk itself for the risk bearers" questions whether there were recent accidents involving the same hazard. (And is it relevant, meaning can people still recall this event?) It can be seen also as previous accidents drawing a lot of media attention or strong protests.

The score that needs to be given by the user of the scoring system for calculating the DF* varies from 0 (which corresponds to hardly any records of accidents of the same type) through 2 (neutral score) to 4 (corresponding to many records of accidents of the same type).

F_I-2 – History of bad advice

The indicator "history of bad advice" is a reflection of whether the company has had a large-scale industrial accident before, which it can be assumed is still in the risk bearers' memory.

The score that needs to be given by the user of the scoring system for calculating the DF* varies from 0 (which corresponds to excellent safety records for the company) through 2 (neutral score) to 4 (corresponding to poor safety records for the company).

F_I-3 – Fair allocation of risks and benefits

The main idea of this indicator "fair distribution of risks and benefits" is to determine whether the risks and benefits are fairly distributed among the risk bearers and the risk beneficiaries. In fact, the aim is to avoid cases of moral hazards when one person is willing to maintain a risky situation because he or she does not bear the consequences and only has the benefits.

The score that needs to be given by the user of the scoring system for calculating the DF* varies from 0 (which corresponds to a fair distribution of risks and benefits) through 2 (neutral score) to 4 (corresponding to an unfair distribution of risks and benefits).

F_I-4 – Common-dread

The indicator "common-dread" questions whether the risk bearers are able to reason in a rational way when facing the hazard/situation. Is it possible that risk bearers have a great sense of dread that would influence their reaction? The idea here is to determine the level of fear of harm or the level of emotional "gut reaction." It can also be seen as a possibility of dramatization of the risk by the public.

The score that needs to be given by the user of the scoring system for calculating the DF* varies from 0 (corresponding to a hazard that is not very frightening) through 2 (neutral score) to 4 (corresponding to a great sense of dread associated with the considered hazard).

F_{II} *(Three Indicators) – "Trust in Risk Bearers" (12 Points Maximum):*

F_{II}-1 – Individual control over the hazard/situation

The indicator "individual control over the hazard/situation" tries to capture the level of per-sonal control over the situation for the person exposed to the risk. It can also be seen as the availability of alternatives when facing the hazard. With this factor, the possible response skills of the risk bearers can be evaluated.

The score that needs to be given by the user of the scoring system for calculating the DF* varies from 0 (which corresponds to total control for the individual) through 2 (neutral score) to 4 (corresponding to no control at all for the individual).

F_{II}-2 – Lack of personal experience

The indicator "lack of personal experience" verifies whether there is a lack of personal experience among risk bearers for the considered risk. In fact, if risk bearers have personal experience with the risk/situation, the risk is more likely to be accepted.

The score that needs to be given by the user of the scoring system for calculating the DF* varies from 0 (which corresponds to a lot of experience for the individual) over 2 (neutral score) to 4 (corresponding to no experience at all for the individual).

F_{II}-3 – Uncertainties: public's point of view

The indicator "uncertainties: public''s point of view' can be seen as the public's knowledge of the considered risk, i.e., how much does the public at large, or more specifically the risk bearers, know about the hazard? Is it well known by the public? Can people easily understand this risk? It can also be seen as the accessibility to reliable information.

The score that needs to be given by the user of the scoring system for calculating the DF* varies from 0 (which corresponds to easily known by the public) through 2 (neutral score) to 4 (corresponding to great uncertainty or a complex hazard that is difficult to understand).

F_{III} (Three indicators) – "Characteristics of the Major Hazard and Its Consequences" (12 Points Maximum):

F_{III}-1 – Uncertainties: science point of view

The indicator "uncertainties: science point of view" can be seen as the experts' knowledge and agreement concerning the studied hazard. How much does science know about the risk? Is it well known by the technical community? Do experts agree on how to manage the risk?

The score that needs to be given by the user of the scoring system for calculating the DF* varies from 0 (which corresponds to perfectly known by the technical community) through 2 (neutral score) to 4 (corresponding to great uncertainty or no agreement at all among scientists).

F_{III}-2 – Environmental impact

The indicator "environmental impact" assesses whether the hazard could have a (major) environmental impact or not. For example, a chemical leak into a river or a radioactive fallout could have a major impact upon perception of the risk.

The score that needs to be given by the user of the scoring system for calculating the DF* varies from 0 (which corresponds to no environmental impact) through 2 (neutral score) to 4 (corresponding to tremendous environmental impact).

F_{III}-3 – Man-made vs natural causes

The main idea of the indicator "man-made versus natural causes" is to determine whether the accident could be caused by human error or natural causes. Human error (score 2) includes, for example, a miscalculation of the resistance of a physical barrier or a mismanagement of the risk. A plane crash as a result of a storm is an example of a neutral score.

The score that needs to be given by the user of the scoring system for calculating the DF* varies from 0 (which corresponds to purely natural causes) through 2 (neutral score) to 4 (corresponding to purely human and/or technical causes).

F_{IV} *(One Indicator) – "Trust in Company's Safety Management" (Four Points Maximum):*
The indicator "trust in the company" (which in this case is also the factor) indicates to what extent it is believed that the people exposed to the risk (or the citizens most likely to criticize company policies) trust the company's risk management abilities? The main idea of this factor is to take into account the perception of risk bearers regarding the company's safety management.

The score that needs to be given by the user of the scoring system for calculating the DF* varies from 0 (which corresponds to a complete trust in management practices) through 2 (neutral score) to 4 (corresponding to a complete mistrust in management practices).

Effective Use of the Scoring System – Calculation of DF*

Different factors can determine whether or not societal concerns might arise from a situation. A value needs to be assigned to each indicator, so that a value is determined for each factor, eventually providing a corresponding weight factor in the equation to calculate the DF*.

However, suitable ranges of values need to be carefully chosen so that the original DF does not change too much. In fact, the idea is to have slight modifications of the value calculated with the economic model developed by Goose [144]. The estimated level of societal concern (low, neutral, or high) should merely be an indication of the value of the weight factors applied to the DF in order to give the modified DF*. This specific point has been verified by a sensitivity analysis that is described in the next section.

In the following model, the user should bear in mind that the advised range of values given by Goose [144] for the DF remains between 3 and 30 (DF ∈ [3;30]). It is possible to imagine a case where high societal concerns could raise the DF over 30, but that should only happen in very specific cases.

An Excel document can be created for this purpose, containing all the necessary calculations on a unique spreadsheet and therefore allowing direct use of the scoring system. The factor table summarizing the layout for each factor and indicator in the Excel document is presented in Figure 4.31. It is important to point out that the model described in what follows should only be considered as a guidance tool. It should therefore be seen as a suggestion and not as the only possible way to do it.

A separate spreadsheet can be created with the weight factors that should be applied depending on the total score for each main factor. An example of how the weight factors are chosen is given in Figure 4.32.

It should be noted that a choice needs to be made concerning the possible negative impact of this kind of assessment. It has been supposed that social impacts are rarely in favor of a company, especially in industry. That is why it is impossible to reduce in this model the value of the DF with a weight factor lower than 1.

The final step is then to multiply the initial value of the DF by the four weight factors (WF$_i$) as described in the following formula:

$$DF^* = DF \times WF_I \times WF_{II} \times WF_{III} \times WF_{IV}$$

Main factors	Indicators	Scores	Score guidance
F_I - 4 items - Awareness of company about the risk and fairness of its management (16 point max)	I-1 - History of the risk itself for the risk bearers		Between 0 and 4
	I-2 - History of bad advice		Between 0 and 4
	I-3 - Fair allocation of risks and benefits		Between 0 and 4
	I-4 - Common-dread		Between 0 and 4
Total for factor I			
F_{II} - 3 items - Trust in risk beakers (12 point max)	II-1 - Individual control over the hazard/situation		Between 0 and 4
	II-2 - Lack of personal experience		Between 0 and 4
	II-3 - Uncertainities: public's point of view		Between 0 and 4
Total for factor II			
F_{III} - 3 items - Characteristics of the major hazard and its consequences (12 points max)	III-1 - Uncertainities: science point of view		Between 0 and 4
	III-2 - Environmental impact		Between 0 and 4
	III-3 - Man-made vs natural causes		Between 0 and 4
Total for factor III			
F_{IV} - 1 items - Trust in company (4 point max)	IV - Trust in company		Between 0 and 4

Figure 4.31 Scoring system for disproportion factor (DF), taking societal acceptability of risks into account: layout of the factors in the Excel file.

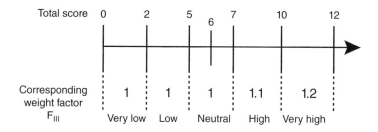

Figure 4.32 Example of factor F_{III} and its corresponding weight factor WF_{III}.

Illustrative Example

This example, where the DF is already calculated, illustrates how the scoring system can be used in practice. The goal is to see whether the model and the results are logical and relevant.

The starting point here is to use one of the examples developed by Goose [144], which was first described in the Health and Safety Executive research report written by Quinn *et al.* [154]. The example concerns a chlorine installation consisting of road tanker deliveries of 2×80 ton vessels to a user location via pipework and supply. Goose [144] calculated the DF value for the plant to be 12.2. However, as it is an illustrative example, the authors did not make assumptions concerning the characteristics of the organization in charge of managing this hazard, which is an important criterion for the assessment of the DF* using the proposed scoring system. That is why in the following example, an existing organization A (notice that we call it A, for confidentiality reasons) was chosen by us in order to have a more concrete case to work with. A is a well-known chemical concern that has a lot of experience in dealing with chemicals, and especially chlorine.

Estimation of the Level of Societal Concern

In this section, the different steps of the scoring system will be detailed briefly and comments will be provided regarding the approach of our proposed model. Disagreements can arise concerning the scores attributed to the indicators. In fact, different sets of results coming from different people would allow one to assess more precisely the repeatability and the robustness of this system.

F_I – Awareness of Company about the Risk and Fairness of its Management

1. A score of 4 has been given here because researchers have shown that there have been a lot of accidents involving chlorine, as is clear from this extract of a *Scientific American* article: "Over the past 10 years, chlorine has been involved in hundreds of accidents nationwide, injuring thousands of workers and townspeople, and killing some, according to federal databases."[1]
2. A score of 1 has been given because, after some research into their accident history, it is clear that, except for a recent explosion in one of their chemical plants, organization A has a relatively good safety record.
3. A score of 3 has been given because the chemical industry is mostly beneficial for the company that sells and uses chemicals, but the workers still receive a wage for their work.
4. A score of 4 has been given because people still have a strong negative image of chemicals and they may have a great sense of dread when facing this kind of event.
   This gives a total score of 12, and a corresponding weight factor $WF_I = 1.2$ (very high-risk area).

F_{II} – Trust in Risk Bearers

1. A score of 4 has been given because once the hazard has occurred, workers have no control over the exposure.
2. A score of 3 has been given because even if the workers are specialized in chemical transport, for example, they will not necessarily have experienced a real chemical exposure before.
3. A score of 3 has been given because exposed workers will not necessarily have a scientific background. They will probably have specific training for that kind of risk, but many uncertainties might remain.

---

[1] http://www.scientificamerican.com/article/chlorine-accidents-take-big-human-toll/

This gives a total score of 10, and a corresponding weight factor $WF_{II} = 1.1$ (high-risk area).

F_{III} – Characteristics of the Major Hazard and its Consequences

1. A score of 0 has been given because the storing and managing of chemicals are well-known subjects. It is therefore possible to suppose that they have long been used to dealing with this kind of hazard.
2. A score of 3 has been given because it would definitely have an environmental impact, but not necessarily a tremendous one.
3. A score of 4 has been given because the hazard is entirely man-made.
   This gives a total score of 7, and a corresponding weight factor $WF_{III} = 1$ (neutral area).

F_{IV} – Trust in Company

The company can be considered as fully trusted, so the score here is 0. This gives a corresponding weight factor $WF_{IV} = 1$ (very low-risk area)

Finally, after multiplying everything together, a modified DF* of 16.1 is found, instead of the original DF of 12.2. The result seems logical and not too impactful. It is reasonable as the chemical industry continues to have a negative image among the public.

Discussion of the Suggested Scoring System

The scoring system described is one that may be used by risk managers so as to include the SAR in economic analyses of industrial risks. It is quite easy to understand and to picture each of the parameters included in the model, and especially how it works as a whole. It is also flexible in the sense that it can be adapted to different situations. Specific parameters can even be chosen as neutral so that they have no impact on the final outcome. Finally, the scoring system is complete in the sense that all the indicators known to be influential for the SAR and relevant within the framework of the prevention of major industrial accidents have been included and weighted according to their importance.

Disadvantages of this model include the subjectivity introduced by the fact that users are free to change some of the parameters and to apply different weight factors depending on the desired sensitivity and the final impact on the DF. Another item of concern is that this model still needs risk managers who are ready to take the leap and include it in economic analyses of their organizations' risks.

The scoring system allows the user to assess the level of societal concerns by identifying situations where public outrage could spill over. It also provides guidance about the main factors influencing societal acceptability along with a means of including it in the previously developed DF model (based solely on the numbers of fatalities and the FN curve) thanks to weight factors. In fact, 11 indicators have been identified, explained and grouped into four main factors.

Regarding further adjustments, it is possible to imagine, for example, the development of actual software with an interface that would make it even more user-friendly and with an extensive database that would allow access to information and good descriptions of the terms or the context. It could provide details about every parameter, but also information on specific cases, and give examples of how to solve certain issues related to this topic (e.g., trust improved with better risk communication, or knowledge of workers increased with awareness campaigns).

4.12   Conclusions

Making decisions generally requires a choice between different alternatives and their possible outcomes. These are frequently associated with risk and uncertainty. The same applies to decisions that companies have to make concerning investments in safety. Not investing in safety entails risk and uncertainty: it can result in loss (in the case of an accident) or in hypothetical gain (if the accident does not occur) [29].

There is no simple recipe for evaluating and managing the risks and uncertainties associated with decision-making [87]. Also, decision-making under uncertainty is not the same as decision-making under risk. Under risk, all outcomes, as well as the likelihood of each outcome, are known. Under uncertainty, some of the alternatives and outcomes, in addition to the likelihood, can be unknown [155]. However, the distinction between risk and uncertainty is not very clear, as risks are very uncertain. After all, likelihoods are only an approximation for predicting uncertain events. So, predictions of risk are characterized by uncertainty [86, 87].

Risk analyses are often used in combination with risk acceptance criteria in order to support the decision-making [86, 88]. Risk acceptance criteria are defined as the upper limits of acceptable risk or tolerable. Based on the risk acceptance criteria, one can decide on the need for risk-reducing measures [88]. The criteria can result from one's own risk appreciation, or they can be legislation-driven or based on corporate guidelines. Defining risk acceptance criteria is, however, difficult. Acceptable risks concerning the safety of assets, personnel, and third parties as external population need to be determined by balancing different concerns [156].

Much debated and complex problems in this regard are the moral aspects of risk, comparison of risks, monetization of risk and, ultimately, the VoSL. These topics have all been discussed in this chapter, and new approaches to deal with them in an organizational context and for the purposes of economic analysis have been suggested.

References

[1] Shafir, E., Simonson, I., Tversky, A. (1993). Reason-based choice. *Cognition*, **49**, 11–36.
[2] Aliev, R.A., Huseynov, O.H., Pedrycz, W. (2012). Decision theory with imprecise probabilities. *International Journal of Information Technology and Decision Making*, **11**(2), 271–306.
[3] Johnson, E.J., Weber, E.U. (2008). Decisions under uncertainty: psychological, economic, and neuroeconomic explanations of risk preference. In: *Neuroeconomics: Decision Making and the Brain* (eds Glimcher, P., Camerer C., Fehr E. & Poldrack R.) Elsevier, New York, pp. 127–144.
[4] Kahneman, D., Tversky, A. (1984). Choices, values, and frames. *American Psychologist*, **39**(4), 341–350.
[5] Levy, J. S. (1992). An introduction to prospect theory. *Political Psychology*, **13**(2) 171–186.
[6] Hayden, B.Y., Platt, M.L. (2009). The mean, the median, and the St. Petersburg paradox. *Judgment Decision Making*, **4**(4) 256–272.

[7] Hey, J.D. (2002). Experimental economics and the theory of decision making under risk and uncertainty. *The Geneva Papers on Risk and Insurance Theory*, **27**, 5–21.

[8] Loewenstein, G., Prelec, D. (1991). Decision making over time and under uncertainty: a common approach. *Management Science*, **37**(7), 770–786.

[9] Weber, M. (1987). Decision making with incomplete information. *European Journal of Operational Research*, **28**, 44–57.

[10] Abdellaoui, M., Bleichrodt, H., Paraschiv, C. (2007). Loss aversion under prospect theory: a parameter-free measurement. *Management Science*, **35**(10), 1659–1674.

[11] Abrams, R.A., Curley, S.P., & Yates, F. (1986). Psychological sources of ambiguity avoidance. *Organizational Behavior and Human Decision Processes*, **38**, 230–256.

[12] Bassen Jr,, A.W. (1987). The St. Petersburg paradox and bounded utility. *History of Political Economy*, **19**(4) 517–523.

[13] Bell, D.E. (1985). Disappointment in decision making under uncertainty. *Operations Research*, **33**(1) 1–27.

[14] Borghans, L., Golsteyn, B.H.H., Heckman, J.J., and Meijers, H. (2009) *Gender Differences in Risk Aversion and Ambiguity Aversion*. IZA Discussion Paper No. 3985.Maastricht University, the Netherlands.

[15] Cosmides, L., Hell, W., Rode, C., & Tooby, J. (1999). When and why do people avoid unknown probabilities in decision under uncertainty? Testing some predicions from optimal foraging theory. *Cognition*, **72**, 269–304.

[16] Fox, C.R., Tversky, A. (1998). A belief-based account of decision under uncertainty. *Management Science*, **44**(7) 879–895.

[17] Gonzalez, R., Wu, G. (1998). Common consequence conditions in decision making under risk. *Journal of Risk and Uncertainty*, **16**, 115–139.

[18] Gonzales, R., Wu, G. (1999). Nonlinear decision weights in choice under uncertainty. *Management Science*, **45**(1), 74–86.

[19] Hey, J.D., Lotito, G., Maffioletti, A. (2010). The descriptive and predictive adequacy of theories of decision making under uncertainty/ambiguity. *Journal of Risk and Uncertainty*, **41**, 81–111.

[20] Tversky, A., Wakker, P. (1995). Risk attitudes and decision weights. *Econometrica*, **63**(6), 1255–1280.

[21] Cooke, A.D.J., Mellers, B.A., Schwartz, A. (1998). Judgment and decision making. *Annual Review of Psychology*, **49**, 447–477.

[22] Bleichrodt, H., Pinto, J.L. (2002). Loss aversion and scale compatibility in two-attribute trade-offs. *Journal of Mathematical Psychology*, **46**, 315–337.

[23] Hsee, C.K., Rottenstreich, Y. (2001). Money, kisses, and electric shocks: on the affective psychology of risk. *Psychological Science*, **12**(3), 185–190.

[24] Kahneman, D. (1991). Judgment and decision making: a personal view. *Psychological Science*, **2**(3) 142–145.

[25] Kahneman, D., Lovallo, D. (1993). Timid choices and bold forecasts: a cognitive perspective on risk taking. *Management Science*, **39**(1), 17–31.

[26] Kahneman, D., Novemsky, N. (2005). The boundaries of loss aversion. *Journal of Marketing Research*, **42**, 119–128.

[27] Kahneman, D., Tversky, A. (1992). Advances in prospect theory: cumulative representation of uncertainty. *Journal of Risk and Uncertainty*, **5**, 297–323.

[28] Köbberling, V., Wakker, P.P. (2005). An index of loss aversion. *Journal of Economic Theory*, **122**, 119–131.

[29] Barkan, R., Erev, I., Zohar, D. (1998). Accidents and decision making under uncertainty: a comparison of four models. *Organizational Behavior and Human Decision Processes*, **74**(2), 118–144.

[30] Idson, L.C., Higgins, T., Liberman, N. (2000). Distinguishing gains from nonlosses and losses from nongains: a regulatory focus perspective on hedonic intensity. *Journal of Experimental Social Psychology*, **36**, 252–274.

[31] Wikipedia (English) (2016). Prospect Theory, https://en.wikipedia.org/wiki/Prospect_theory.

[32] LessWrong (2011) Prospect Theory: a Framework for Understanding Cognitive Biases, http://lesswrong.com/lw/6kf/prospect_theory_a_framework_for_understanding/ (accessed 24 February 2016).

[33] Ariely, D., Huber, J., Wertenbroch, K. (2005). When do losses loom larger than gains?, *Journal of Marketing Research*, **42**, 134–138.

[34] Driver-Linn, E., Gilbert, D.T., Kremer, D.A., Wilson, T.D. (2006). Loss aversion is an effective forecasting error. *Psychological Science*, **17**(8), 649–653.

[35] Gächter, S., Herrmann, A., Johnson, E.J. (2010) individual-level loss aversion in riskless and risky choices. *CeDEx Discussion Paper Series*. University of Nothingham, Nothingham.

[36] Kahneman, D., Knetsch, J.L., Thaler, R.H. (1991). Anomalies: the endowment effect, loss aversion, and status quo bias. *Journal of Economic Perspectives*, **5**(1), 193–206.

[37] Jovanovic, P. (1999). Application of sensitivity analysis in investment project evaluation under uncertainty and risk. *International Journal of Project Management*, **17**(4), 217–222.

[38] Gollier, C., Treich, N. (2003). Decision-making under scientific uncertainty: the economics of the precautionary principle. *The Journal of Risk and Uncertainty*, **27**(1), 77–103.

[39] March, J.G., Shapira, Z. (1987). Managerial perspectives on risk and risk taking. *Management Science*, **33**(11) 1404–1418.

[40] Finucane, M.L., MacGregor, D.G., Peters, E., Slovic, P. (2004). Risk as analysis and risk as feelings: some thoughts about affect, reason, risk and rationality. *Risk Analysis*, **24**(2) 311–322.

[41] Finucane, M.L., MacGregor, D.G., Peters, E., Slovic, P. (2005). Affect, risk, and decision-making. *Health Psychology*, **24**(4) S35–S40.

[42] Cabanis, M., Herrlich, J., Kircher, T., Klingberg, S., Krug, A., Landsberg, M., Wölwer, W. (2014). Investigation of decision-making under uncertainty in healthy subjects: a multi-centric fMRI study. *Behavioural Brain Research*, **261** 89–96.

[43] Duncan, R.B. (1972). Characteristics of organizational environments and perceived environmental uncertainty. *Administrative Science Quarterly*, **17**(3) 313-327.

[44] Durbach, I.N., Stewart, T.J. (2012). Modeling uncertainty in multi-criteria decision analysis. *European Journal of Operational Research*, **223**, 1–14.

[45] Etner, J., Jeleva, M., Tallon, J. (2012). Decision theory under ambiguity. *Journal of Economic Surveys*, **26**(2), 234–270.

[46] Hastie, R. (2001). Problems for judgment and decision making. *Annual Reviews Psychology*, **52**, 653–683.

[47] Ho, T., Huynh, V., Nakamori, Y., Ryoke, M. (2007). Decision making under uncertainty with fuzzy targets. *Fuzzy Optimization and Decision Making*, **6**, 255–278.

[48] Lipshitz, R., Strauss, O. (1997). Coping with uncertainty: a naturalistic decision-making analysis. *Organizational Behavior and Human Decision Processes*, **69**(2), 149–163.

[49] Liu, X. (2004). On the methods of decision making under uncertainty with probability information. *International Journal of Intelligent Systems*, **19**, 1217–1238.

[50] Najjaran, H., Sadiq, R., Tesfamariam, S. (2010). Decision making under uncertainty – an example for seismic risk management. *Risk Analysis*, **30**(1) 78–94.

[51] Chichilnisky, G. (2000). An axiomatic approach to choice under uncertainty with catastrophic risks. *Resource and Energy Economics*, **22**, 221–231.

[52] Huettel, S.A., McCarthy, G., Song, A.W. (2005). Decisions under uncertainty: probabilistic context influences activation of prefrontal and parietal cortices. *The Journal of Neuroscience*, **25**(13), 3304–3311.

[53] Troffaes, M.C.M. (2007). Decision making under uncertainty using imprecise probabilities. *International Journal of Approximate Reasoning*, **45**, 17–29.

[54] Simon, H.A. (1989) *Bounded Rationality and Organizational Learning. The Artificial Intelligence and Psychology Project*. Technical Report AIP – 107, Carnegie Mellon University, Pittsburgh, PA.

[55] Eisenhardt, K.M., Zbaracki, M.J. (1992). Strategic decision making. *Strategic Management Journal*, **13**, 17–37.

[56] Whyte, G. (1993). Escalating commitment in individual and group decision making: a prospect theory approach. *Organizational Behavior and Human Decision Processes*, **54**, 430–455.

[57] Crisan, L.G., Heilman, R.M., Houser, D. (2010). Emotion regulation and decision making under risk and uncertainty. *American Psychological Association*, **10**(2) 257–265.

[58] Bechara, A., Naqvi, N., Shiv, B. (2006). The role of emotion in decision making. *Current Directions in Psychological Science*, **45**(5) 260–264.

[59] Harinck, F., Mersmann, P., Van Beest, I., Van Dijk, E. (2007). When gains loom larger than losses. *Psychological Science*, **18**(12), 1099–1105.

[60] Weber, E.U. (2006). Experience-based and description-based perceptions of long-term risk: why global warming does not scare us (yet). *Climatic Change*, **77**, 103–120.

[61] Epstein, S., Pacini, R. (1999). The relation of rational and experiential information processing styles to personality, basic beliefs, and the ratio-bias phenomenon. *Journal of Personality and Social Psychology*, **76**(6) 972–987.

[62] Dane, E., Pratt, M.G. (2007). Exploring intuition and its role in managerial decision making. *Academy of Management Review*, **32**(1) 33–54.

[63] Beattie, J., de Vries, N., van der Pligt, J., Zeelenberg, M. (1996). Consequences of regret aversion: effects of expected feedback on risky decision making. *Organizational Behavior and Human Decision Processes*, **65**(2) 148–158.

[64] Riabacke, A. (2006). Managerial decision making under risk and uncertainty. *IAENG International Journal of Computer Science*, **32**(4) 1–7.

[65] Donkers, B., Melenberg, B., Van Soest, A. (2001). Estimating risk attitudes using lotteries: a large sample approach, *Journal of Risk and Uncertainty*, **22**(2) 165–195.

[66] Allman, J.M., Camerer, C.F., Grether, D.M., Kovalchik, S., Plott, C. R. (2005). Aging and decision making: a comparison between neurologically healthy elderly and young individuals. *Journal of Economic Behavior & Organization*, **58**, 79–94.

[67] Frederick, S. (2005). Cognitive reflection and decision making. *Journal of Economic Perspectives*, **19**(4), 25–42.

[68] Booth, A.L. and Nolen, P.J. (2009) *Gender Differences in Risk Behavior: Does Nurture Matter?* IZA Discussion Paper No. 4026. Institute for the Study of Labour, Bonn

[69] Eckel, C., Grossman, P.J. (2008) Men, women and risk aversion: experimental evidence, Chapter 113, in *Handbook of Experimental Economics Results*, vol. **1**, Part 7.Elsevier, Amsterdam

[70] Fuller, C.W., Vassie, L.H. (2004). *Health and Safety Management. Principles and Best Practice*. Prentice Hall, Essex.

[71] Kletz, T. (1998). *Process Plants: A Handbook for Inherently Safer Design*. Braun-Brumfield, Ann Arbor, MI.

[72] Kletz, T., Amyotte, P. (2010). *Process Plants. A Handbook for Inherently Safer Design*, 2nd edn. CRC Press, Boca Raton, FL.

[73] Meyer, T., Reniers, G. (2013). *Engineering Risk Management*. De Gruyter, Berlin.

[74] Wilson, R., Crouch, E.A.C. (2001). *Risk-Benefit Analysis*, 2nd edn. Harvard University Press, Newton, MA.

[75] Busenitz, L.W., Barney, J.B. (1997). Differences between entrepreneurs and managers in large organizations: biases and heuristics in decision-making. *Journal of Business Venturing*, **12**, 9–30.

[76] Krueger, N. Jr.,, Dickson, P.R. (1994). How believing in ourselves increases risk taking: perceived self-efficacy and opportunity recognition. *Decision Sciences*, **25**(3), 385–400.

[77] Riabacke, A. (2006). Managerial decision making under risk and uncertainty. *IAENG International Journal of Computer Science*, **32**, 453–459.

[78] Panopoulos, G.D., Booth, R.T. (2007). An analysis of the business case for safety: the costs of safety-related failures and the costs of their prevention. *Policy and Practice in Health and Safety*, **1**, 61–73.

[79] Booth, R.T. (2015) http://www.hastam.co.uk/is-zero-accidents-a-valid-safety-aim-part-2-the-case-in-favour-with-reservations-by-professor-richard-booth/ (accessed 6 February 2016).

[80] Heinrich H.W. (1950). *Industrial Accident Prevention*, 3rd edn. McGraw-Hill Book Company, New York.

[81] Bird, E., Germain, G.L. (1985). *Practical Loss Control Leadership, The Conservation of People, Property, Process and Profits*. Institute Publishing, Loganville, GA.

[82] James, B., Fullman, P. (1994). *Construction Safety, Security and Loss Prevention*, Wiley-Interscience, New York.

[83] Krause T. R. (2011). *New Findings on Serious Injuries and Fatalities*. BST (Behavioural Science Technology), Ojai, CA.

[84] Hopkins, A. (2010), *Failure to Learn. The BP Texas City Refinery Disaster*. CCH Australia Ltd., Sydney.

[85] Van Nunen, K., Reniers, G., Ponnet, K., Cozzani, V. (2016). Major accident prevention decision-making: a large-scale survey-based analysis. *Safety Science*, **88**, 242–250.

[86] Aven, T., Kristensen, V. (2005). Perspectives on risk: review and discussion of the basis for establishing a unified and holistic approach. *Reliability Engineering and System Safety*, **90**, 1–14.

[87] Klinke, A., Renn, O. (2002). A new approach to risk evaluation and management: risk-based, precaution-based, and discourse-based strategies. *Risk Analysis*, **22**(6), 1071–1094.

[88] Rodrigues, M.A., Arezes, P., Leão, C.P. (2014). Risk criteria in occupational environments: critical overview and discussion. *Procedia – Social and Behavioral Sciences*, **109**, 257–262.

[89] Chorus, C.G. (2010). A new model of random regret minimization. *European Journal of Transport and Infrastructure Research*, **10**(2), 181–196.

[90] Dohmen, T., Falk, A., Huffman, D., Sunde, U., Schupp, J., Wagner, G.G. (2011). Individual risk attitudes: measurement, determinants, and behavioral consequences. *Journal of the European Economic Association*, **9**(3) 522–550.

[91] Butler, J.V., Guiso, L., Jappelli, T. (2014). The role of intuition and reasoning in driving aversion to risk and ambiguity. *Theory and Decision*, **77**(4) 455–484.

[92] Stanovich, K.E., West, R.F. (2000) Individual differences in reasoning: implications for the rationality debate? *Behavioral and Brain Sciences*, **23** 645–665.

[93] Klein, G. (1998) *Sources of Power: How People Make Decisions*. MIT Press, Cambridge, MA.

[94] Klein, G. (2003) *Intuition at Work: Why Developing Your Gut Instincts Will Make You Better at What You Do.* Doubleday, New York.

[95] Dijksterhuis, A. (2004). Think different: the merits of unconscious thought in preference development and decision making. *Journal of Personality and Social Psychology*, **87**(5) 586–598.

[96] Lee, L., Amir, O., Ariely, D. (2009). In search of Homo economicus: cognitive noise and the role of emotion in preference consistency. *Journal of Consumer Research*, **36**, 173–187.

[97] Zuckerman, M., Kuhlman, D.M. (2000). Personality and risk taking: common biosocial factors. *Journal of Personality*, **68**, 999–1029.

[98] Rolison, M.R., Scherman, A. (2003). College student risk-taking from three perspectives. *Adolescence* **38**, 689–704.

[99] Boyer, T.W. (2006). The development of risk-taking: a multiperspective review. *Developmental Review* **26**, 291–345.

[100] Ponnet, K., Reniers, G., Kempeneers, A. (2015). The association between students' characteristics and their reading and following safety instructions. *Safety Science*, **71**, 56–60.

[101] Lang, S. (1973). *Calculus of Several Variables*. Addison-Wesley, Reading, MA.

[102] Simon, C. P., Blume, L.E. (1994). *Mathematics for Economists*. W.W. Norton and Company, New York.

[103] Adams, J.G.U. (1995). *Risk*. UCL Press, London.

[104] Slovic, P. (Ed.) (2000). *The Perception of Risk*. Earthscan, Virginia.

[105] Fife, I., Machin, E.A. (1976). *Redgrave's Health and Safety in Factories*. Butterworth, London.

[106] Melnick, E.L., Everitt, B.S. (2008). *Encyclopedia of Quantitative Risk Analysis and Assessment*. John Wiley & Sons, Ltd, Chichester.

[107] Ale, B.J.M., Hartford, D.N.D., Slater, D. (2015). ALARP and CBA all in the same game. *Safety Science*, **76**, 90–100.

[108] Ale, B.J.M. (2009). *Risk: An Introduction. The Concepts of Risk, Danger and Chance*. Routledge, Abingdon.

[109] Duffey, R.B., Saull, J.W. (2008). *Managing Risk: the Human Element*. John Wiley & Sons, Ltd, Chichester.

[110] Ball, D.J. and Floyd, P.J. (1998) *Societal Risks. Final Report*. HSE, London.

[111] Reniers, G., Ale, B., Dullaert, W., Foubert, B. (2006). Decision support systems for major accident prevention in the chemical process industry: a developers' survey. *Journal of Loss Prevention in the Process Industries*, **19**, 604–662.

[112] Sunstein, C.R. (2002). *Risk and Reason. Safety, Law and the Environment*. Cambridge University Press, Cambridge.

[113] Evans, A.W., Verlander, N.Q. (1993). What is wrong with criterion FN-lines for judging the tolerability of risk? *Risk Analysis*, **17**(2) 157–168.

[114] Bedford, T. (2013). Decision-making for group risk reduction: dealing with epistemic uncertainty, *Risk Analysis*, **33**(10), 1884–1898.

[115] Tweeddale, M. (2003). *Managing Risk and Reliability of Process Plants*. Gulf Professional Publishing (Elsevier Science), Burlington, MA.

[116] van Danzig, D. (1956). Economic decision problems for flood prevention, *Econometrica*, **24**, 276–287,

[117] Jonkman, S.N., Vrijling, J.K., van Gelder, P.H.A.J.M., Arends, B. (2003). Evaluation of tunnel safety and cost effectiveness of measures. Safety and Reliability 2003, Proceedings of ESREL 2003, Swets and Zeitlinger, Lisse.

[118] van de Poel, I., Fahlquist, J.N. (2013). Risk and responsibility. In: *Essentials of Risk Theory* (eds Roeser, S., Hillerbrand, R., Sandin, P. & Peterson, M.). Springer, Dordrecht.

[119] Roeser, S. (2006). The role of emotions in the moral acceptability of risk. *Safety Science*, **44**, 689–700.

[120] Roeser, S. (2007). Ethical intuitions about risks. *Safety Science Monitor*, **11**(3) 1–13.

[121] Sunstein, C.R. (2005). *Laws of Fear. Beyond the Precautionary Principle*. Cambridge University Press, Cambridge.

[122] Slovic, P. (2002). The affect heuristic. In: *Heuristics and Biases: the Psychology of Intuitive Judgment* (eds Gilovitch, T., Griffin, D., & Kahneman, D.), Cambridge University Press, Cambridge.

[123] Wilde, G. (1976). The risk compensation theory of accident causation and its practical consequences for accident prevention. Paper Presented at the Annual Meeting of the Österreichische Gesellschaft für Unfallchirurgies, Salzburg, Austria.

[124] Adams, J.G.U. (1985). *Risk and Freedom: the Record of Road Safety Regulation*. Transport Publishing Projects, London.

[125] Adams, J.G.U. (1988). Risk homeostasis and the purpose of safety regulation, *Ergonomics*, **31**(4) 407–428.

[126] Kunreuther, H. (1992). A conceptual framework for managing low-probability events. In: *Social Theories of Risk* (eds Krimsky, S. & Golding, D.), Praeger, London.

[127] Harris, C.E., Pritchard, M.S., Rabins, M.J. (2008). *Engineering Ethics. Concepts and Cases*, 4th edn. Wadsworth, Belmont, CA.

[128] Asveld, L., Roeser, S. (eds) (2009). *The Ethics of Technological Risk*. Earthscan, London.

[129] Hansson, S.O. (2009). Risk and safety in technology. In: *Handbook of the Philosophy of Science* (ed. Meijers, A.). Philosophy of Technology and Engineering Sciences, vol. **9**. Elsevier, Oxford, pp. 1069–1102.

[130] Van de Poel, I., Royakkers, L. (2011). *Ethics, Technology and Engineering*. Wiley-Blackwell, London.

[131] Rayner, S. (1992). Cultural theory and risk analysis. In: *Social Theories of Risk* (ed. Krimsky, G.). Praeger, London.

[132] Kelman, S. (1981). *What Price Incentives? Economists and the Environment*. Auburn House Publishing Company, Boston, MA.

[133] Dorman, P. (2009). *Markets and Mortality. Economics, Dangerous Work, and the Value of Human Life*. Cambridge University Press, Cambridge.

[134] Leonard, N. (1969). Future economic value in wrongful death litigation. *Ohio State Law Journal*, **30**, 502–514.

[135] Dehez, P., Drèze, J. (1982). State-dependent utility, the demand for insurance and the value of safety. In: *The Value of Life and Safety: Proceedings of a Conference held by the "Geneva Association"* (ed. Jones-Lee M.W.). North-Holland, Amsterdam.

[136] Jones-Lee, M.W., Poncelet, A.M. (1982). The value of marginal and non-marginal multiperiod variations in physical risk. In: *The Value of Life and Safety: Proceedings of a Conference held by the "Geneva Association"* (ed. Jones-Lee M.W.). North-Holland, Amsterdam.

[137] Australian Government, Department of Prime Minister and Cabinet (December 2014) Best Practice Regulation Guidance Note – Value of Statistical Life.

[138] IENM (2013) Letter of 26 April 2013 to the Parliament, Ministerie van Infrastructuur IENM/BSK 2013/19920.

[139] HSE (2001) Reducing Risk, Protecting People. Her Majesty's Stationery Office, Norwich NR3 1BQ, p. 36.

[140] DoT (2013) Guidance on Treatment of the Economic Value of a Statistical Life in U.S. department of Transportation Analysis. Department of Transportation, United States. Memorandum to secretarial officers and modal administrators.

[141] Button, K. (1993) Overview of Internalizing the Social Costs of Transport. OECD ECMT Report.

[142] Miller, T.R. (2000). Variations between countries in values of statistical life. *Journal of Transport Economics and Policy*, **34**(2) 169–188.

[143] Sunstein, C.R. (2013) The Value of a Statistical Life: Some Clarifications and Puzzles. Regulatory Policy Program, RPP-2013-18.

[144] Goose, M. (2006). *Gross Disproportion, Step by Step – A Possible Approach to Evaluating Additional Measures at COMAH Sites*. In: Institution of Chemical Engineers Symposium Series, vol. **151**, p. 952. Institution of Chemical Engineers, Rugby.

[145] Fischhoff, B., Slovic, P., Lichtenstein, S. (1978), *How Safe is Safe Enough? A Psychometric Study of Attitudes Towards Technological Risks and Benefits*, Elsevier, Amsterdam.

[146] Huang, L., Ban, J., Sun, K., Han, Y., Yuan, Z., Bi, J. (2013). The influence of public perception on risk acceptance of the chemical industry and the assistance for risk communication. *Safety Science* **51**, 232–240.

[147] Gurian, P.L. (2008). *Risk Perception, Risk Communication, and Risk Management*. Drexel University Libraries.

[148] Adams, J. (2009). Risk management: the economics and morality of safety revisited, In: *Safety-Critical Systems: Problems, Process and Practice*. Springer, Berlin.

[149] Baker, E., Dunbar, I. (2004) Modelling Societal Concern for the Rail Industry. Research Programme for the RSSB (Rail Safety & Standards Board).

[150] Mansfield D (2003) Gauging societal concerns, Symposium Series, vol. **149**, In: *Hazard XVII: Process Safety – Fulfilling Our Responsibilities*, pp. 15–29, IChemE, London.

[151] Sandman, P.M. (2012) *Responding to Community Outrage: Strategies for Effective Risk Communication*.[First published in 1993 by the American Industrial Hygiene Association, copyright transferred to the author, Peter M. Sandman, in 2012.]

[152] Ottway, H.J. (1982), *Beyond Acceptable Risk: On the Social Acceptability of Technologies*. Elsevier, Amsterdam.

[153] Achille, J., Ponnet, K., Reniers, G. (2015) Calculation of an adjusted disproportion factor (DF*) which takes the societal acceptability of risks into account. *Safety Science* DOI: 10.1016/j.ssci.2015.12.007, In Press.

[154] Quinn, D.J., Davies, P.A. (2004) *Development of an Intermediate Societal Risk Methodology*. HSE Books, Research Report RR283. HSE, London, http://www.hse.gov.uk/research/rrhtm/rr283.htm (accessed 24 February 2016).

[155] Mousavi, S., Gigerenzer, G. (2014). Risk, uncertainty, and heuristics. *Journal of Business Research*, **67** 1671–1678.

[156] Abrahamsen, E.B., Aven, T. (2008). On the consistency of risk acceptance criteria with normative theories for decision-making. *Reliability Engineering and System Safety*, **93** 1906–1910.

5

Cost-Benefit Analysis

5.1 An Introduction to Cost-Benefit Analysis

A cost-benefit analysis (CBA) is an economic evaluation in which all costs and consequences of a certain decision are expressed in the same units, usually money [1]. Such an analysis may be employed in relation to operational safety, to aid normative decisions about safety investments. One should keep in mind, however, that CBAs cannot demonstrate whether one safety investment is intrinsically better than another. Nevertheless, a CBA allows decision-makers to improve their decisions by adding appropriate information on costs and benefits to certain prevention or mitigation investment decisions. As decisions on safety investments involve choices between different possible risk management options, CBAs can be very useful. Moreover, decisions may be straightforward in some cases, but this may not always be true. Especially in the process industries or the nuclear sector, for instance, where there are obviously important type I as well as type II risks that should be managed and controlled, and a variety of both of them, risk management options may be difficult and not at all obvious. Moreover, CBA is not a pure science and sometimes needs to employ debatable concepts, such as the value of human life (see also Chapter 4), the value of body parts, the question of who pays the prevention costs, and who receives the safety benefits.

There are two types of CBA. On the one hand, it is possible to carry out a so-called "ex ante CBA," in order to verify whether a certain investment is worthwhile. On the other hand, a so-called "ex post CBA" can be carried out at the end of a project to evaluate the financial achievements accompanying the project [2]. Although, in this chapter, we are mainly interested in an ex ante CBA, whereby safety managers may carry out an economic exercise and thus simulate and assess the financial soundness of the implementation of safety and prevention measures, the technique described further in this chapter can also be used to carry out an ex post CBA. There are, however, several reasons why we focus on ex ante analysis in this book. An important reason why companies should consider executing CBAs for the evaluation of prevention investment decisions is that such analyses can help employees to convince managers of the importance of safety measures from an economic point of view. In addition, a CBA can also help managers in the efficient allocation of a safety budget, as some safety measures may turn out to be more efficient than others, when dealing with identical, similar or comparable risks.

Operational Safety Economics: A Practical Approach Focused on the Chemical and Process Industries,
First Edition. Genserik L.L. Reniers and H.R. Noël Van Erp.
© 2016 John Wiley & Sons, Ltd. Published 2016 by John Wiley & Sons, Ltd.

In any case, it should be borne in mind that the result of any CBA, being a recommendation for a prevention investment decision, is meant merely to assist the decision-maker in the decision process by making costs and benefits more transparent and more objective.

The decision-maker is thus recommended to use this approach with caution, as the available information is subject to varying levels of quality, detail, variability, and uncertainty. Nevertheless, the tool is far from unusable and can provide meaningful information for aiding decision-making, especially if it takes the levels of variability and uncertainty into account and thus avoids misleading results.

From the previous, it is clear that CBAs can be used to determine whether an investment represents an efficient use of resources. A safety investment project represents an allocation of means (money, time, etc.) in the present that will result in a particular stream of non-events, or expected hypothetical benefits, in the future. The role of a CBA is to provide information to the decision-maker, in this case an employee or a manager who will appraise the safety investment project. The main purpose of the analysis is to obtain relevant information about the level and distribution of benefits and costs of the safety investment. Through this information, an investment decision within the company can be guided and made more objective. The role of the analysis is thus to provide the possibility of a more objective evaluation and not to adopt an advocacy position either in favor of or against the safety investment, as there are also many other aspects that should be taken into account when deciding about safety investments, such as, for example, social acceptability and regulatory affairs.

The costs and benefits of a safety investment project within an organization include those affecting the financial position of the company in the broadest sense possible. This means that all claims resulting from damage to society and others are included, but damage to society and others for which there will be no claims whatsoever from the company are excluded. There is no doubt that type II accidents, for example, have wider implications on society, but if these would (theoretically) not affect the firm's financial position in any way, they are omitted from the CBA helping the company's safety decision-making process.

A safety investment project makes a difference, as the future will be different depending on whether the company decides to invest or not, or to invest in an alternative investment option. Thus, in the CBA, two hypothetical worlds are envisaged: one without the safety investment, and another with the safety investment. During a CBA, a monetary value is put on the difference between the two hypothetical worlds. This process is shown in Figure 5.1.

As a safety investment project involves costs in the present and both costs and benefits in the future, the net benefit stream will be negative for a certain period of time and will then become positive at a certain point in time. This should be the case for both type I and type II risks, but the manner in which the calculations are performed differ. Therefore, a distinction between both types of risks is made later in this chapter.

5.2 Economic Concepts Related to Cost-Benefit Analyses

5.2.1 Opportunity Cost

An important concept in economics is the existence of "opportunity costs." In this book on operational safety economics, this concept dictates that if a budget is spent on a certain safety investment, it is no longer available and thus cannot be spent on another safety investment, or a financial investment for that matter. As the resources will have been used, they are now

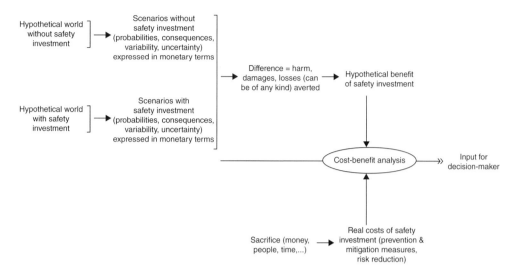

Figure 5.1   Cost-benefit analysis process for safety investments.

unavailable for use in other safety tasks, measures or any kind of investment. This implies that any hypothetical benefits arising from the use of other (e.g., safety) investments, besides the one that the company decides to invest in, have to be forgone. This is called the "opportunity cost" of the investment decision. In economic language, an opportunity cost can be defined as "the amount lost by not using the resource (labor or capital) in its best alternative use" [3]. Regarding safety, it is obviously very difficult to measure all the forgone hypothetical benefits in reality, but the concept of opportunity costs offers a useful guide in thinking through decisions to the very end, as it explicitly shows the trade-offs that have to be made when allocating resources in operational safety.

As an example, assume a €100 000 safety investment (all possible costs included) leading to a hypothetical benefit of €120 000 (all possible direct and indirect avoided losses are considered). Hence, the profit of this safety investment amounts to €20 000. However, if another – smaller – safety investment of €75 000 had been made, a hypothetical benefit of €94 000 could have been realized. If the €25 000 that was still available after this smaller investment (€75 000 + €25 000 = €100 000) had been invested with a return of 5%, there would be a return of €1250. The smaller investment, in this illustrative case, would lead to a profit of €19 000 + €1250 = €20 250. Hence, there would be an opportunity cost of €250.

5.2.2   Implicit Value of Safety

One of the trade-offs that have to be made is a budget allocation between prevention of type I risks and prevention of type II risks.

The following illustrative example shows that by not thinking economic analyses through, wide variations in the criteria that may be used by different people in the same situation, and by the same people in different situations, can lead to inconsistencies in the allocation of implicit values of safety, in turn leading to inefficient allocation of resources [4]. Assume two safety investment projects that have the same costs, e.g., €1 000 000. The first safety investment project (SIP1) concerns type I risks and, according to the accident scenarios, will reduce the number of fatalities by two in 10 years, as well as avoiding other non-fatal accidents for at least €10 000 000 (the expected value, taking probabilities into account) in 10 years. The second safety investment project (SIP2) is concerned with type II risks and, according to the accident scenarios considered, will reduce the number of fatalities by 10 in 10 years, and will also lead to €500 000 (expected value, taking probabilities into account) worth of non-fatal hypothetical benefits in a decade. Table 5.1 summarizes the data of this illustrative example.

Table 5.1 The implicit value of safety.

	Safety investment costs (€ million)	Number of fatalities avoided in 10 years	Non-fatal hypothetical benefits
Safety Investment Project 1 (SIP1)	1	2	€10 million
Safety Investment Project 2 (SIP2)	1	10	€500 000

In this illustrative example, choosing between these two possible projects implicitly determines the value attached to a life by the company's decision-maker. A decision-maker who prefers SIP2 over SIP1 implicitly decides that an additional €9.5 million of non-fatal benefits are considered to be worth less than the additional eight lives saved under SIP2. In other words, the choice for SIP2 indicates that the value of life for this decision-maker is higher than (€10 million − €0.5 million)/(10 − 2) = €1 187 500. Conversely, choosing for SIP1 over SIP2 indicates that the value of life is lower than €1 187 500 for this decision-maker.

In fact, making a budget allocation choice for type I risks and type II risks unavoidably leads to placing a value on safety, human life or a related parameter. So even if one objects on principle to the valuation of safety, or human life, it is often unavoidable when business decisions have to be made [4].

5.2.3   Consistency and Uniformity of Safety Investment Decisions

A difficulty of safety investment decision-making has to do with establishing consistent safety performance across different plant units or facilities: the implementation of uniform operational safety standards in every case may not lead to an efficient method of resource allocation. In fact, using uniform operational safety standards may give rise to differences in the implicit value of safety within different organizational contexts and could lead to suboptimal safety investments.

An illustrative example can throw some light on to this consistency and uniformity problem. Assume you have an organization where the two types of risks (type I and type II) are present. On the one hand, there are 200 000 hours of exposure to type I risks per year, and there is an accident rate of one fatality per 50 000 hours' exposure to such risks. On the other hand, safety management estimates that workers are exposed to type II risks for 100 000 hours/year, and, using type II scenarios, it is estimated that the accident rate is three fatalities per 50 000 hours exposed (expected and accumulated value). Assume that the organization sets a uniform standard of safety performance for both types of risk at one fatality per 100 000 hours exposed. What will be the consequence?

Table 5.2 provides an overview of the estimated annual costs of safety investments to reduce accident rates to various values for both types of risk.

Table 5.2 Costs of reducing fatal accident rates for type I and type II risks.

Fatality rate per 50 000 hours exposed to risk	Type I risks		Type II risks	
	Number of fatalities annually	Annual safety investments k€	Number of fatalities annually	Annual safety investments k€
3	–	–	6	–
2	–	–	4	50
1	4	–	2	200
0.75	3	20	1.5	300
0.5	2	50	1	450
0.25	1	100	0.5	800
0.125	0.5	150	–	–
0	0	400	0	1400

From Table 5.2, if the organization sets a uniform standard for both types of risk at one fatality per 100 000 hours exposed (or 0.5 per 50 000 hours exposed), the number of fatalities due to type I risks will be reduced from four to two, and due to type II risks from six to one. Hence, in theory, seven lives in total will have been gained per year. Now consider the costs/year to achieve this goal, and investigate whether this would be the best possible allocation of resources.

To achieve the uniform standard of 0.5 lives per 50 000 hours exposed, the cost would be €50 000 for type I risks and €450 000 for type II risks. Hence, in total, the cost would be €500 000. However, this uniformity reasoning leads to a different implicit value of safety: avoiding one more fatality related to type I risks would cost the organization €50 000, while avoiding a type II related extra fatality would cost the company €950 000. There is obviously a huge difference between the implicit values of life for both types of risk, and the uniformity approach has thus created an inconsistency within the company over the valuation of human life.

An alternative option for this organization would therefore be to use the same implicit value of life for both types of risk, and not necessarily to pursue uniform operational safety standards. If such an approach were followed, the budget of €500 000 that was used to save seven lives in the previous reasoning should be used in another way. If this budget is allocated differently between type I and type II risks, an optimization in the number of lives saved can be achieved, at a lower total cost. If €150 000 is invested in safety for type I risks, there would be a reduction from four to 0.5 fatalities. If a further €300 000 were invested in type II safety investments, the number of fatalities would be reduced from six to 1.5. Hence, a total reduction of (3.5 + 4.5 =) 8 fatalities would have been realized at a budget of €450 000. Moreover, using this approach, the cost of avoiding one extra fatality is €500 000 for both types of risk [calculated as follows: type I risks, €250 000 for 0.5 fatalities; type II risks, (€800 000 − €300 000)/(1.5 − 0.5) = €500 000 for one fatality].

In summary, the implementation of uniform operational safety standards for both types of risk in every case may not create an optimal method of resource allocation and safety managers therefore need to calculate optimal budget allocations for type I and type II risks based on organization-specific characteristics.

5.2.4 Decision Rule, Present Values, and Discount Rate

If a company uses a CBA, the recommendation of whether to accept or reject an investment project is based on the following process:

1. identification of costs and benefits;
2. calculation of the present values of all costs and benefits;
3. comparison of the total present value of costs and the total present value of benefits.

In order to compare the total costs and the total benefits, comprising the costs and benefits occurring at different points in time, one needs to take a discount rate into account in the calculation to obtain the present values. Thus, during a CBA, all cash flows, from both costs and benefits in the future, need to be converted to values in the present. This conversion is carried out by discounting the cash flows by a discount rate. The discount rate represents the rate at which people (or companies) are willing to give up consumption in the present in exchange for additional consumption in the future. Another definition is that in a multi-period model, people value future experiences to a lesser degree than present ones, as they are sure about present events and unsure about future events, which are subject to the environment. Thus the higher the discount rate they choose, the lower the present values of the future cash flows [5].

An investment project is recommended when the total net present value (NPV) of all cash flows is positive, and an investment project is usually rejected when the NPV is negative. To calculate the NPV related to project management, all cash flows are determined, and future cash flows are recalculated to today's value of money by discounting them by the discount rate. The formula usually quoted to calculate the NPV is:

$$\text{NPV} = \sum_{t=0}^{T} \frac{X_t}{(1+r)^t}, \tag{5.1}$$

where X_t represents the cash flow in year t, T is the time period considered (usually expressed in years), and r is the discount rate.

Applied to operational safety, the NPV of a project expresses the difference between the total discounted present value of the benefits and the total discounted present value of the costs. A positive NPV for a given safety investment indicates that the project benefits are larger than its costs.

$$\text{NPV} = \text{present value(benefits)} - \text{present value(costs)},$$

where

If NPV ≥ 0, recommend safety investment;
If NPV < 0, recommend to reject safety investment.

It is evident that the cash flows, i.e., prevention costs and certainly expected hypothetical benefits (due to non-events), may be uncertain. Different approaches can be used in this regard. The cash flows can, for example, be expressed as expected values, taking the uncertainties in the form of probabilities into consideration, and also increasing the discount rate to outweigh the possibilities for unfavorable outcomes. In case of type II risks, it is recommended to use scenario analyses, determining expected cash flows for different scenarios (e.g., worst-case and most credible case) and using a disproportion factor (DF) (see Sections 4.11.2.1, 5.4.2.2, and 8.12).

There can be different categories of costs related to a safety investment, e.g., initial costs, installation costs, operating costs, maintenance costs, inspection costs . These costs are clearly represented by negative cash flows. Some costs (e.g., initial costs and installation costs) occur in the present and thus do not have to be discounted, while other costs (e.g., operating, maintenance, and inspection costs) occur throughout the remaining lifetime of the facility and will thus have to be discounted to the present. There may also be different categories of benefits linked to a safety investment, such as supply chain benefits, damage benefits, legal benefits, insurance benefits, human and environmental benefits, intervention benefits, and reputation benefits, among others. The benefits represent positive cash flows, which all occur throughout the remaining lifetime of the facility and thus will all have to be discounted to the present.

In order to clarify the discount rate principle, all cash flows (for both costs and benefits) are assumed to occur on an arbitrarily chosen date, which can, for example, be chosen to be the last day of the calendar year in which they occur. This assumption converts the continuous cash flows to a discrete range of cash flows, occurring at the end of each year. The cash flows at the end of each year then have to be discounted to a present value, using a discount factor. As stated before, cash flows occurring in the current year do not have to be discounted. Therefore the current year is called "year 0," and the following years "year 1," "year 2," ... , "year n." Costs and benefits occurring in year 1 are discounted back one period, those occurring in year 2 are discounted back two periods, and those occurring in year n are discounted back n periods. The implicit assumption is made that the discount rate remains the same throughout the remaining lifetime of the facility [5].

Thus, to calculate the present value of a benefit occurring in year 1, it needs to be discounted for one period to come to a present value in year 0. Similar to the calculation of a benefit occurring in year 1, the present value of benefits occurring in years 2 and 3 are obtained by discounting them two and three periods, respectively. Similar to the previous calculations, the present value (PV) of a benefit occurring in year n is obtained by discounting it over n periods.

These calculations can be found in the following range:

$$\text{PV of a benefit in year 1} = \frac{\text{benefit}}{(1+r)},$$

$$\text{PV of a benefit in year 2} = \frac{\text{benefit}}{(1+r)^2},$$

$$\vdots$$

$$\text{PV of a benefit in year } n = \frac{\text{benefit}}{(1+r)^n}.$$

Now that the concept and method of discounting future cash flows have been clarified, suppose a safety investment project has a cost in year 0 and then the same level of costs and benefits at the end of each and every subsequent year for the remaining lifetime of the facility. This means that the costs in year i are the same for all i, i.e., $C_i = C$; likewise, the benefits in year i are the same for all i, i.e., $B_i = B$. This concept is called an "annuity." The present value (PV) of such an annuity is given by the following formula, with n being the remaining lifetime of the facility:

$$\text{PV(Annuity of a cost)} = C + \frac{C}{(1+r)} + \frac{C}{(1+r)^2} + \ldots + \frac{C}{(1+r)^n} \qquad (5.2)$$

$$\text{PV(Annuity of a benefit)} = B + \frac{B}{(1+r)} + \frac{B}{(1+r)^2} + \ldots + \frac{B}{(1+r)^n}. \qquad (5.3)$$

C and B are the equal annual costs (cost categories where costs are made in the future) and benefits (all benefits categories), respectively, that occur at the end of each year, and are assumed to remain constant. This assumption is valid as long as inflation is omitted from the calculations and as long as the annual costs are assumed not to increase over time due to aging. These assumptions keep the caluculations simple while explaining the cost-benefit approach. Each term in the formula is formed by multiplying the previous term by $1/(1+r)$. As the above formulas can become very long, the formula for calculating the present value of annuities can be rewritten, by way of the series solution

$$1 + \frac{1}{(1+r)} + \frac{1}{(1+r)^2} + \ldots + \frac{1}{(1+r)^n} = 1 + \frac{1}{r} - \frac{1}{r(1+r)^n},$$

as

$$\text{PV(Annuity)} = A + \frac{A}{r} - \frac{A}{r(1+r)^n},$$

where A is the yearly cost or benefit of a cost-benefit category. Note that this general annuity goes to $(n+1)A$ as the discount ratio r goes to zero. The term

$$\frac{1}{r} - \frac{1}{r(1+r)^n} = \frac{(1+r)^n - 1}{r(1+r)^n} \qquad (5.4)$$

of the series solution is called "the annuity (discount) factor" and is applicable whenever the annuity starts from year 1 [5].

Using this model, the benefits and costs in the future are assumed to be constant, and inflation is not included into the future costs and benefits, as already mentioned. Inflation is the process that results in a rise of the nominal prices of goods and services over time. Therefore in this (simplified) model, the real rate of interest[1] should be used as the discount rate instead of the money rate of interest.[2] As the money rate of interest, m, includes two components, the real rate of interest, r, and the anticipated rate of inflation, i, the anticipated rate of inflation is built into the money rate of interest:

$$m = r + i.$$

As inflation is not included in the numerator of the formula for calculating the present value of annuities (as the costs and benefits are constant throughout the whole remaining lifetime), it cannot be included into the denominator.

5.2.5   Different Cost-Benefit Ratios

Several approaches are possible for presenting the cost-benefit principle, and different cost-benefit ratios can be calculated. Notice that sometimes benefits are divided by costs, in which case a benefit–cost ratio is obtained, and sometimes costs are divided by benefits, giving a cost-benefit ratio. In the case of a benefit–cost ratio, the ratio should ideally be higher than 1, and as high as possible, while in the case of a cost-benefit ratio, it should ideally be lower than 1, and as low as possible. The following ratios are mentioned by Fuller and Vassie [4]:

Value of an averted loss:

> Benefit–cost ratio = value of averted losses (= hypothetical benefits)/
>
> cost of safety measures over their lifetime

Value of equivalent life:

> Benefit–cost ratio = value of equivalent lives saved over the lifetime
>
> of the safety measures/cost of safety measures over their lifetime

Value of risk reduction:

> Benefit–cost ratio = [(liability of the original risk) – (liability of the residual risk)]/
>
> costs of safety measures over their lifetime

To illustrate the calculation of a benefit–cost ratio based on the value of risk reduction, consider the following example. Assume that an organization has 500 employees, and that the probability of a fatality in the organization is 1 in 10 000 per year. If a safety investment X were carried out, the fatality rate would be decreased to 0.8 in 10 000 per year. The safety investment X has a cost of €1 000 000 over its lifetime of 10 years.

---

[1] Real rate of interest (r): does not include the anticipated rate of inflation (i).
[2] Money rate of interest (m): includes two components, the real rate of interest (r) and the anticipated rate of inflation (i): $m = r + i$.

Assume that the value of statistical life (VoSL) is €8 000 000, consistent with previous chapter figures. Then, the benefit–cost ratio can be calculated as:

$$\text{Benefit-cost ratio} = \frac{500 \times 10^{-4} \times 8\,000\,000 \times 10 - 500 \times (0.8 \times 10^{-4}) \times 8\,000\,000 \times 10}{1\,000\,000}$$

$$= \frac{800\,000}{1\,000\,000} = 0.8 < 1$$

Therefore, subject to the need to refine the calculation to take into account NPVs and opportunity costs associated with the €1 000 000 safety investment, it can be deduced from the very rough benefit–cost ratio that the safety investment X would not be cost-beneficial.

5.3   Calculating Costs

The purpose of implementing operational safety measures is to reduce present and future operational risks. By "reducing the risk," the prevention of accidents is indicated, as well as the mitigation of consequences if an accident were to occur after all. Safety measures can be costly. Before discussing the different cost factors of implementing safety measures, different types of measures are briefly discussed.

5.3.1   Safety Measures

There are four different classifications into which safety measures can be divided. First, risk reduction measures can be classified into protection and prevention measures, depending on their characteristics. Protection measures (including mitigation measures) lower the consequences, while prevention measures decrease the probability of an accident (see also Meyer and Reniers [6]). Second, safety measures can also be classified into active or passive systems according to what is necessary to be able to perform their function. Third, a classification can be made according to their impact on the severity and probability of occurrence; of the many safety measures, only some of them will, for example, play a role in the prevention of catastrophic or disastrous events. There may sometimes be a need, therefore, to focus the efforts and identify priority safety elements. The third classification of safety measures is thus into safety measures and safety-critical measures. Fourth, safety measures can be looked at from three dimensions: people, procedures, and technology. The interplay among people, procedures and technology safety measures defines the observable part of the safety culture in an organization [see also the safety culture egg model (The Egg Aggregated Model) from Chapter 2].

5.3.2   Costs of Safety Measures

As stated earlier, in order to be able to implement new safety measures and upgrade existing safety systems, a company has to reserve substantial funding. In this section, the various costs related to new safety measures that a company may decide to implement are discussed. Table 5.3 provides a clear overview of the different kinds of costs of safety measures.

For each of the costs mentioned in Table 5.3, formulas are elaborated to calculate every subcategory of costs.

5.3.2.1 Initiation Safety Costs

Under the initiation costs of safety measures, five different kinds of safety costs can be grouped:

1. Investigation safety costs
2. Selection and design safety costs
3. Material safety costs
4. Training safety costs
5. Changing guidelines and informing safety costs.

Table 5.3 Cost categories of safety measures.

Type of safety cost	Subcategory of safety cost
Initiation (Section 5.3.2.1)	Investigation Selection and design Material Training Changing guidelines and informing
Installation (Section 5.3.2.2)	Production loss Start-up Equipment Installation team
Operation (Section 5.3.2.3)	Utilities
Maintenance (Section 5.3.2.4)	Material Maintenance team Production loss Start-up
Inspection (Section 5.3.2.5)	Inspection team
Logistics and transport safety (Section 5.3.2.6)	Transport and loading/unloading of hazardous materials Storage of hazardous materials Drafting control lists Safety documents
Contractor safety (Section 5.3.2.7)	Contractor selection Training
Other safety (Section 5.3.2.8)	Other prevention measures

These costs will not have to be discounted to present values, as they will occur in the present (hence in the basic year, year 0). Each of the different types of costs is explained more in depth in the following sections.

Investigation Safety Costs

The investigation, carried out by the so-called "investigation team" studying the potential of a safety project, brings with it costs related to the investigation and audit activities, internally (the health and safety department of the company) or externally, or both. The purpose of this effort is to check whether additional safety measures or upgrades to the existing safety system are possible and necessary. The costs can be estimated and/or calculated by multiplying the hourly wage of an employee by the number of hours the investigation/audit takes, and again by the number of employees participating in the investigation or audit. If, however, employees with significantly varying wage levels participate, the investigation team costs can be calculated separately for each category of employees. Another possibility is to take the average wage level of all employees participating, in order to simplify the work and only have to work with one category.

Investigation team safety costs:

$$w_1 \times h_1 \times n_1 \text{ (employee category 1)} + w_2 \times h_2 \times n_2 \text{ (employee category 2)} + \dots$$
$$+ w_t \times h_t \times n_t \text{ (employee category } t)$$

where

w_i = the hourly wage of category i (€/hour per person);
h_i = the number of hours of category i (hours);
n_i = the number of employees of category i (number of people).

This calculation is represented by the following formula:

$$\text{Investigation team safety costs} = \sum_{i=1}^{t} w_i h_i n_i$$

where t is the number of employee categories.

Selection and Design Safety Costs

If the investigation or audit suggests that upgrades in the safety system are possible or necessary, a prevention and/or mitigation measure will have to be selected and designed. Such a measure is, of course, accompanied by costs, which can be calculated by multiplying the hourly wage of all employees involved by the number of hours they work on the design, and then again multiplying by the number of employees participating. They can also be calculated separately for categories of employees with varying wage levels.

Selection and design safety costs:

$$w_1 \times h_1 \times n_1 \text{ (employee category 1)} + w_2 \times h_2 \times n_2 \text{ (employee category 2)} + \dots$$
$$+ w_t \times h_t \times n_t \text{ (employee category } t)$$

where

w_i = the hourly wage of category i (€/hour per person);
h_i = the number of hours of category i (hours);
n_i = the number of employees of category i (number of people).

This calculation is represented by the following formula:

$$\text{Selection and design safety costs} = \sum_{i=1}^{t} w_i h_i n_i$$

where t is the number of employee categories.

Material Safety Costs

The actual safety measure, and the components that constitute it, also sometimes requires a budget (e.g., in case of a dyke that needs to be built around a storage tank). The material costs and costs related to the creation of the safety measure can be calculated by multiplying the price per unit of the necessary materials by the number of units the company requires to create the safety measure.

Material safety costs:

$$u_1 \times n_1 \text{ (material 1)} + u_2 \times n_2 \text{ (material 2)} + \dots + u_s \times n_s \text{ (material } s\text{)}$$
where

u_i = the price per unit for material i (€/unit);
n_i = the amount of units for material i (number of units).

This calculation is represented by the following formula:

$$\text{Material safety costs} = \sum_{i=1}^{s} u_i n_i$$

where s is the number of different materials.

Training Safety Costs

In order to calculate these safety costs, the assumption is made that the company provides training to its employees working in the facility related to the new safety measure. It is assumed that some employees or external consultants or coaches will be given the task of disseminating the necessary information and explaining how to work with the new safety measure and how to handle it properly in the case of an emergency. The costs arising because of this assignment to some employees or external consultants or coaches can be calculated and estimated by multiplying the hourly wage of an employee by the number of hours this process takes, and again by the number of employees participating in the assignment. If, however, employees with significantly varying wage levels participate, the training costs can also be calculated separately for each category of employees. Another possibility is to estimate the costs by taking the average wage level of all employees participating, so you only have to work with one category.

Training safety costs:

$$w_1 \times h_1 \times n_1 \text{ (employee category 1)} + w_2 \times h_2 \times n_2 \text{ (employee category 2)} + \dots$$
$$+ w_t \times h_t \times n_t \text{ (employee category } t\text{)}$$

where

w_i = the hourly wage of category i (€/hour per person);
h_i = the number of hours of category i (hours);
n_i = the number of employees of category i (number of people).

This calculation is represented by the following formula:

$$\text{Training safety costs} = \sum_{i=1}^{t} w_i h_i n_i$$

where t is the number of employee categories.

Changing Guidelines and Informing Safety Costs

In order to calculate the costs resulting from the required changes to guidelines and the necessary disseminating activities, the assumption is made that in addition to training, the company will inform the personnel of the new safety measure through some kind of brochure, newsletter, or guide. This brochure will also contain the altered guidelines and safety instructions. These costs can be calculated by multiplying the price per unit of brochures/guides by the number of them needed. One unit can, in this case, either represent one brochure or a pack of brochures that may contain, for example, 100 brochures. This will depend on which price is used, the price per brochure or the price per batch of brochures (procedures).

Changing guidelines and informing safety costs:

$$u_1 \times n_1 \text{ (brochure 1)} + u_2 \times n_2 \text{ (brochure 2)} + \dots + u_s \times n_s \text{ (brochure } s\text{)}$$

where

u_i = the price per unit for brochure i (€/unit);
n_i = the amount of units for brochure i (number of units).

This calculation is represented by the following formula:

$$\text{Changing guidelines and informing safety costs} = \sum_{i=1}^{s} u_i n_i$$

where s is the number of different brochures.

5.3.2.2   Installation Safety Costs

The installation safety costs are made up of different sub-costs:

1. Production loss safety costs
2. Start-up safety costs
3. Equipment safety costs
4. Installation team safety costs.

Similar to the initiation safety costs, the installation safety costs will not have to be discounted to present values, as they will occur only in the present (hence in the basic year, year 0). Each of the different types of costs is explained in more depth in the following sections.

Production Loss Safety Costs

When a safety measure is implemented, in some cases the production has to be stopped temporarily, resulting in a production loss. This production loss is accompanied by costs because of the non-producing status of the facility or installation. Production loss safety costs can be calculated by multiplying the production capacity/rate of the facility by the duration of the stop, and again by the profit per unit sold [7].

This calculation is represented by the following formula.

> Production loss safety costs:
>
> $$\text{ProdRate} \times t \times \text{Profit}$$
>
> where
>
> ProdRate = the production capacity/rate of the factory or installation (number of units/hour);
>
> t = the duration of the stop in production (hours);
>
> Profit = the profit per unit sold (€/unit).

Start-up Safety Costs

The implementation of a new safety measure can cause a temporary slowdown in production due to the required restart of the facility (because of the required production halt due to safety measure implementation). The costs related to the temporary slowdown in production for safety-related reasons are called start-up safety costs and can be calculated by multiplying the difference in production rate before and after the halt in production by the duration from the time the production line is reactivated after the implementation of the new measure to the time it returns to the initial production rate, and again by the profit per unit sold [7].

This calculation is represented by the following formula.

> Start-up safety costs:
>
> $$[\text{ProdRate(old)} - \text{ProdRate(new)}] \times t \times \text{Profit}$$
>
> where
>
> ProdRate(old) = production capacity/rate of the facility before the halt in production due to safety (number of units/hour);

ProdRate(new) = production capacity/rate of the facility after the halt in production for safety reasons (number of units/hour);
t = the duration from the time the production line is reactivated to the time it returns to the initial production capacity/rate (hours);
Profit = the profit per unit sold (€/unit).

Note that if the production rate at the time of the start-up is exactly the same as the production rate before the halt in production, the start-up safety costs will be zero.

Equipment Safety Costs

The installation of a new safety measure usually requires equipment (to be bought or rented). Equipment indicates all kinds of working tools, but also, for example, machinery and modes of transportation. These equipment costs can be calculated by multiplying the price per unit of the equipment by the units needed to install the safety measure.

Equipment safety costs:

$u_1 \times n_1$ (equipment category 1) + $u_2 \times n_2$ (equipment category 2) + ... + $u_s \times n_s$ (equipment category s)

where

u_i = the price per unit for brochure i (€/unit);
n_i = the amount of units for equipment i (number of units).

This calculation is represented by the following formula:

$$\text{Equipment safety costs} = \sum_{i=1}^{s} u_i n_i$$

where s is the number of different equipment categories.

Installation Team Safety Costs

The installation team safety costs are related to the employees who are installing the new safety measure in the facility. These can be calculated and estimated by multiplying the hourly wage of the participating employees by the number of hours the installation takes, and again by the number of employees involved. If employees with significantly varying wage levels are involved, the installation team costs can be calculated separately for each category of employees. Another possibility is to take the average wage level of all employees participating, and thus work with just one category.

Installation team safety costs:

$w_1 \times h_1 \times n_1$ (employee category 1) + $w_2 \times h_2 \times n_2$ (employee category 2) + ...
+ $w_t \times h_t \times n_t$ (employee category t)

where

w_i = the hourly wage of category i (€/hour per person);
h_i = the number of hours of category i (hours);
n_i = the number of employees of category i (number of people).

This calculation is represented by the following formula:

$$\text{Installation team safety costs} = \sum_{i=1}^{t} w_i h_i n_i$$

where t is the number of employee categories.

5.3.2.3 Operation Safety Costs (Utilities)

Utility safety costs will have to be discounted to present values, as they will not only occur in the present (i.e., in the basic year, year 0), but throughout the remaining lifetime of the facility. Active safety systems, for example, need energy sources and other utilities external or internal to the system to perform their function. Without these utilities, the active safety system will not be able to function. Examples of external energy sources include electric power, hydraulic power, manpower, and system pressure. In a CBA, one may choose to calculate the annual utility safety costs by multiplying the price per unit of a utility by the units needed per year.

Annual utility safety costs:

$u_1 \times n_1$ (utility category 1) $+ u_2 \times n_2$ (utility category 2) $+ \ldots + u_s \times n_s$ (utility category s)

where

u_i = the price per unit for utility i (€/unit);
n_i = the amount of units for utility i (number of units).

This calculation is represented by the following formula:

$$\text{Yearly utility safety costs} = C_u = \sum_{i=1}^{s} u_i n_i$$

where s is the number of different utility categories.

The assumption is made that the utility safety costs represent the same level of costs at the end of each year for a specific time interval. As mentioned earlier, the cost stream C_1, $C_2, \ldots, C_n$, where n the remaining life span of the facility in years, and $C_i = C$ for all i, is termed an annuity. The total present value is not just the sum of the utilities' costs for each year, such as was calculated in the previous cost sections, because the utilities' costs occur throughout the remaining lifetime of the facility and thus have to be calculated taking into account a discount factor. The total present value is given by the formula in the following box, as discussed in Section 5.2.4, (Eq. 5.4).

Total present value of utilities cost:

$$\frac{(1+r)^n - 1}{r(1+r)^n} C_u$$

where

C_u = the yearly cost of utility use for safety purposes (€/year);
n = the remaining life span of the facility (years);
r = the discount rate (%).

5.3.2.4 Maintenance Safety Costs

1. Material safety costs
2. Maintenance team safety costs
3. Production loss safety costs
4. Start-up safety costs.

These safety costs will have to be discounted to present values, as they will not only occur in the present (in the basic year, year 0), but throughout the remaining life span of the facility. Each of the different types of safety costs is explained in greater detail in the following sections.

Material Safety Costs

Maintenance of safety measures requires replacements for decrepit materials. The material costs of the replacement materials can be calculated by multiplying the price per unit of the materials by the units needed for the maintenance of the safety measure per year.

Yearly maintenance material safety costs:

$$u_1 \times n_1 \text{ (material 1)} + u_2 \times n_2 \text{ (material 2)} + \ldots + u_s \times n_s \text{ (material } s)$$

where

u_i = the price per unit for material i (€/unit);
n_i = the amount of units for material i (number of units).

This calculation is represented by the following formula:

$$\text{Yearly maintenance material safety costs} = C_{mm} = \sum_{i=1}^{s} u_i n_i$$

where s is the number of different materials.

These costs represent the maintenance material costs of one maintenance period, which is defined as one year. Thus if it is assumed that maintenance occurs on a yearly basis and the yearly cost is always the same, termed C_{mm}, the total present value of all maintenance materials needed during the lifetime of the safety measure can be calculated by taking into account a discount factor, because the maintenance material costs occur throughout the remaining lifetime of the facility. The total present value is then given by the formula in the following box for annuities, as discussed in Section 5.2.4, (Eq. 5.4).

Total present value of maintenance material safety costs:

$$\frac{(1+r)^n - 1}{r(1+r)^n} C_{mm}$$

where

C_{mm} = the yearly cost of maintenance material (€/year);
n = the remaining life span of the facility (years)
r = the discount rate (%).

Maintenance Team Safety Costs

The maintenance team safety costs are related to the maintenance activities of employees for the installed safety measure(s). These can be calculated and estimated by multiplying the hourly wage of such an employee by the number of hours the maintenance takes, and again by the number of employees participating. If employees with significantly varying wage levels are participating, the maintenance team costs can be calculated separately for each category of employees. Another possibility is to take the average wage level of all employees participating, and thus work with just one category.

Yearly maintenance team safety costs:

$$w_1 \times h_1 \times n_1 \text{ (employee category 1)} + w_2 \times h_2 \times n_2 \text{ (employee category 2)} + \dots$$
$$+ w_t \times h_t \times n_t \text{ (employee category } t)$$

where

w_i = the hourly wage of category i (€/hour per person);
h_i = the number of hours of category i (hours);
n_i = the number of employees of category i (number of people).

This calculation is represented by the following formula:

$$\text{Yearly maintenance team safety costs} = C_{mt} = \sum_{i=1}^{t} w_i h_i n_i$$

where t is the number of employee categories.

These costs represent the maintenance team costs for one maintenance period, which is defined as 1 year. Thus, if we assume that maintenance occurs on a yearly basis and the yearly cost is always the same, termed C_{mt}, the total present value of all maintenance teams needed during the lifetime of the safety measure can be calculated by taking into account a discount factor, because the maintenance team costs occur throughout the remaining lifetime of the facility. The total present value is given by the formula in the following box for annuities, as discussed in Section 5.2.4, (Eq. 5.4).

Total present value of maintenance team safety costs:

$$\frac{(1+r)^n - 1}{r(1+r)^n} C_{mt}$$

where

C_{mt} = the yearly cost of maintenance team for safety measures (€/year);
n = the remaining life span of the facility (years);
r = the discount rate (%).

Production Loss Safety Costs

When maintenance is periodically necessary for the optimal functioning of the safety measure, sometimes the production has to be stopped temporarily, resulting in a production loss. This production loss is accompanied by costs arising from the non-producing status of the facility. Production loss costs per maintenance period can be calculated by multiplying the production rate of the factory/refinery by the duration of the stop, and again by the profit per unit sold [7].

This calculation is represented by the following formula.

Production loss safety costs:

$$C_{pl} = \text{ProdRate} \times t \times \text{Profit}$$

where

ProdRate = the production capacity/rate of the facility (number of units/hour);
t = the duration of the stop in production (hours);
Profit = the profit per unit sold (€/unit).

These safety costs represent the maintenance production loss safety costs of one maintenance period, which is defined as 1 year. Thus, if it is assumed that maintenance occurs on a yearly basis and the yearly cost is always the same, termed C_{pl}, the total present value of all maintenance production loss during the lifetime of the safety measure can be calculated by taking into account a discount factor, because the maintenance production loss costs occur throughout the remaining lifetime of the facility. The total present value is given by the formula in the following box, (Eq. 5.4).

Total present value of maintenance production loss safety costs:

$$\frac{(1 + r)^n - 1}{r(1 + r)^n} C_{pl}$$

where

C_{pl} = the yearly cost due to production loss caused by safety measures (€/year);
n = the remaining life span of the facility (years);
r = the discount rate (%).

Start-up Safety Costs (after Maintenance)

Maintenance of a new safety measure can cause a temporary slowdown in production due to the restart of the facility after halting production for necessary maintenance. The costs arising from the temporary slowdown in production are called start-up costs, and can be calculated by multiplying the difference in production rate before and after the halt in production by the duration from the time the production line is reactivated after the maintenance period of the safety measure to the time it returns to the initial production rate, and again by the profit per unit sold [7].

This calculation is represented by the following formula.

Start-up safety costs (after maintenance):

$$C_{sum} = [\text{ProdRate(old)} - \text{ProdRate(new)}] \times t \times \text{Profit}$$

where

ProdRate(old) = production rate of the factory before the halt in production (number of units/hour);

ProdRate(new) = production rate of the factory after the halt in production (number of units/hour);

t = the duration from the time the production line is reactivated to the time it returns to the initial production rate (hours);

Profit = the profit per unit sold (€/unit).

Notice that if the production rate at the time of the start-up is exactly the same as the rate before the halt in production, the start-up costs will be zero.

Also notice that the safety costs above represent the maintenance start-up safety costs of one maintenance period, which is defined as 1 year. Thus, if it is assumed that maintenance occurs on a yearly basis and the yearly cost is always the same, termed C_{sum}, the total present value of all maintenance start-ups during the lifetime of the safety measure can be calculated by taking into account a discount factor, because the maintenance start-up costs occur throughout the remaining lifetime of the facility. The total present value is given by the formula in the following box, (Eq. 5.4).

Total present value of maintenance start-up costs:

$$\frac{(1 + r)^n - 1}{r(1 + r)^n} C_{sum}$$

where

C_{sum} = the yearly cost due to start-ups (€/year);

n = the remaining life span of the facility (years);

r = the discount rate (%).

5.3.2.5 Inspection Safety Costs (Inspection Team Costs)

This cost will have to be discounted to a present value, as it will not only occur in the present (in the basic year, year 0), but throughout the remaining life span of the facility.

The inspection team safety costs are related to the periodic inspection and audit activities of the safety department of the company or an external auditing company, to check whether the safety measures are effective [8]. Carrying out periodic risk assessments can also be considered part of these safety costs. These inspection team costs can be calculated and estimated by multiplying the hourly wage of an employee by the number of hours the inspection takes, and again by the number of employees participating. If employees with significantly varying wage levels are involved, the inspection team safety costs can be calculated separately for each category of employees. Another possibility is to take the average wage level of all employees participating, and thus only work with one category.

Yearly inspection team safety costs:

$$w_1 \times h_1 \times n_1 \text{ (employee category 1)} + w_2 \times h_2 \times n_2 \text{ (employee category 2)} + \ldots$$
$$+ w_t \times h_t \times n_t \text{ (employee category } t)$$

where

w_i = the hourly wage of category i (€/hour per person);
h_i = the number of hours of category i (hours);
n_i = the number of employees of category i (number of people).

This calculation is represented by the following formula:

$$\text{Yearly inspection team safety costs} = C_{\text{insp}} = \sum_{i=1}^{t} w_i h_i n_i$$

where t is the number of employee categories.

These costs, however, represent the inspection team's safety costs for one inspection period, which is defined as 1 year. Thus if it is assumed that these costs occur on a yearly basis and the yearly cost is always the same, termed C_{insp}, the total present value of all teams needed during the lifetime of the safety measure is calculated by considering a discount factor, because the inspection team costs occur throughout the remaining lifetime of the facility. The total present value is given by the formula in the following box, (Eq. 5.4).

Total present value of inspection team safety costs:

$$\frac{(1 + r)^n - 1}{r(1 + r)^n} C_{\text{insp}}$$

where

C_{insp} = the yearly cost of the inspection team (€/year);
n = the remaining life span of the facility (years);
r = the discount rate (%).

5.3.2.6 Logistics and Transport Safety Costs

Materials need to be transported and stored in a safe way. Control lists as well as safety documents need to be drawn up, filled in, and updated. The sub-categories of this cost category are as follows:

1. Transport and loading/unloading of hazardous materials safety costs
2. Storage of hazardous materials safety costs
3. Drafting control lists safety costs
4. Safety documents safety costs.

Transport and Loading/Unloading of Hazardous Materials

The transportation of materials and substances, such as the transport of gas cylinders, entails costs due to existing legislation and to extra measures for safety. Transport indeed requires compliance with existing regulations (e.g., ADR18) during transportation and during loading and unloading of goods, and sometimes extra safety measures are needed.

The transport costs of materials can be calculated by multiplying the transport price per material unit or good with the number of units or goods transported, for all materials that need to be transported.

$$\text{Transport costs related to safety} = \sum_{i=1}^{n} u_i x_i$$

where

u_i = transport cost per good or material unit i (€/unit)
x_i = number of goods or material units i (number of units)
n = total number of materials (number of materials)

These costs represent the transport safety costs of all materials transported during 1 year. Thus, if it is assumed that these costs occur on a yearly basis and the yearly cost is always the same, termed C_{transp}, the total present value of all transport costs can be determined by the formula in the following box, (Eq. 5.4).

Total present value of transport safety costs:

$$\frac{(1+r)^n - 1}{(1+r)^n r} C_{\text{transp}}$$

where

C_{transp} = the yearly cost of transportation of materials (€/year);
n = the remaining life span of the facility (years);
r = the discount rate (%).

Storage of Hazardous Materials Safety Costs

The storage costs can be determined by multiplying the storage price per material unit or good by the number of units or goods stored, for all materials that needs to be stored:

$$\text{Storage costs} = \sum_{i=1}^{n} u_i x_i$$

where

u_i = storage cost per good or material unit i (€/unit)
x_i = number of goods or material units i (number of units)
n = total number of materials (number of materials)

These costs represent the storage safety costs of all materials stored during 14 years. Thus, if it is assumed that these costs occur on a yearly basis and the yearly cost is always the same, termed C_{storage}, the total present value of all storage costs can be determined by the formula in the following box.

Total present value of storage safety costs:

$$\frac{(1+r)^n - 1}{(1+r)^n r} C_{\text{storage}}$$

where

C_{storage} = the yearly cost of stored materials (€/year);
$\phantom{C_{\text{storage}}}n$ = the remaining life span of the facility (years);
$\phantom{C_{\text{storage}}}r$ = the discount rate (%).

Drafting Control Lists Safety Costs

It is necessary to draft control lists for transportation and storage [9]. The total cost of drafting such lists is obtained by multiplying the hourly wage of persons responsible for the drafting by the number of hours needed to draft the lists per category of person, and summing all categories and persons. This cost only needs to be considered once, as once the lists have been drafted, they can be reused afterwards.

Drafting control lists safety costs:

$$w_1 \times h_1 \times n_1 \text{ (employee category 1)} + w_2 \times h_2 \times n_2 \text{ (employee category 2)} + \ldots$$
$$+ w_t \times h_t \times n_t \text{ (employee category } t)$$

where

w_i = the hourly wage of category i (€/hour per person);
h_i = the number of hours of category i for drafting the lists (hours);
n_i = the number of employees of category i (number of people).

This calculation is represented by the following formula:

$$\text{Drafting control lists safety costs} = C_{\text{control lists}} = \sum_{i=1}^{t} w_i h_i n_i$$

where t is the number of employee categories.

Safety Documents Safety Costs

Safety documents periodically need to be filled in by employee(s) [9]. The cost arising from the filling in of safety documents can be determined by multiplying the hourly wage of those employees by the number of hours needed to fill in the documents and again by the number of employees participating. If employees with significantly varying wage levels participate, the safety documents safety costs can be calculated separately for each category of employees. Another possibility is to take the average wage level of all employees participating, and thus work with just one category.

Safety documents safety costs:

$$w_1 \times h_1 \times n_1 \text{ (employee category 1)} + w_2 \times h_2 \times n_2 \text{ (employee category 2)} + \ldots$$
$$+ w_t \times h_t \times n_t \text{ (employee category } t)$$

where

> w_i = the hourly wage of category i (€/hour per person);
> h_i = the number of hours of category i for filling in the safety documents (hours);
> n_i = the number of employees of category i filling in safety documents (number of people).

This calculation is represented by the following formula:

$$\text{Yearly safety document safety costs} = C_{SD} = \sum_{i=1}^{t} w_i h_i n_i$$

where t = number of employee categories.

These costs represent the safety document safety costs of one period, which is defined as, say, 1 year. Thus, if it is assumed that these costs occur on a yearly basis and the yearly cost is always the same, termed C_{SD}, the total present value of all teams needed during the filling in of safety documents is calculated by considering a discount factor, because the safety document costs occur yearly for the remaining lifetime of the facility under consideration. The total present value is given by the formula in the following box.

Total present value of safety documents safety costs:

$$\frac{(1 + r)^n - 1}{(1 + r)^n r} C_{SD}$$

where

> C_{SD} = the yearly cost of filling in the safety documents (€/year);
> n = the remaining life span of the facility (years);
> r = the discount rate (%).

5.3.2.7 Contractor Safety Costs

If a company works with contractors, they need to be selected, taking safety into account. The selection process, as well as the contractor training aimed at company safety, represents a safety cost. Moreover, a loss of working time as a result of training the contractors should also be considered. Therefore, contractor safety costs include:

1. Contractor selection safety costs
2. Training safety costs.

Contractor Selection Safety Costs
Contractor firms need to be chosen with "safety" as one of the most important selection parameters. The selection is conducted by employees of the company, and the costs can thus be determined by taking these employee costs into consideration.

Contractor selection safety costs:

$w_1 \times h_1 \times n_1$ (employee category 1) $+ w_2 \times h_2 \times n_2$ (employee category 2) $+ \ldots$
$+ w_t \times h_t \times n_t$ (employee category t)

where

$w_i =$ the hourly wage of category i (€/hour per person);
$h_i =$ the number of hours of category i for selecting the contractors (hours);
$n_i =$ the number of employees of category i selecting the contractors (number of people).

This calculation is represented by the following formula:

$$\text{Contractor selection safety costs} = C_{Sel} = \sum_{i=1}^{t} w_i h_i n_i$$

where t is the number of employee categories.

These costs represent the contractor selection safety costs for one period, which is defined as, say, 1 year. Thus, if it is assumed that these costs occur on a yearly basis and the yearly cost is always the same, termed C_{Sel}, the total present value of all contractor selection procedures is calculated by considering a discount factor, because the contractor selection costs may occur yearly for the remaining lifetime of the facility under consideration. If the contractor selection costs only need to be incurred once, there is evidently no need to use a discount factor and calculate a NPV. If another time period is used for the selection, e.g., a 5-year period, the formula needs to be adjusted for this. The total present value for a 1-year period is given by the formula in the following box.

Total present value of safety documents safety costs:

$$\frac{(1+r)^n - 1}{(1+r)^n r} C_{Sel}$$

where

$C_{Sel} =$ the yearly cost of filling in the safety documents (€/year);
$n =$ the remaining lifespan of the facility (years);
$r =$ the discount rate (%).

Contractor Training Safety Costs

Contractor employees, when selected by a company, often need to receive safety training within the company, as well as receiving instructions and guidelines for working at or with certain installations. These extra costs, which are related to safety, should also be taken into consideration.

Contractor training safety costs:

$w_1 \times h_1 \times n_1$ (employee category 1) $+ w_2 \times h_2 \times n_2$ (employee category 2) $+ \ldots$
$+ w_t \times h_t \times n_t$ (employee category t)

where

> w_i = the hourly wage of category i (€/hour per person);
> h_i = the number of hours of category i for training the contractors (hours);
> n_i = the number of employees of category i training the contractors (number of people).

This calculation is represented by the following formula:

$$\text{Contractor training safety costs} = C_{ST} = \sum_{i=1}^{t} w_i h_i n_i$$

where t is the number of employee categories.

These costs represent the training safety costs of one period, which is defined as, say, 1 year. Thus, if it is assumed that these costs occur on a yearly basis and the yearly cost is always the same, termed C_{ST}, the total present value of all safety training needs is calculated by considering a discount factor, because the safety document costs occur yearly for the remaining lifetime of the facility under consideration. The total present value is given by the formula in the following box.

Total present value of safety documents safety costs:

$$\frac{(1+r)^n - 1}{(1+r)^n r} C_{ST}$$

where

> C_{ST} = the yearly cost of filling in the safety documents (€/year);
> n = the remaining lifespan of the facility (years);
> r = the discount rate (%).

5.3.2.8   Other Safety Costs

Safety costs that cannot be assigned to one of the categories already discussed in the previous sections are listed under "other safety costs" and can/should be mentioned in this category.

5.4   Calculating Benefits (Avoided Accident Costs)

The purpose of implementing safety measures is to reduce present and future risks. By "reducing the risk," the prevention of accidents is indicated, as well as the mitigation of the consequences of an accident should it occur after all. Thus, the benefits of a safety investment/measure can be regarded as the difference in consequences without and with a safety investment/measure, taking into account the difference in likelihood of an accident occurring. The "consequences without safety measure" can be seen as the potential (hypothetical) consequences of accident scenarios. The "consequences with safety measure" are those consequences that are still possible after taking a specific safety measure for the accident scenario. In this section, the various financial aspects of consequences related to an accident (scenario) will be discussed, as well as the formulas for calculating these aspects.

5.4.1 Distinction between Various Accident Costs

The literature mentions a number of accident cost categories and taxonomies, the most used and well-known accident cost categories being direct and indirect costs, and insured and uninsured costs.

5.4.1.1 Direct and Indirect Accident Costs

Direct accident costs represent costs that are immediately visible and tangible. They can be seen as "logical, common-sense consequences of the accident." Conversely, indirect costs are those accident costs that are difficult to assess, and they are often intangible and invisible.

The costs of accidents are often much higher than merely the sum of the direct and visible costs. In fact, indirect costs usually represent a multiple of the direct costs and they are therefore a very important factor when analyzing accidents and making decisions on safety investments for dealing with both type I and type II risks. A number of researchers have tried to draft ratios of direct over indirect costs, and a variety of ratios can be found in the literature, depending on the nature of the study (e.g., depending on the industrial sector in which the research was conducted). A well-known and much used ratio for type I accident costs is that of Heinrich, the father of industrial safety discussed in Sections 1.1. Based on a study of 75 000 type I accidents, Heinrich [10] concluded that indirect costs are four times higher than direct costs. But, as already mentioned, other studies have found different ratios.

Gavious *et al.* [7], for example, define the total cost of an accident as the sum of the direct costs, the indirect costs, the so-called payment costs and the immeasurable costs. In this taxonomy, the direct costs imply all costs due to installation damages, product losses, and equipment detriment. Medical costs, fines and insurance costs are considered to be direct costs by these authors. Indirect costs are caused by production delays and/or halts. Due to production problems and halts, delivery to clients could be problematic and contractual costs could result, as well as having to pay third parties, and so on. Payment costs include having to pay employees despite them being at home due to injury. The immeasurable costs result from reputation decrease and moral or psychological damage to employees, possibly leading to lower quality of life and lower productivity. An overview of the taxonomy of accident costs as proposed by Gavious *et al.* [7] can be found in Table 5.4.

5.4.1.2 Insured and Uninsured Accident Costs

The division of accident costs into insured and uninsured costs was conceptualized by Simmonds [11]. The insured costs comprise employees' wages, insured medical expenses and damage to property. Uninsured costs are composed of uninsured medical costs, wage costs due to lower output of employees upon return, cost of education or training of new employees, and so on. It should be noted that the categorization of insured and uninsured costs depends on the type of accident and the insurance premium. An illustrative scheme of a possible taxonomy is provided by Sun *et al.* [12] (see Figure 5.2).

A distinction needs to be made between avoided type I accident costs and avoided type II accident costs. Some costs may apply in both cases, i.e., in both type I and type II accidents, but others might only apply to one type of accident. The next section therefore discusses all kinds of avoided costs (in other words, benefits), but clearly highlights which types of accident the avoided accident costs are applicable to.

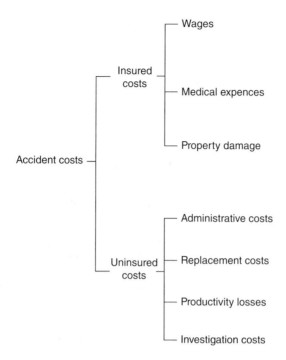

Figure 5.2   Insured and uninsured costs. (Source: Sun *et al.* [12]. Reproduced with permission from Taylor & Francis.)

Table 5.4   Taxonomy of accident costs.

Direct costs	Damage	Damage to installations, products and equipment
	Medical	Evacuation to hospital, used materials for first aid, hospitalization
	Juridical	Paying fines
	Insurance	Insurance premium rise
Indirect costs	Capacity losses	Production delays/stops, production decrease and problems
	Production scheme	Planning problems with clients and suppliers
	Recruitment	Costs of replacement of employees
	Wage costs	Costs of employees carrying out the accident investigation
Payment costs	Wage costs	Wage costs of injured employees being at home
Immeasurable costs	Moral damages	
	Reputation decrease	

Gavious *et al.* [7]. Reproduced with permission from Elsevier.

5.4.2   Avoided Accident Costs

For each of the avoided accident cost categories mentioned in Table 5.5, formulas were developed that can be used in a CBA and/or a cost-effectiveness analysis exercise or tool in order to calculate the benefits linked to type I and type II risks. The table highlights what types of risk the formulas can be used for. If there are no accurate variables available for use in some of the subcategories, information derived from previously executed projects within the company

Table 5.5   Avoided accident cost categories.

Type of avoided accident cost	Subcategory of avoided accident cost
Supply chain (Section 5.4.2.1)	Production-related (type I + type II) Start-up (type I + type II) Schedule-related (type I + type II)
Damage (Section 5.4.2.2)	Damage to own material/property (type I + type II) Damage to other companies' material/property (type II) Damage to surrounding living areas (type II) Damage to public material property (type II)
Legal (Section 5.4.2.3)	Fines (type I + type II) Interim lawyers (type II) Specialized lawyers (type II) Internal research team (type II) Experts at hearings (type II) Legislation (type II) Permit and license (type II)
Insurance (Section 5.4.2.4)	Insurance premium (type I + type II)
Human and environmental (Section 5.4.2.5)	Compensation victims (type I + type II) Injured employees (type I + type II) Recruitment (type I + type II) Environmental damage (type I + type II)
Personnel (Section 5.4.2.6)	Productivity of personnel (type I + type II) Training of new or temporary employees (type I + type II) Wages (type I + type II)
Medical (Section 5.4.2.7)	Medical treatment at location (type I + type II) Medical treatment in hospitals and revalidation (type I + type II) Using medical equipment and devices (type I + type II) Medical transport (type I + type II)
Intervention (Section 5.4.2.8)	Intervention (type I + type II)
Reputation (Section 5.4.2.9)	Share price (type II)
Other (Section 5.4.2.10)	Accident investigation (type I + type II) Manager working time (type I + type II) Clean-up (type I + type II)

can be used; another option is to use estimated consequences from an independent partner company. If needed, this information can then eventually be used to determine one or more flat-rate amounts representing one or more of the avoided cost subcategories in Table 5.5.

As the consequences of an accident only become a reality when the accident actually occurs, the probability of occurrence should be taken into account in the calculation of the expected hypothetical benefits in some way. Therefore the consequences, calculated using either the appropriate formula or a flat-rate amount, will have to be multiplied by the probability of occurrence in order to obtain the expected hypothetical benefits due to an accident scenario. Thus, if the different kinds of hypothetical consequences are considered to be spread out on a yearly basis, and the yearly cost arising from these consequences is considered to be always the same, $C_i = C$ for all i, then the total present value of all hypothetical costs of an accident scenario during the remaining lifetime of the facility can be calculated by taking into account both the remaining lifetime and a discount factor. The total present value of each of the consequence subcategories is given by the formula for annuities, as discussed in Section 5.2.4.

This calculation has to be executed for the cases both with and without the implementation of the safety measure. The difference between the two present values of consequence costs represents the maxmax hypothetical benefit (see also Section 4.2.4) resulting from the implementation of the new safety investment:

$$\text{Maxmax hypothetical benefits} = (\text{Total present value of consequence costs}$$

$$\text{without safety investment})$$

$$- (\text{Total present value of consequence costs}$$

$$\text{with safety investment})$$

This calculation is identical for all of the consequences discussed in the following sections.

If the probabilities of accident scenarios are used, the expected hypothetical benefits can be determined. Indeed, the value that is generated by the various formulas explained further in this section needs to be multiplied by the frequency of occurrence or the probability of the event, in order to obtain the expected annual avoided costs. Afterwards, this annual avoided cost is used in the formula for annuities in order to obtain the present value of the consequences for the remaining lifetime of the facility. This needs to be carried out for both the situations without and with the new safety measure, and the difference between the two present values represents the expected hypothetical benefit with regard to that specific subcategory. This calculation is identical for all defined subcategories of consequences and has to be carried out for every one of them.

5.4.2.1   Supply Chain Avoided Costs

Production-related Avoided Costs (Type I + Type II)
When an accident occurs, it is possible that (a part of) the production will be halted, resulting in a production loss. This production loss is accompanied by costs because of the non-producing status of (a part of) the factory or plant. Production loss costs can be calculated by multiplying the production capacity/rate of the facility by the estimated duration of the halt, and again by the profit per unit sold [7].

This calculation is represented by the following formula.

Avoided production loss costs:

$$\text{ProdRate} \times t \times \text{Profit}$$

where

$\text{ProdRate} = $ the production rate of the factory (number of units/hour);
$t = $ the duration of the halt in production due to an accident (hours);
$\text{Profit} = $ the profit per unit sold (€/unit).

Start-up Avoided Costs (Type I + Type II)

When the production is re-started after an accident, a temporary slowdown in production due to the restart of the facility can occur. The costs accompanied by the temporary slowdown in production are called start-up costs, and these can be calculated by multiplying the difference in the production rate before and after the halt in production by the duration from the time the production line is reactivated after the accident occurred to the time when the production line returns to the initial production rate, and again by the profit per unit sold [7].

This calculation is represented by the following formula.

Avoided start-up costs:

$$[\text{ProdRate(old)}) - \text{ProdRate(new)}] \times t \times \text{Profit}$$

where

$\text{ProdRate(old)} = $ production rate of the factory before the halt in production due to an accident (number of units/hour);
$\text{ProdRate(new)} = $ production rate of the factory after the halt in production due to an accident (number of units/hour);
$t = $ the duration from the time the production line is reactivated to the time when it returns to the initial production rate (hours);
$\text{Profit} = $ the profit per unit sold (€/unit).

Note that if the production rate at the time of the start-up is exactly the same as the that before the halt in production, the start-up costs will be zero.

Schedule-related Avoided Costs (Type I + Type II)

If an accident occurs, this will also affect the production timetable of the factory, which can cause problems with clients. The possibility exists that clients may cancel one or more contracts, or may demand a lower price due to the delay. One solution may be to hire a contractor who can help the company to provide the necessary products to meet the company's time schedule. However, a scheduling problem will not only affect clients and customers, but also partners and suppliers. If the company produces part of a product and a partner company finishes the partly completed product, the partner company will also face supply chain costs, as the company will have to wait longer for the partly completed products to arrive. In the case of suppliers, the problem is that, if the company cannot produce, its inventory stays the same. Because of the latter the company will not need fresh suppliers at the normal rate, and thus the agreements with suppliers will have to be changed/canceled, which will also cause the suppliers some scheduling problems.

The costs arising from these scheduling problems can be arrived at by adding up three aspects. The first is obtained by multiplying the fine for a canceled order/contract by the number of canceled orders/contracts. The second is obtained by multiplying the fine due to delays in deliveries per day by the number of days the orders are late, and again by the number of orders that have a delay. Finally, the third cost is calculated by multiplying the number of units supplied by the contractor, by the difference between the cost per unit charged by the contractor and the in-house cost per unit [7].

This calculation is represented by the following formula.

Avoided schedule costs:

$$\sum_{i=1}^{n}(F^i_{\text{canceled}}n^i_{\text{canceled}}) + \sum_{i=1}^{n}(F^i_{\text{delay}}d^i n^i_{\text{delay}}) + \sum_{p=1}^{q}[n^p_{\text{contractor}}(C^p_{\text{contractor}} - C^p_{\text{in-house}})]$$

where

F^i_{canceled} = the fine for a canceled order/contract of type i (€/contract);

n^i_{canceled} = the number of canceled orders/contracts of type i (number of contracts);

F^i_{delay} = the fine due to delays in deliveries per day of contractual agreements of type i (€/delay per day);

d^i = the number of days late of orders of type i (days);

n^i_{delay} = the number of orders of type i that have a delay (number of orders that have a delay);

$n^p_{\text{contractor}}$ = the number of units of product p supplied by the contractor due to problems with the company's own production (number of units);

$C^p_{\text{contractor}}$ = the cost per unit of product p charged by the contractor (€/unit);

$C^p_{\text{in-house}}$ = the in-house cost per unit of product p (€/unit);

n = number of types of contracts;

q = number of products that need to be contracted out.

5.4.2.2 Damage Avoided Costs

Avoided Costs with Respect to Damage to Own Material/Property (Type I + Type II)
An accident may lead to damage to buildings, infrastructure, products, machines, and so on. These costs are labeled as "damage costs" and are usually taken into account in any CBA.

These avoided costs are represented by the following formula.

Avoided damage costs:

$A + B + C$

where

A = damage to the company's equipment and machines (€);

B = damage to the company's buildings and other infrastructure (€);

C = damage to the company's raw materials and finished goods (€).

Avoided Costs with Respect to Damage to Other Companies' Material/Property (Type II)
An type II accident might cause damage to other companies' material and property, in addition to damage to the company's own assets. The company needs to pay for the damage incurred by other companies, as they will probably file claims against the company that caused the damage [8]. These costs are also labeled as damage costs, and should be taken into account in an economic analysis.

These avoided costs are represented by the following formula.

Avoided damage costs:

$$A + B + C$$

where

A = damage to other companies' equipment and machines (€);
B = damage to other companies' buildings and other infrastructure (€);
C = damage to other companies' raw materials and finished goods (€).

Avoided Costs with Respect to Damage to Surrounding Living Areas (Type II)
An type II accident sometimes causes damage to residential properties. The company will have to pay for this damage, as the inhabitants are likely to file claims against the company that caused the accident. These costs are also labeled as damage costs and should be taken into account in the CBA. For example, in the case of the major accident that occurred in Buncefield in Hertfordshire, UK, in 2005, there was widespread damage to both commercial and residential properties near the site. Some properties close to the depot were destroyed, while others suffered damage. As a consequence, about 2000 people had to be evacuated from their homes, some for long periods of time. In addition, properties as far as 8 km away from the Buncefield site suffered minor damage, such as broken windows [13].

Another example is the Deepwater Horizon disaster that took place in the Gulf of Mexico in 2010. The consequences to the flora and fauna in the Gulf of Mexico and along the Louisiana coastline were truly disastrous and, as a result, the local fishermen's economy simply ceased to exist for a certain period, which led to huge compensation claims.

Such avoided costs can be represented by the following formula.

Avoided damage costs:

$$A$$

where

A = damage to surrounding living areas(€)

Avoided Costs with Respect to Damage to Public Material/Property (Type II)
In some cases, a type II accident will cause damage to public material and property. The company needs to pay for that damage as well, as the local government will probably file claims against the company that caused the accident [8]. These costs are also labeled as damage costs.

These avoided costs are represented by the following formula.

Avoided damage costs:

$$A + B + C$$

where

$A =$ damage to public equipment and public machines (€);
$B =$ damage to public buildings and other public infrastructure (€);
$C =$ damage to public materials and public goods (€).

5.4.2.3 Legal Consequences Avoided Costs

The different types of legal consequences of an accident are explained in greater detail in this section. The legal aspects turn out to be an important part of the hypothetical benefits, especially in the case of type II risks. One can imagine that whenever an accident occurs, especially a major one, the legal department of a company will be placed under a lot of stress. In the case of a type II accident, a lot of financial resources will have to be spent to handle that pressure by hiring additional staff members and experts to deal with the complexity of such an accident. In addition, the legal environment in which the company operates will change according to the occurrence of catastrophes, and the company will need to make sure that it complies with these changes.

Fines-related Avoided Costs (Type I + Type II)
If an accident occurs, the government and other organizations will try to identify responsible individuals or a responsible group of individuals. In some cases the company as a whole will be held responsible for the accident, while in other cases employees, managers or other persons may be held responsible [14].

Responsible persons or the responsible organization may be exposed to civil liability, administrative liability and/or criminal liability for the major accident. Having both administrative and criminal liability carries with it the obligation of paying fines. The difference between the two types of liability is explained here. Criminal liability arises when someone or some organization has hurt or killed people, or the property has been destroyed on purpose, whereas administrative liability comes into play when one has not operated and behaved according to the law and to prescribed procedures and methods. Thus, if an accident is caused due to violations of safety procedures and/or through breaking the law, the organization may be exposed to fines and claims by the authorities due to the administrative liability [7, 14]. Another difference lies in the importance and weight of the sentence, which are significantly different. On top of the fine, criminal liability will result in a criminal record and there may be serious punishment as a consequence, such as a custodial sentence. This is not the case with administrative liability, which only results in a fine.

Consider, as an example, the major accident that took place at Total's ammonium nitrate warehouse in the AZF fertilizer factory in an industrial zone in Toulouse, France. As a result of this accident, 30 people were killed and about 3000 were injured. Serge Biechlin, the former CEO of the Total subsidiary, Grande Paroisse, received a criminal liability fine of US $58 000 and was sentenced to 1 year in prison. The court decided on a 3-year jail term, 2 years suspended, and a US $58 000 fine for manslaughter. Grande Paroisse also received a criminal liability fine of €225 000, which is the maximum amount possible. The court ruled that Biechlin had contributed to creating the situation that resulted in the damage and did not take steps to avoid it, and had also exposed employees and the surrounding population to a serious risk [15].

However, these fines are usually negligible compared to the compensation one has to pay for being exposed to civil liability after a major accident. Civil liability means compensating for every form of damage that arises as a consequence of the major accident. Thus the responsible party has to compensate for lost lives, injured people, materials, buildings, and so on [14]. In the case of the major accident at AZF, as a result of civil liability, Grande Paroisse has paid out more than €2 billion (US $2.7 billion) in compensation to more than 16 000 victims, according to Total's figures [15].

In former times, governments and other organizations always held individuals responsible for major accidents, and they therefore paid the price for having the criminal, administrative, and civil liability. They did this because organizations and companies could not be held liable for major accidents. This has now changed, and companies can be sentenced and forced to pay fines. Because companies cannot go to jail, managers and other employees are still being pursued over fines and possible jail time [14].

Another example of managers of industrial facilities being held responsible for a major accident is the case of the 1984 gas disaster at Union Carbide's pesticide plant in Bhopal, India. Even 30 years after the accident, India was still pushing the US to extradite Warren Anderson, the former boss of the facility. Right up until he died in 2014, he faced charges of criminal negligence, and thus probably would have been sentenced to prison if the US had agreed to send him to India. Anderson was initially arrested in India after the catastrophe, but then managed to flee the country. From 1984, the US authorities turned down frequent extradition requests for Anderson. Seven local managers who were working at the Union Carbide plant in 1984 were convicted [16].

These avoided costs are represented by the following formula.

Avoided fines-related costs:

$$A + B + C$$

where

$A =$ civil liability fines (€);
$B =$ criminal liability fines (€);
$C =$ administrative liability fines (€).

Interim Lawyers Avoided Costs (Type II)

If a major accident occurs, the government will assemble a research team to discover what caused the accident and what the consequences are for the country, society, and environment. A company lawyer will also be assigned to help carry out this research and this person will therefore be unavailable in his or her usual company role. Therefore the company will need to hire an interim lawyer for the full duration of the research.

The costs related to the hiring of interim lawyers due to the occurrence of a major accident can be calculated and estimated by multiplying the daily wage of such a lawyer by the number of days he or she will be hired, and again by the number of lawyers that the company wishes to hire. If, however, the company decides to hire both junior and senior interim lawyers, the user may want to calculate the costs separately for both categories of lawyers.

These avoided costs are represented by the following formula.

Interim lawyers avoided costs:

$$w_1 \times d_1 \times n_1 \text{ (junior lawyers)} + w_2 \times d_2 \times n_2 \text{ (senior lawyers)}$$

where

w_1, w_2 = the hourly wage of junior lawyers and senior lawyers (€/day per lawyer);
d_1, d_2 = the number of days of junior lawyers and senior lawyers per lawyer (days);
n_1, n_2 = the number of junior lawyers and senior lawyers (number of lawyers).

Specialized Lawyers Avoided Costs (Type II)

In the event of a trial regarding major accidents, companies will hire lawyers who are specialized in these types of disaster. These lawyers will require substantial salaries and are expensive for any organization, even for a large multinational, as trials surrounding major accidents can take several years. However, the costs of specialized lawyers vary widely depending on the country in which the accident occurs or where the trial takes place. Often a deal is made between the two parties or a flat-rate amount is used. In any case, trials and lawsuits surrounding major accidents that take several years can easily cost the company hiring specialized lawyers several millions of euros [14].

The costs related to the hiring of specialized lawyers after a major accident can be calculated and estimated by multiplying the hourly wage of such a lawyer by the number of hours he/she will be hired, and again by the number of lawyers that the company wishes to hire. If, however, the company decides to hire specialized lawyers with widely varying wage levels, the user may want to calculate the avoided costs separately for those categories of lawyers.

Specialized lawyers avoided costs:

$$w_1 \times h_1 \times n_1 \text{ (lawyer category 1)} + w_2 \times h_2 \times n_2 \text{ (lawyer category 2)} + \ldots$$
$$+ w_t \times h_t \times n_t \text{ (lawyer category } t)$$

where

w_i = the hourly wage of category i (€/hour per person);
h_i = the number of hours of category i (hours);
n_i = the number of lawyers of category i (number of lawyers).

This calculation is represented by the following formula:

$$\text{Specialized lawyers avoided costs} = \sum_{i=1}^{t} w_i h_i n_i$$

where t is the number of lawyer categories

Internal Research Team Avoided Costs (Type II)

Independent of the research team assembled by the government and of any investigation carried out by other organizations involved in the major accident, a research team will always also be assembled by the company itself. This research team will mainly consist of health and safety experts and other specialists, and its purpose is to analyze the available information to identify the possible causes of the accident and make recommendations aimed at preventing similar accidents in the company's plants in future [14, 17].

An example of such a team is BP's investigation team after the Deepwater Horizon accident. The team began its work immediately in the aftermath of the incident, working independently of other accident response activities and organizations. When the investigation was being conducted, numerous similar investigations were ongoing. These included investigations by the US Coast Guard, the Bureau of Energy Management, the Regulation and Enforcement Joint Investigation, Transocean, the President of the United States' National Commission, and the Norwegian Institute for Water Research. More than 50 specialists, both internal and external, participated in the investigation of BP itself. The people involved were specialists in safety, drilling, exploration, well control, subsea engineering, and many more related fields. The outcomes of the research done by the government and the company will serve as evidence in determining the liable party [14, 17].

The costs related to the internal team investigating a major accident can be calculated and estimated by multiplying the daily wage of people participating by the number of days they will be hired, and again by the number of people the company wishes to assemble. If employees with significantly varying wage levels participate, the user may want to calculate the internal investigation team costs separately. Another possibility is to take the average wage level of all employees participating, meaning the user of the tool will only have to work with one category.

Internal research team avoided costs:

$$w_1 \times h_1 \times n_1 \text{ (employee category 1)} + w_2 \times h_2 \times n_2 \text{ (employee category 2)} + \ldots$$
$$+ w_t \times h_t \times n_t \text{ (employee category } t)$$

where

w_i = the hourly wage of category i (€/hour per person);
h_i = the number of hours of category i (hours);
n_i = the number of employees of category i (number of people).

This calculation is represented by the following formula:

$$\text{Internal research team avoided costs} = \sum_{i=1}^{t} w_i h_i n_i$$

where t is the number of employee categories

Experts at Hearings Avoided Costs (Type II)

In addition, experts in their field will sometimes be invited to testify and state their opinion in court. The party that is held accountable for a type II accident will pay the salary of these experts. The company will also have the possibility of hiring additional experts, to challenge the findings of the initial experts.

The costs related to the hiring of experts due to the occurrence of a major accident can be calculated and estimated by multiplying the hourly wage of experts participating by the number of hours they will be hired, and again by the number of experts that the company wishes to hire. If experts with significantly varying wage levels participate, the user may want to calculate the experts' costs separately. Another possibility is to take the average wage level of all experts participating, meaning the user of the tool will only have to work with one category.

Experts avoided costs:

$$w_1 \times h_1 \times n_1 \text{ (expert category 1)} + w_2 \times h_2 \times n_2 \text{ (expert category 2)} + \dots$$
$$+ w_t \times h_t \times n_t \text{ (expert category } t)$$

where

w_i = the hourly wage of category i per employee (€/hour per person);
h_i = the number of hours of category i (hours);
n_i = the number of experts of category i (number of experts).

This calculation is represented by the following formula:

$$\text{Experts avoided costs} = \sum_{i=1}^{t} w_i h_i n_i$$

where t is the number of expert categories

Avoided Costs Related to Legislation Changes (Type II)

Companies all over the world have invested resources in the field of safety as a response to legislation and directives. For instance, in Europe the major industrial accident prevention legislation is called the Seveso Directive, the first version of which was issued in 1982 as a result of the major accident that occurred in Seveso, Italy, 6 years earlier, in 1976. Vierendeels *et al.* [18] indicate that there are two drivers for legislation changes: (i) scientific progress and societal changes; and (ii) a shock effect (i.e., major accidents). However, the exact relationship between the occurrence of a major accident and a change in legislation is not unambiguous, and this relationship is discussed here. First, the history of major risk regulation is explained, and the connection between the occurrence of major accidents and legislation is clearly underlined.

At the beginning of the nineteenth century, more specifically the year 1810, the first regulation regarding major risks was born, as a result of an accident in 1794 in Grenelle, France, in which about 1000 people died. Over the following decades, similar catastrophes occurred in Europe, triggering the enactment of similar legislation regarding major industrial risks. In 1982 the first European Directive concerning major risks, Seveso I, was issued as a response to major accidents in Flixborough in the UK (in 1974) and Seveso in Italy (in 1976). The legislation has been changed several times since then, mainly because of subsequent major industrial accidents. For example, the accident that occurred at Bhopal in India in 1984 and the Rhine pollution at Basel in Switzerland in 1986 resulted directly in new amendments in 1987 and 1988. Together with major accidents in Mexico City in 1984 and on the Piper Alpha oil rig in 1987, these accidents resulted in changes in legislation, and in 1996 the Seveso II Directive was approved. However, major accidents continued to happen and mainly because of the shock effects, such as caused by the accidents in Baia Mare in Romania in 2000, Enschede in the Netherlands in the same year, and Toulouse in France in 2001, the legislation was changed and amended again in 2003. On June 1, 2015, a new version of the legislation for major risks, the Seveso III Directive, entered into force in Europe. It is clear, then, that legislation is subject to frequent changes as new accidents and new challenges arise. This is a time-consuming and costly process not only for governments, but also for private companies, as they have to analyze and implement the altered regulations in order to comply [18].

There is, therefore, clearly a correlation between the occurrence of major accidents and legislation changes, and, as already indicated, it is a result of the so-called shock effect. An accident can have a shock effect on regulation if the consequences are severe enough. A study by Vierendeels *et al.* [18] indicates that an accident causing 20 casualties, or even, in some cases, eight to 10 casualties, can have such a shock effect. The difference in the number of victims that can produce such an effect is explained by the fact that the media and the public, and thus also the politicians, often perceive an event as less significant if the casualties are employees or rescue workers, and more significant if they are civilians from the surrounding community. The number of fatalities, however, is not the only cause of a significant shock effect that can lead to a change in regulation. Others include, for example, a visible and traceable disposal of chemical substances in air, water, or soil that could have long-term health effects as well as the cost of an accident. If the accident costs exceed €1 billion, it is highly probable that the resulting shock effect will be large enough to bring about a change in legislation, even if there are no casualties involved [18].

Future changes in major accident legislation will be accompanied by a large financial, administrative, and operational burden. However, costs related to legislation changes as a result of a major accident are difficult to quantify directly. Nonetheless, they can be calculated indirectly by multiplying the total safety budget of the type of facility under consideration by the estimated increase (in %) of the safety budget (due to the occurrence of an accident scenario that will cause the legislation to change and become more complex).

These costs are represented by the following formula.

Legislation-related avoided costs:

$$S \times I$$

where

S = the total safety budget of this facility (€);
I = increase of the safety budget for this facility after the type II accident occurs (%).

Permits and Licenses Avoided Costs (Type II)

It will become harder for a company to obtain the necessary exploitation permits and operating licenses in the country where a major accident involving the company might occur. In general, companies (e.g., those with activities in the fields of petrochemicals, energy, and chemicals) need to be rigorous in obeying the rules for obtaining permits and licenses. They cannot afford to lose operating and exploitation permits through reckless behavior, as this could be a financial disaster for the company. For instance, if a company has had a relatively large number of minor accidents and is close to not obtaining a renewal of its permits and licenses, a major accident might well mean the end of their operating permit.

The costs related to obtaining new permits and licenses due to the occurrence of a major accident are difficult to quantify. However, they can be estimated by multiplying the total costs of having to close down the facility by the likelihood that the company will lose its operating permit due to a major accident.

These avoided costs are represented by the following formula.

Permit- and license-related avoided costs:

$$C \times L$$

where

C = the costs of having to close down the facility (€);
L = the likelihood of losing the operating permit (%).

Lawsuits Avoided Costs (Type II)

The costs accompanying lawsuits and trials can become substantial and can pose a significant threat to the liquidity of the company. This is primarily caused by the fact that such lawsuits and trials can last a number of years, sometimes more than a decade. Thus, it would be in the best interests of any company to avoid such time- and money-consuming trials. But how can this be achieved? One possible solution is by acting in a proactive way, as Fina did in 1995 after the explosion in Eynatten, Belgium. On the June 18, 1995 an explosion destroyed the Fina-owned road restaurant and gas station next to the E40 highway. This explosion was caused by a gas leak in the kitchen of the restaurant, and it killed 16 people and injured 20 more who had to be transported by helicopter to a hospital in Aken, Germany [19, 20]. Immediately after the explosion, the emergency procedure was set in motion. During this procedure, a helicopter, the police, the fire department of three nearby villages, the Red Cross, and the civil protection unit were present at the explosion site. In total, between 100 and 150 people were present [21]. Instead of acting in the traditional way, whereby a company waits to compensate the victims until it is clear who is liable and responsible for the accident, Fina took the initiative and compensated the injured and the families of the victims straight away. It would become clear later on after investigation which company was liable and thus responsible for the explosion, and once that decision was taken, the compensation that Fina had already paid could be recouped from their insurance company or the insurance company of the responsible party. The most important fact for Fina at that moment was that the people who suffered from this accident received compensation immediately. By acting in this proactive way, Fina possibly reduced the amount of expensive lawsuits and trials and therefore the number of lawyers required. None of the victims sued Fina for injuries or deaths. This approach also reduced the number of times that Fina appeared in the media in a negative way, as they surely would have through numerous trials and continued denial of their responsibility. In addition, this further reduced the negative impact of the explosion on Fina's image and reputation, as they had sent a clear message to the world by compensating other parties immediately, proving that their company takes full responsibility and that they care about others [14].

As noted, the reduction in the number of trials would have had a positive effect on the image and reputation of the company, in turn possibly having a positive effect on its sales, which might have been reduced as a result of the negative media attention. In turn, by acting proactively, the company took care of a possible market-value decrease, represented by a decrease in its share price on the stock market. Companies operating in certain areas, such as, for example, the chemical, oil, and gas industries, should always remember that it takes years to build a good image. By contrast, a good reputation can be destroyed in just one brief moment, which it will take several years to rebuild.

It should be noted that insurance companies will usually not let the company act in such a proactive way (i.e., compensating losses before the court decision), as they will want to keep the amount spent in the aftermath of an accident as low as possible, because they will have to bear the majority of those costs. Instead of taking into account the reduction of the legal costs, they will be more concerned with the possibility that the other party will ask for more money as compensation than they deserve. However, this turns out to be negligible compared with the major legal costs and indirect costs resulting from reputational risk to their client in the case of a major (type II) accident.

5.4.2.4 *Insurance Avoided Costs (Type* **I** *+* **Type II***)*

In order to understand insurance premium increases, insurances as a whole are discussed briefly before discussing the premium. Consider the following definitions of insurance [22]:

- The pooling of fortuitous losses by transfer of the risks to insurers, who agree to indemnify insured parties for such losses, to provide other pecuniary benefits on their occurrence or to render services connected with the risk. Pooling is the sharing of total losses among a group and is thus the spreading of losses incurred by a few over an entire group.
- A system to protect persons, groups, or businesses against certain risks of financial loss by transferring the risks to a large group who agree to share the financial losses in exchange for premium payments.
- A written contract between two parties providing a promise of reimbursement in the case of loss; purchased by people or companies so concerned about hazards that they have made prepayments to an insurance company for such risk treatment.

In summary, insurances arrange the transferring of risks and the sharing of losses. It is always a contract between two parties, the insured party and the insurance company. Table 5.6 lists some important characteristics about insurances. These features are vital to understand how insurances work and how they affect the financial position of a company.

Insurances need an individual approach; they are different according to insurer, facility, and country. Therefore no insurance in any industry is exactly the same. There are all kinds of insurances, such as business interruption insurances and property damage insurances, which all assist the company with the financial burden it would face were an accident to occur. For example, in the process industry, the most important insurances are [23]:

- Property and physical damage
- Machinery breakdown
- Business interruption
- Liability: public liability, product liability, recall
- Employers' liability (for accidents in which employees are injured or killed)
- Environmental insurances: own premises and third liability
- Transport (marine cargo/goods in transit/storage risks)
- Project insurances
- D&O (directors' and officers') insurances.

Table 5.6 Insurance characteristics.

What do insurances actually do?
- Insurance aims at restoring/indemnifying/compensating the insured should the risk translate into an incident.
- It does not put the insured in a better position than they were before the loss (that would invite fraud).
- It is based on probabilities – the incident(s) may or may not occur.

What are insurances not?
- The total solution.
- A guarantee that an incident will not occur.
- A guarantee that all costs of damages will be repaid.
- A protection against reputation and market losses or image destruction.
- An alternative for risk management.

Why does a company need insurances?
- Because operational risk management cannot eliminate all operational risks.
- Because it is an acceptable choice to transfer operational risk (as one possible risk treatment option).
- Because bad luck does exist.
- To receive financial compensation from a third party for losses companies perhaps cannot bear themselves.
- Because it contributes to a stable working environment.
- Because it may be a regulatory or contractual requirement.
- Because through insurances the long-term future of the company will not be compromised.

Depré [22].

By paying a yearly insurance premium, companies aim to ensure that the insurance company will take responsibility for (part of) the financial burden if an accident would occur. Many minor accidents or some major accidents will affect the insurance premium that a company has to pay. In order to understand how this premium will change, companies should first understand how premiums are calculated. An insurance company takes into account both the value of the assets that need to be insured and the accidents and damage that have occurred in the past. Because these data are from the past, insurance companies extrapolate them to the future to arrive at the required insurance premium. The premium also depends on the size of a company (e.g., a multinational or a national company), and on the geographical locations of it plants. In addition, the insurance company will add a profit margin to the insurance premium.

In the specific case of type II accidents, one problem is that there is too little information available, what makes it very difficult, if not impossible, to accurately extrapolate into the future. In addition, the intrinsic value of the assets to be insured is very large in some cases (e.g., in the process industry). This results in insurance premiums being very high. The combination of the lack of data and magnitude of insurance premiums means that premiums are usually determined during negotiations between the company and an insurance broker, representing the insurance company. Therefore, there is no formula or rule of thumb for determining the premium increase after a major accident, as this will be discussed during negotiations between the company and the insurance broker. However, because (among other things) all accidents and damage that occurred in the past are taken into account in the calculation of the premium, we know that the premium will definitely increase after the occurrence of a major accident.

Notice that in case of a number of minor accidents, it is much easier to determine the rise of the insurance premium, as a lot more information is available in these cases.

In the case of type II major accidents, there is an additional factor that can influence the insurance premium. This additional factor is the reinsurance company, which reinsures the insurance company. If, for example, there have been a larger than expected number of major industrial accidents, the reinsurance company will have a hard time dealing with those claims, resulting in more difficult negotiations between the reinsurance and insurance companies. This, in turn, results in more difficult negotiations between the insurance company and the companies where the major accidents occurred, resulting in larger premium increases [24].

This all means that it is very difficult to predict by what percentage the insurance premium will increase after a number of minor or major accidents, but the insurance broker or people working in the company's insurance department will estimate this percentage increase by looking at the magnitude of the consequences predicted by the health and safety department. Although the insurance premium consequences are hard to predict, they are calculated by multiplying the current total insurance premium cost of the facility by the expected increase in the premium.

The insurance premium avoided costs can thus be represented by the following formula.

Insurance premium avoided costs:

$$P \times I$$

where

$P =$ the current total premium cost of the facility (€);
$I =$ the expected increase of the premium (%).

5.4.2.5 Human and Environmental Avoided Costs

Compensation Victims Avoided Costs (Type I + Type II)
Whenever an accident causes casualties, the company will have to compensate the families for their losses [8]. These costs are also labeled as compensation victims costs, and should be taken into account in any CBA. These consequences can be calculated by multiplying the VoSL in the specific country or region (see Chapter 4 for a discussion on calculating the VoSL considering the probability of the accident scenario) by the expected number of fatalities.

These costs are then represented by the following formula.

Compensation victims costs:

$$\text{VoSL} \times n$$

where

$\text{VoSL} =$ the value of one human life (€/person);
$n =$ the number of fatalities (number of persons).

Injured Employees Avoided Costs (Type I + Type II)

Whenever an accident causes injuries, both minor and major injuries, the company will have to compensate the injured people for their injuries [8]. These consequences are calculated by multiplying the cost of lightly and seriously injured workers in that specific region by the expected number of lightly and heavily injured people.

These avoided costs are represented by the following formula.

Injured employees avoided costs:

$$C_1 \times n_1 \text{ (lightly injured)} + C_2 \times n_2 \text{ (seriously injured)}.$$

where

C_1, C_2 = the cost of one lightly injured and seriously injured worker (€/person), respectively;

n_1, n_2 = the number of lightly injured and seriously injured workers (number of injured persons), respectively.

Avoided Recruitment Costs (Type I + Type II)

As some employees may be injured or killed or leave the company due to an accident, new employees will have to be recruited. The recruiting cost is thus the cost of hiring new workers, which includes the time invested in recruiting and training the new workers. The recruitment consequences are calculated by multiplying the sum of the hiring and training costs by the number of newly recruited employees. Hiring costs include advertising, interviews and assessments, and other costs [7].

These avoided costs are represented by the following formula.

Recruitment avoided costs:

$$(H_1 + T_1) \times n_1 \text{ (employee category 1)} + (H_2 + T_2) \times n_2 \text{ (employee category 2)}$$
$$+ \ldots + (H_t + T_t) \times n_t \text{ (employee category } t)$$

where

H_i = the hiring cost per employee of category i (€/person);
T_i = the training cost per employee of category i (€/person);
n_i = the number of newly recruited employees of category i (number of persons).

These avoided costs are represented by the following formula:

$$\text{Recruit costs} = \sum_{i=1}^{t} (H_i + T_i) n_i$$

where t is the number of employee categories.

Avoided Costs with Respect to Environmental Damage (Type I + Type II)
Whenever an accident causes environmental damage, the company will receive claims to compensate for the damage. These consequences are calculated by multiplying the mass of the spill by the expected cost per mass unit spilled.

These avoided costs are represented by the following formula.

Environmental damage avoided costs:

$$S \times C$$

where

S = the mass of the spill (e.g., tons);
C = the cost per mass spilled (€/mass unit).

5.4.2.6   Personnel-related Avoided Costs

Accidents often result in situations where employees are temporarily unable to carry out their job and daily activities, for short or long periods of time, or sometimes even indefinitely. Alternatively, employees may be obliged to perform activities other than the ones they are used to. Such situations entail avoided accident costs related to personnel and their productivity.

Lowered/Lost Productivity Avoided Costs (Type I + Type II)
Productivity of employees often decreases due to an incident or accident. This productivity loss is not merely the result of the employee who is actively involved in the accident, but can also result from other employees displaying lower productivity patterns. Furthermore, irrespective of the fact that due to an accident an employee can be incapable to work for a certain period of time, when that person returns to work he or she often displays lower productivity. It is sometimes possible that physical problems and restrictions and/or an altered risk perception (the so-called Hawthorne effect, cf. [12, 25]) can lead to different behaviors resulting in lower productivity. If "adapted work" is foreseen for the employee, productivity will most likely be lower as well. If the employee needs to be replaced, productivity levels will also be lower, especially initially, owing to a lack of experience and expertise [11, 26].

Formulas to deal with lost productivity are as follows

Productivity loss of employees:

$$\sum_{i=1}^{n}(T_{1i} \times W_i \times N_i) + \sum_{i=1}^{n}(E_{1i} \times T_{2i} \times W_i \times N_i)$$

where

T_{1i} = period of time in which employees of category i do not work due to an accident (hours)
T_{2i} = period of time in which employees of category i work less efficiently due to the Hawthorne effect (hours);
W_i = wage of employees of category i involved (€/hour per person);
N_i = number of employees of category i involved (number of persons);
E_{1i} = efficiency loss of employees immediately after the accident due to the Hawthorne effect (%);
n = number of employees categories (dimensionless).

---

and

Productivity loss of employees after return:

$$\sum_{i=1}^{n} (E_{2i} \times T_i \times W_i \times N_i)$$

where

E_{2i} = the efficiency loss of employees of category i doing "adapted work" (%);

W_i = wage of employee of category i involved (€/hour per person);

T_i = period of time during which adapted work needs to be carried out by employees of category i (hours);

N_i = number of employees of category i involved (number of persons);

n = number of employees categories (dimensionless).

Training of Temporary Workforce Avoided Costs

Training people who are to replace those employees who suffered an accident also represents a cost. The most important part of this cost is time. The time needed by the trainer having to train the person who replaces the injured employee, as well as the training time of the substitute, needs to be counted. The latter can also be seen as lowered productivity, as, during the training period, the substitute does not attain his/her optimal productivity [11, 26, 27].

Training cost of temporary employees:

$$\sum_{i=1}^{n} (T_{1i} \times L_{1i} \times N_i) + \sum_{j=1}^{m} (E_j \times T_{2j} \times L_{2j} \times M_j)$$

where

E_j = the efficiency loss of employees of category j doing "adapted work" (%)s;

L_{1i} = wage of trainer of category j (€/hour per person);

T_{1i} = period of time devoted *for* the training *by* trainers of category j (hour);

L_{2j} = wage of substitute employee of category j (€/hour per person);

T_{2j} = period of time needed for the substitute to reach the optimal productivity (hour);

N_i = number of trainers of category i (number of persons);

M_j = number of substitute employees of category j (number of persons);

n = number of trainer categories (dimensionless);

m = number of substitute employee categories (dimensionless).

Wage Avoided Costs

Accidents always go hand in hand with a lot of loss of time. The "wage cost" represents the amount of time that company employees cannot devote to their regular tasks as a result of an accident. This may be the result of necessary medical treatment at the company's first aid department, in which case the corresponding wage cost may be negligible. However, when the accident leads to a longer period of lost work, the wage cost can be significant [7, 11, 25, 26, 28]. There may also be a wage cost due to employee colleagues having to work extra hours in the event of an accident.

Total wage cost:

$$\sum_{i=1}^{t} T_i \times L_i \times N_i + X$$

where

  T_i = period of time lost due to the accident of an employee of category i (hours);
  L_i = wage of the involved employee of category i (€/hour per employee of category i);
  N_i = number of employees involved in the accident (number of employees of category i);
  t = number of employee categories;
  X = Costs due to extra work of colleague employees (€).

5.4.2.7 Medical-related Avoided Costs

This category of avoided costs only applies to accidents involving one or more injured persons. Medical expenses are often an important part of the total cost of an accident, but they are mostly considered as insured costs. The degree to which such costs are, in fact, insured depends on the insurance policy.

Medical Treatment at Location Avoided Costs (Type I + Type II)
Large companies usually have their own medical service department, so that in the case of occupational accidents, medical personnel of the organization may offer first aid. Sometimes, it is necessary for the medical service to travel to the site of an accident, leading to a possible cost. It should be noted that due to legislation and/or precaution, not all medical services should and/or could be canceled within an organization, even if the number of accidents were zero [11].

Avoided Costs Related to Medical Treatment in Hospitals and Revalidation (Type I + Type II)
Some of the more severe occupational accidents need to be treated in hospitals by specialized personnel. This may also represent a substantial avoided cost:

Total cost medical treatment:

  X

where

  X = total expenses for all medical treatments and revalidation in hospitals and transports to hospitals in relation to employees involved in an accident (€).

Avoided Costs Related to Using Medical Equipment and Devices (Type I + Type II)
Avoided costs related to used medical equipment and devices is mainly applicable to companies having their own medical services. Depending on the nature and severity of the

accident, employees can be treated in the medical facilities of the organization, and medical equipment, devices, and material can be consumed in such cases. First, well-educated medical personnel need to be present in the case of certain equipment. Such personnel and their training and education represent an avoided cost. Second, medical material may include bandages, painkillers, and so on.

Medical Transport Avoided Costs (Type I + Type II)
If an accident requires employees to be treated in hospital instead of by the organization's medical services, they need to be transported to the hospital. This transportation cost should be taken into consideration in a CBA.

5.4.2.8   Intervention Avoided Costs (Type I + Type II)

Whenever an accident occurs, and certainly in case of a major one, different types of intervention personnel will be needed. Intervention types can range widely and include fire services, police services, ambulance services, and special unit services if toxic material is involved in the accident. The option to include fire and police department costs should at least be considered, as in some cases the company will have to pay an amount of money for their services, although these interventions by the fire and police departments are public services [8]. The intervention avoided costs can be calculated by taking the sum of the avoided costs for the specified intervention types.

Intervention avoided costs:

$$F + P + A + S$$

where

F = the fire department costs that may be charged to the company (€);
P = the police department costs that may be charged to the company (€);
A = the ambulance services costs that may be charged to the company (€).
S = the special units costs that may be charged to the company (€).

5.4.2.9   Reputation Avoided Costs (Type II)

Consequences related to the reputation of the company subject to a major accident are hard to quantify. One possible way to do so is by considering the share price consequences, as share prices display the investors' image of the current performance and future expectations of the company, which can be seen as the "reputation."

   Consider the example of the BP share price drop due to the Deepwater Horizon drilling rig major accident in April 2010. Following the oil rig disaster, the BP share price dropped more than 50% in value. On April 20, the day of the major accident, the share price was £655.40. As information regarding the severity and consequences of the disaster became widely known, the share price plunged, reaching a low of £302.90 on June 29, a decline of 53.78% in comparison to the April 20 share price. From this day on, the share price gradually increased again. The price seems to have stabilized around £450.00, a recovery of some 50% of its loss, although still some 30% below the pre-accident share price and pre-accident market value of about $190 billion.

The share price avoided costs can be calculated by multiplying the current total market value of the company by the expected drop in the share price.

Share price avoided costs:

$$M \times D$$

where

> M = the current total market value of the company (€);
> D = the expected drop in the share price (%).

A company can use a rule of thumb for the expected drop in share price and anticipate an expected decrease of share price (expressed as a percentage), depending on the consequences of a disaster scenario.

5.4.2.10 Other Avoided Costs

Accident Investigation Avoided Costs (Type I + Type II)
When an accident occurs, a person or a team is assigned to investigate it (not necessarily related to any legal affairs; see Section 5.4.2.3). Depending on local legislation and the type of accident, organizations are obliged to report it to the authorities through an accident investigation report [12]. Organizations also often want to determine and map the causes of accidents in order to take the necessary preventive measures to avoid future similar incidents. The literature thus mentions a variety of accident investigation approaches, each with pros and cons. The costs of accident analyses arise from the time that employees have to devote to the investigation, and sometimes also from technical studies. In the case of type II accidents, in particular, technical investigations may be an important avoided cost.

The time-related costs are composed of the wages of people carrying out the accident investigation. This cost can be determined by using the wage per employee category involved in the investigation [26]. Sometimes, certain additional employee-related costs are present that should be added to the costs. Such additional costs involve, for instance, the further processing of the accident investigation file, and sending the report to all concerned parties.

Accident investigation avoided costs:

$$\sum_{i=1}^{n} (t_i \times L_i \times N_i) + X + \text{TS}$$

where

> t_i = Time in total spent on the investigation (hours);
> L_i = employee wages, possible different categories of employees (€ per hour per employee category i);
> N_i = number of employees involved in the accident investigation of category i (number of people);
> n = number of employee categories;
> X = additional employee–related accident investigation costs (€);
> TS = cost related to the technical studies needed for the accident investigation (€).

Manager Work Time Avoided Costs (Type I + Type II)

Managers of all levels (middle management, higher management and board of directors) will be forced to invest time if an accident occurs. They will have to investigate the accident, guide the employees, possibly deal with press attention and, in certain cases, attend lawsuits and other legal processes [7]. The manager work time consequences can be calculated and estimated by multiplying the total number of hours lost by all managers of a certain manager category by the cost per hour of the lost work time of managers of that category. As the work of managers with significantly varying wage levels will be affected, the manager work time consequences can be calculated separately for each category of managers.

Manager work-time costs:

$h_1 c_1$ (manager category 1) + $h_2 c_2$ (manager category 2) + ...
   + $h_n c_n$ (manager category n)

where

h_i = the total number of hours lost by all managers of category i (hours);
c_i = the cost per hour of the lost work time of managers of category i (€/hour).

These costs are represented by the following formula:

$$\text{Manager work time costs} = \sum_{i=1}^{n} h_i c_i$$

where n is the number of manager categories.

Clean-up Avoided Costs (Type I + Type II)

An avoided cost that is often forgotten is the clean-up cost resulting from an accident. Before rebuilding and restoring the initial situation, the whole accident area needs to be cleaned up. Besides the employees, an independent cleaning company may sometimes need to be hired to execute this clean-up assignment. The avoided clean-up costs can be calculated and estimated by multiplying the hourly wage of an employee by the number of hours the cleaning will take, and again by the number of employees participating.

Avoided clean-up costs:

$w \times h \times n$

where

w = the hourly wage of a cleaning employee per employee (€/hour per employee);
h = the number of hours worked by a cleaning employee (hours);
n = the number of employees cleaning up (number of employees).

5.4.3 Investment Analysis (Economic Concepts Related to Type I Risks)

In the case of type I risks, certain economic concepts exist that are linked to the costs and benefits and help to make an investment analysis to steer a recommendation for the safety investment. The economic concepts are "internal rate of return" (IRR) and "payback period" (PBP).

5.4.3.1 Internal Rate of Return

The IRR can be defined as the discount rate at which the present value of all future cash flows (or monetized expected hypothetical benefits) is equal to the initial investment or, in other words, it is the rate at which an investment breaks even. Generally speaking, the higher an investment's IRR, the more desirable it is to carry on with the investment. As such, the IRR can be used to rank several possible investment options an organization is considering. Assuming all other factors are equal among the various investments, the safety investment with the highest IRR would then be recommended to have priority. Note that the IRR is sometimes referred to as "economic rate of return" (ERR).

An organization should, in theory, undertake all safety investments available with IRRs that exceed a minimum acceptable rate of return pre-determined by the company. Investments may, of course, be limited by availability of funds or safety budget to the company.

Because the IRR is a rate quantity, it is an indicator of the efficiency, quality, or yield of an investment. This is in contrast to the NPV, which is an indicator of the value or magnitude of an investment.

A rate of return for which the NPV, expressed as a function of the rate of return, is zero, is the IRR, r^*. This can be expressed as follows, (Eq. 5.1):

$$NPV(r^*) = \sum_{n=0}^{N} \frac{C_n}{(1+r^*)^n} = 0 \tag{5.5}$$

In cases where a first safety investment displays a lower IRR but a higher NPV over a second safety investment, the first investment may be recommended over the second investment. Furthermore, note that the IRR should not be used to compare investments of different duration. For example, the NPV added by an investment with longer duration but lower IRR could be greater than that of an investment of similar size, in terms of total net cash flows, but with shorter duration and higher IRR.

As a simple illustrative example, the following problem can be given for calculating an IRR. Assume that a safety investment is given by the sequence of cash flows (initial investment costs and yearly hypothetical benefits) shown in Table 5.7.

From Table 5.7 and the above equation, the IRR r^* can be determined by solving the following equation, (Eq. 5.5):

$$NPV(r^*) = -123\,400 + \frac{36\,200}{(1+r^*)^1} + \frac{54\,800}{(1+r^*)^2} + \frac{48\,100}{(1+r^*)^3} = 0$$

In this case, the answer is 5.96% (in the calculation, i.e., $r^* = 0.0596$).

Table 5.7 Sequence of cash flows for
a safety investment.

Year (n)	Cash flow (C_n) (€)
0	−123 400
1	36 200
2	54 800
3	48 100

5.4.3.2 Payback Period

The PBP is calculated by counting the time (usually expressed as a number of years) it will take to recover an investment. Hence, a break-even point of investment is determined in terms of time. The PBP of a certain safety investment for type I risks is a possible determinant of whether to go ahead with the safety project or not, as longer PBPs are typically not desirable for some companies. It should be noted that the PBP ignores any benefits that occur after the determined PBP and, therefore, does not measure profitability. Moreover, the time value of money is not taken into account in the concept, and nor is the opportunity cost considered. The PBP may be calculated as the cost of safety investment divided by the annual benefit inflows.

> As an illustrative example, let's assume that a company invests €400 000 in safety equipment. The expected (hypothetical) benefit from the new equipment is expected to be €100 000 per year for 10 years. The payback period is 4 years (€400 000 divided by €100 000 per year).
>
> A second safety project requires an investment of €200 000 and it generates expected hypothetical benefits as follows: €20 000 in year 1; €60 000 in year 2; €80 000 in year 3; €100 000 in year 4; and €70 000 in year 5. The payback period is 3.4 years (€20 000 + €60 000 + €80 000 = €160 000 in the first 3 years + €40 000 of the €100 000 occurring in year 4).

Note that the payback calculation uses cash flows, not net income. The PBP simply computes how fast a company will recover its cash investment.

5.5 The Cost of Carrying Out Cost-Benefit Analyses

Although the demand for CBAs over time is increasing in organizations (as it is in society), it should be clear that it takes many resources (efforts of all kinds, such as time, skill, and money) to carry out a CBA in a good way, certainly also in case of safety investment studies. The reason for these resources for CBA is that, if carried out well, the technique is used to evaluate the ratio between all benefits and all costs of certain investments. Hence, financial terms need to be assigned to all categories of costs and benefits as displayed in the previous sections, to be able to calculate the so-called "cost-benefit ratio" as accurately as possible. If only partial information is used, or inaccurate information, the ratios will be incorrect and safety investment decisions may be influenced. So, carrying out a CBA in itself can be an important part of the expenditure of a safety investment study.

5.6   Cost-Benefit Analysis for Type I Safety Investments

It is possible to determine whether the costs of a safety measure outweighs – or not – its benefits. In general, the idea is simple, as previously explained: compare the costs of the safety measure with its benefits. As seen in the previous sections, the costs of a safety measure are relatively easy to determine, but the benefits are much more difficult to calculate. In the approach explained in this section, the benefits are expressed as the "reduced risk," taking into account the costs of accidents with and without the safety measure implementation. The following equation may be used for this exercise [29]:

$$\{(C_{without} \times F_{without}) - (C_{with} \times F_{with})\} \times Pr_{control} > safety\ measure\ cost.$$

Alternatively, if there is not enough information regarding the initiating events' frequencies to use this equation, the following equation can be used:

$$(C_{without} - C_{with}) \times F_{accident} \times Pr_{control} > safety\ measure\ cost,$$

where

$C_{without}$ = cost of accident without safety measure;
C_{with} = cost of accident with safety measure;
$F_{without}$ = statistical frequency of initiating event if the safety measure is not implemented;
F_{with} = statistical frequency of initiating event if the safety measure is implemented;
$F_{accident}$ = statistical frequency of the accident;
$Pr_{control}$ = probability that the safety measure will perform as required.

These formulas show immediately why this approach may only be carried out for (type I) risks where sufficient data are available: if not, the required "statistical frequencies" are unknown, the probabilities may not be known, and rough estimates (more or less informed *guesses*) will have to be used. If sufficient information is available, the results from these equations for determining the cost-benefit of a safety measure will be reliable.

5.7   Cost-Benefit Analysis for Type II Safety Investments

5.7.1   Introduction

In this section, a systematic approach to analyzing and evaluating safety investments aimed at preventing and mitigating type II risks, and based on the DF, is described and explained .

Type II accidents are related to extremely low frequencies and a high level of uncertainty. To take this into account, the CBA preferably involves a DF (see also Chapter 4 and Section 8.12) in order to reflect an intended bias in favor of safety over costs. This safety mechanism is vital

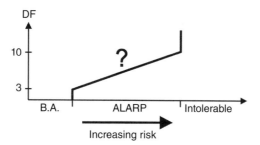

Figure 5.3   Disproportion factor (DF). ALARP, as low as reasonably practicable; B.A., broadly accept-able. (Source: Rushton [31].)

in the calculation to determine the adequate level of investment in prevention measures, as, on the one hand, the probability influences the hypothetical benefits substantially through the number of years over which the total accident costs can be spread out, and on the other, the uncertainty regarding the consequences is high [31].

Usually, CBAs state that the investment is not encouraged if the costs are higher than the benefits. If, however, a DF is included, an investment in safety is reasonably practicable unless its costs are grossly disproportionate to the benefits. If the following equation is true, then the safety measure under consideration is not reasonably practicable, as the costs of the safety measure are disproportionate to its benefits:

$$\frac{\text{Costs}}{\text{Benefits}} > \text{DF} \Rightarrow \text{Costs} > \text{DF} \times \text{Benefits}. \tag{5.6}$$

In order to give an idea of the size of the DF, some guidelines and rules of thumb are avail-able. They state that DFs are rarely greater than 10, and that the higher the risk, the higher the DF must be in order to stress the magnitude of those risks in the CBA. This means that in cases where the risk is very high, it might be acceptable to use a DF > 10 [31]. Although a value > 10 is allowed, Rushton [31] strongly advises against using a DF > 30.

This increase of the DF according to the risks is shown in Figure 5.3 (from Rushton [31]). This figure again indicates that CBAs should be used to evaluate safety measures for risks in the "as low as reasonably practicable" (ALARP) region. The more the risks leans toward the "intolerable" part of the ALARP region, the larger the DF needs to be.

Figure 5.3 highlights the principle that the higher the risk, the higher the DF must be, and that in cases where the risk is very high and tends toward the unacceptable region, it might be acceptable to use high DFs.

In brief, companies can demonstrate to governments and other people that additional mea-sures are not reasonably practicable, based on CBAs taking a DF into account. An advantage of using a DF in the analysis is that the company can claim to be biased in favor of safety above costs. Note that in theory it would also be possible to use the DF for type I risks, if company safety management wished to pursue certain safety investments for this type of risk.

However, operational safety-related investment decisions should indeed be weighted in favor of safety, especially in case of high-impact, low-probability (HILP) risks. The process in practice is preferably not one of simply balancing the costs and benefits of measures, but rather of always implementing the safety measures (due to the high Maxmax hypothetical benefits in case of HILP risks; see Section 4.2.4), except where they are ruled out because they involve so-called "grossly disproportionate" sacrifices. As several factors come into play, there is no simple formula for computing which risks are situated in the "ALARP" region (see Section 4.10.1). Nonetheless, safety-related decisions need to be justified based on some form of economic analysis. Moreover, when comparing the sacrifice (investment cost of safety measure) and the risk reduction (hypothetical benefit of the safety measure), the usual rule applied by a CBA model is that the investment should be made if the benefit outweighs the costs. However, for making decisions in the tolerable part of the ALARP region, the rule is that the safety measure should be implemented unless the sacrifice is "grossly disproportionate" to the risk. By using this successful practice, the investment costs are allowed to outweigh the benefits, and the safety investment is pursued. However, the question remains as to what extent costs can outweigh benefits before being judged "grossly disproportionate." The answer to this question depends on factors which are summarized by the DF. In Thomas and Jones [32], the concept of maximum justifiable spend (MJS) is introduced as follows:

$$\text{MJS} = \text{cost of failure} \times \text{probability of failure} \times \text{proportionality factor}$$

Maximum justifiable spend is the amount of money it is worth spending, based on the magnitude of the risk, to reduce the risk to zero. It gives a criterion for identifying what additional risk reduction measures are worth considering. We suggest that the measures being more costly than the MJS can be discarded immediately. Those that are lower than the MJS should be implemented in order to reduce the risk and contribute to achieving acceptable ALARP. There is also a gray area where additional assessment efforts are required. The formula can be used to rank and classify alternative safety measures, but it relies on preliminary computations and risk assessments about the level of the DF which should be employed and which can be considered realistic. The approach proposed in this section can be used to evaluate ALARP safety investments embedded within an economic ex ante risk.

In the following sections, a formula is presented to derive the value of the DF which makes the NPV of a safety investment equal to zero. Using the results of such a simulation exercise, it is possible to compare alternative safety measures. Moreover, given the limitations and restricted understanding of the three "How" factors (see Section 4.11.2.2), it is important to judge whether the DF associated with each safety measure "behaves" in a reasonable way.

5.7.2 Quantitative Assessment Using the Disproportion Factor

Three main features are associated with every safety investment, as shown in Table 5.8.

The cost of a safety investment can be divided into M (corresponding to the initial investments, e.g., the purchasing cost of new equipment and materials directly related to the intervention) and m (the yearly recurring costs due to maintenance, energy costs, yearly equipment, depreciation and interest expenses, material and training costs). A safety investment is evaluated considering a time horizon n that should be defined by the investor. More specifically, the time horizon should be compatible with the asset life of the safety measure to be analyzed. Hence, within n years, the safety investment is supposed to maintain its effectiveness, without any significant deterioration of its performance.

In order to assess the financial impact of a safety investment, the NPV equation should be adapted to the evaluation of the cost-effectiveness of a safety measure by explicitly including the DF. More specifically, the investment is represented by the cost of the safety measure to be evaluated (i.e., M). This cost is supposed to be entirely sustained in the initial year (year 0) when the investment needs to be evaluated. Owing to the characteristics of the measure, yearly recurring costs (e.g., due to maintenance activities) might be required over the time horizon in which the investment is evaluated. These recurring costs are needed to maintain the functionality of a safety measure and keep its effectiveness at its initial level. These costs are expressed as a percentage of the measure's initial cost and are assumed to be sustained starting from year 1 until n, where n represents the time horizon in which the investment is evaluated. Therefore the cost value C_t to be considered in the formula of the NPV assumes the following form:

$$C_t = \begin{cases} M, & \text{if } t = 0, \\ M \times m, & \text{otherwise.} \end{cases} \tag{5.7}$$

On the other side, consistent with what was previously explained in this chapter, the benefits are quantified as the monetary savings that can be achieved if the disruptions caused by an accident, which might happen with probability p, are avoided or mitigated thanks to the safety investment that has been pursued. To quantify the savings, the basic notion of risk is used, as explained in Section 2.2. As defined by the Center for Chemical Process Safety [33], risk can be seen as an index of potential economic loss, human injury, or environmental damage, which is measured in terms of both the incident probability and the magnitude of the loss, injury, or damage. The risk associated with a specific (unwanted) event is thus expressed as the product of two factors: the likelihood that the event will occur (p) and its consequences (V) considering both financial and human aspects. A risk is therefore an index of the "expected consequence" of the unwanted event. Two types of losses (financial loss f and human loss h) are considered to quantify the value of V (see Talarico *et al.* [34] for more details):

$$V = f + ch,$$

where c represents a factor translating human loss into financial terms.

Table 5.8 Features associated with a safety investment to be evaluated.

Symbol	Description
V	Total loss which results from the adverse event
f	Financial loss
h	Loss of human life
c	Factor translating loss of human life into financial terms
p	Likelihood that the adverse event will occur
n	Time horizon in years
a	Risk aversion factor
e	Effectiveness of the safety investment expressed as a percentage
M	Initial cost of the safety investment, including the installation, expressed in a specific currency
m	Yearly recurring cost expressed as a percentage of the initial investment's cost, M

Moreover, the risk aversion of a decision-maker toward a high-consequence accident scenario is also considered by using the risk aversion factor a. Together with the DF, the risk aversion factor a can be used as a parameter to balance the risk awareness of the decision-maker as a way to incentivize investments in safety.

Assuming further that a safety investment, whose effectiveness is represented by the letter e, is adopted, the risk of an accident can be decreased. In fact, the probability of an accident that might trigger consequences estimated to be equal to V can be lowered due to the safety investment as in the following formula:

$$R_{\text{with measure}} = (1 - e)pV^a.$$

Assuming that no safety investment is pursued to prevent a potential accident, the expected risk is calculated by $R = p{\cdot}V^a$. Therefore the profit of having a safety measure can be measured as the marginal savings that can be obtained compared with a case in which no investments in safety are made. More specifically, the marginal savings are represented by the avoided expected losses in the case of accident due to a lower overall risk. In the following equation, the marginal gain is shown:

$$R_{\text{no safety investment}} - R_{\text{with safety investment}} = pV^a - (1 - e)pV^a = epV^a.$$

Finally, assuming that the safety investment allows one to decrease the risk of accidents during a fixed time horizon whose length is n, the total effect of the safety measure on the risk reduction can be estimated by multiplying the probability p by n. As a result, the potential benefits quantified in the year 0 can be assumed to have the following form:

$$B_t = \begin{cases} epnV^a, & \text{if } t = 0, \\ 0, & \text{otherwise.} \end{cases} \tag{5.8}$$

5.7.3   Decision Model

During the risk assessment phase, risk experts analyze the features of a system that might potentially be affected by a HILP accident. Accident types are investigated and possible consequences are calculated. These consequences can also be estimated from a financial point of view, as described earlier. The probabilities of the accident scenarios are also estimated. Using this information, the expected risks associated with accident scenarios are quantified. In Figure 5.4, a decision model is presented that can be followed within an organization to assess, evaluate and decide about a safety investment regarding a type II accident scenario.

As can be seen in Figure 5.4, this risk assessment approach serves as an input for both a technical and a financial assessment. These phases can be executed in parallel, based on the findings of the previous step. The technical assessment is focused on the definition of the most suitable safety investments, based on the risks that might affect the system. A list of possible safety investments is drafted and for every possible investment some basic features are determined (e.g., installation cost, maintenance, effectiveness, duration). Moreover, some of the available investments might be not compatible, from a technical point of view, with the system that needs to be protected. For this reason, the incompatible measures should be discarded from the following steps of the analysis. Furthermore, the goal of the financial assessment is to define the safety budget, for example, discarding some safety investments as too expensive.

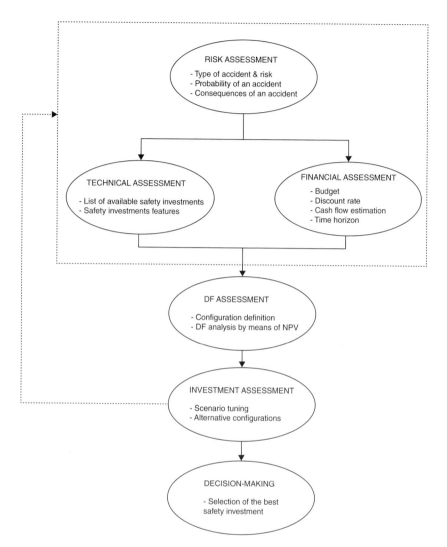

Figure 5.4   General schema of the decision model for evaluating safety investments based on the disproportion factor (DF).

Furthermore, some of the parameters required to estimate the benefits of the safety investments need to be defined, such as the discount rate and the time horizon to analyze the investment.

For every feasible safety investment, an evaluation is subsequently performed to analyze its financial impact and to determine the DF to make the safety investment profitable. A specific configuration might be required by the financial and technical assessment phase to evaluate the investment. This configuration provides specific information in terms of the features of the selected safety investment, time horizon, discount rate, and so on, which can be used to evaluate the investment. Moreover, several scenarios can be analyzed by carrying out a sensitivity analysis whereby the consequences and probabilities of accident scenarios are used as test parameters. Therefore, the main goal of the investment assessment is to assess the robustness of the choice under different scenarios and assumptions. More specifically, different configurations might be tested to explore how the DF, which is associated with the safety investment to be evaluated, is affected by changing one of the test parameters. Sometimes, the financial and technical assessments need to be reiterated in order to realign the technical and financial elements included in the decision model. For every possible safety investment, the proposed process can be repeated. In addition, other investments can be analyzed and compared with each other. In some cases, especially for HILP (type II) accident scenarios, where there is no consensus between risk experts, alternative scenarios, presenting, for example, higher or lower accident probabilities (cf. difference among the worst-case scenario, worst-credible scenario, most-credible scenario, etc.), might be considered. Afterwards, the whole decision approach is repeated to assess the impact on the safety investments to be selected.

The final step is represented by the decision-making in which alternative investments are evaluated on the basis of elements such as the DF for which the NPV is equal to zero; substituting Eqs. (5.2) and (5.3) into Eq. (5.6) and going from an inequality to an equality, we obtain the formula for the break-even DF:

$$DF^0 = \frac{PV(C)}{PV(B)} = \frac{\sum_{t=0}^n C_t/(1+r)^t}{\sum_{t=0}^n B_t/(1+r)^t} = \frac{M + \frac{(1+r)^n - 1}{(1+r)^n r} Mm}{epnV^a}, \quad (5.9)$$

where Eqs. (5.7) and (5.8) are substituted into Eqs. (5.2) and (5.3), respectively.

The lower the DF, the better the investment from a financial point of view. Different simulations can be carried out, such as comparing safety investments given an accident scenario (consequence and likelihood) and/or configuration, such as the time horizon.

5.7.4 Simulation on Illustrative Case Studies

In this section, some sensitivity analyses are carried out to simulate the decision process of evaluating safety investments in a realistic scenario. Section 5.7.4.1 describes the features of the safety measures and the values of the main technical and financial parameters which are considered in the illustrative case studies, while Section 5.7.4.2 provides some recommendations and explanations for the outcomes of the simulation exercise.

5.7.4.1 Input Information

The decision model described in Section 5.7.3 has been widely tested on realistic data representing some safety measures used in chemical plants to prevent and/or mitigate domino effects (see Janssens et al. [35] for more details). More specifically, the safety measures listed in Table 5.9 have been used to validate the decision model described earlier.

A full factorial experiment has been carried out, testing the effects on the selected safety measures for different scenarios (summarized in Table 5.10) and technical and financial configurations (summarized in Table 5.11).

5.7.4.2 Results and Recommendations

A sensitivity assessment was performed using both the outcomes of the risk assessment (see Section 5.7.4.1) and the test parameters (including the consequences and the probabilities of a potential accident) to evaluate possible safety investments. The analysis is carried out by fixing some of the parameters decided during the risk, financial, and technical assessments and analyzing the influence of the parameters which are not fixed on the DF. In the remainder of this section, if not explicitly mentioned, the basic parameters summarized in Table 5.12 have been used to generate the graphs.

In Figure 5.5, five safety investments are assessed based on the calculated DF whereby the NPV equals zero and considering different scenarios in which three accidents scenarios, and hence three hypothetical benefits (i.e., €2 million, €5 million, and €10 million), are considered during the risk assessment phase. From now on the value of the DF that makes the NPV equal to zero is denoted by DF^0 to avoid confusion and simplify the reading. As expected, measure 5 (i.e., "fire resistant coating"; see Table 5.9) presents a lower DF^0 because the investment it requires is lower compared with the other measures. Therefore, from the viewpoint of the

Table 5.9 Features of the safety investments.

Safety investment number	Description	M (€ millions)	e (%)
1	Concrete wall surrounding tank of 25 m + sprinkler without additional foam	15	95
2	Automatic sprinkler installation with additional foam	10	93
3	Automatic sprinkler installation without additional foam	8	90
4	Deluge system (water spray system opened as signaled by a fire alarm system)	4	86
5	Fire-resistant coating	2	81

Table 5.10 Possible scenarios after the risk assessment.

Scenario characteristic	Value
Types of accident scenario	Vapor cloud explosions and fires BLEVE without chemical reactions BLEVE with chemical reactions
Maximum estimated damages potentially triggered by the accident scenario (V)	€0.5 million, €1 million, €2 million, €5 million, €10 million
Probability of the accident scenario (p)	10^{-4}, 10^{-5}, 10^{-6}, 10^{-7}, 10^{-8}
Risk aversion factor (a)	1.3, 1.35, 1.4, 1.45, 1.5, 1.55, 1.6

BLEVE, boiling liquid expanding vapor explosion.

Table 5.11 Technical and financial parameters.

Parameter	Value
Discount rate (i)	1%, 2%, 3%
Yearly recurrent cost (m)	1%, 3%, 5%, 10%
Time horizon (n)	5, 10, 15, 25 years

Table 5.12 Basic parameters setting used in the simulation.

Parameter	Value
Risk aversion factor (a)	1.45
Yearly recurring cost (m)	3%
Time horizon (n)	25
Discount rate (i)	2%
Probability of accident (p)	10^{-5}

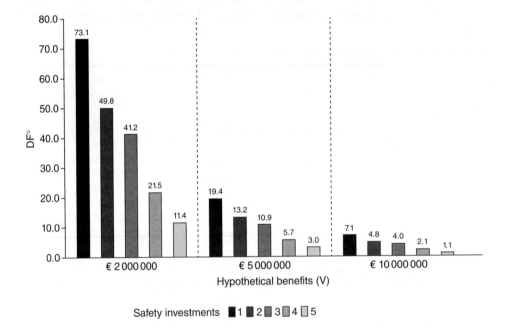

Safety investments ■1 ■2 ■3 ■4 □5

Figure 5.5 DF^0 [the disproportion factor (DF) that makes the net present value (NPV) equal to zero] associated with alternative safety investments, while evaluating different accident scenarios.

safety investor, the benefits are more proportionate to the costs. With reference to measure 1, a higher DF^0 needs to be used to justify the selection of this measure that nevertheless has a higher effectiveness to prevent and mitigate major accidents. As shown in Figure 5.5, when the potential benefits of preventing major accidents are higher, the DF^0 values associated with the different safety investments decrease. This result is justifiable from a decision-maker's perspective because investments are more proportional to potential savings.

It should be noted that sometimes low safety investments imply the use of simple safety measures (e.g., measures 4 and 5 in Figure 5.5), which might not be a viable option for major accidents due to technical and safety reasons (e.g., a reduced effectiveness to prevent major accidents). Therefore, these investment options should be discarded during the risk assessment phase, while considering catastrophic accident scenarios.

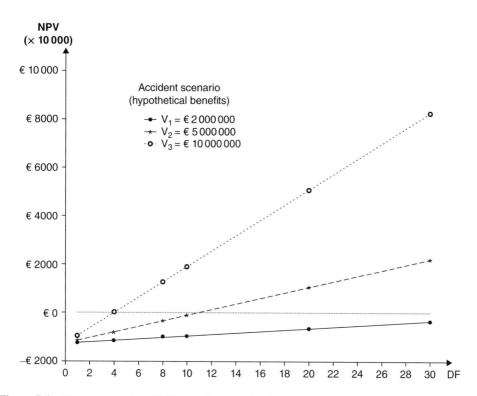

Figure 5.6 Net present value (NPV) and disproportion factor (DF) associated with different accident scenarios.

Impact of Different Accident Scenarios on DF⁰

After having shortlisted the investments to be analyzed in the decision process, one might consider focusing on a single safety investment to assess its robustness by studying the relationship between its main features and the DF^0. In this way, alternative options can be compared more adequately and can be assessed using a multidimensional decision-making approach. For example, in Figure 5.6 a specific investment (investment 3, "Automatic sprinkler installation without additional foam") with a fixed parameter setting (see Table 5.12) is investigated. Assuming three different accident scenarios (and thus three different values of hypothetical benefits), a simulation on the relationship between the DF^0 and the NPV is shown. As expected, the higher the potential benefit, the lower the DF^0 required to make the investment profitable. Analyzing Figure 5.6, one can see that the safety investment can be justified from a financial point of view only if the hypothetical benefits are greater than €2 million. In fact, for minor accident scenarios (in this case hypothetical benefits of €2 million or lower), the total costs of the investment can compensate the potential benefits, only when assuming a DF > 30. As values of DF > 30 are generally considered by investors to be very high or even too high, the measure may not represent a reasonable investment from an economic perspective and thus should be ruled out when the accident scenarios that might affect an organization are triggering "minor" consequences. For accidents with a significant impact, triggering damages of, say, €5 million or €10 million, the measure presents a zero NPV for DFs equal to 10.9 and 4.0,

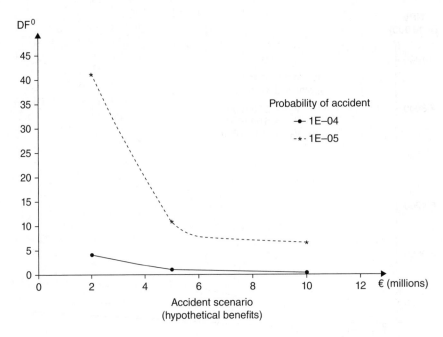

Figure 5.7   Relationship between the probability of an accident and the DF^0 [the disproportion factor (DF) that makes the net present value equal to zero] associated with different scenarios.

respectively. Obviously, these DFs are < 30 and can thus be considered reasonable according to Goose [30], implying that the measure represents a viable safety investment in the case of major accidents.

In Figure 5.7, the impact of the probability of a potential accident is assessed. In particular, the higher the probability, the lower the value of the DF^0 that is needed to make the investment profitable from a financial point of view. If the probability of the accident is very low, the decision-maker will tend to avoid safety investments, as the hypothetical benefits are more uncertain and more disproportionate than the costs. For this reason, a higher DF is required to stress the importance of safety in the case of HILP accidents.

Impact of Technical and Financial Parameters on DF^0

Focusing on a specific safety investment, the relationships between financial and/or technical parameters and the DF^0 can be explored further. In Figure 5.8 the relationship between the yearly recurring costs and the DF^0 is shown for a specific safety investment (measure number 2, "Automatic sprinkler installation with additional foam" in this case), while considering hypothetical benefits of €5 million. The other technical and financial parameters are set to the values reported in Table 5.12. As expected, the greater the yearly recurring costs, the higher the DF^0 due to increased sacrifices in safety required by a firm (see Figure 5.8). A simulation also considering different time horizons, in which the safety measure can maintain its effectiveness, has been performed. As shown in Figure 5.9, the longer the time over which the safety investment can maintain its effectiveness before becoming obsolete, the more attractive the investment will be and thus the lower the DF^0. In fact, the shorter the time horizon, the

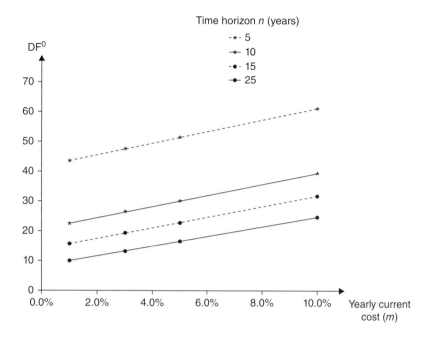

Figure 5.8   Relationship between DF^0 [the disproportion factor (DF) that makes the net present value equal to zero] and the yearly recurring costs for different time horizons.

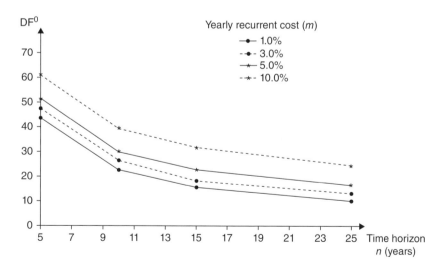

Figure 5.9   Relationship between the disproportion factor (DF) and the asset life for different maintenance costs.

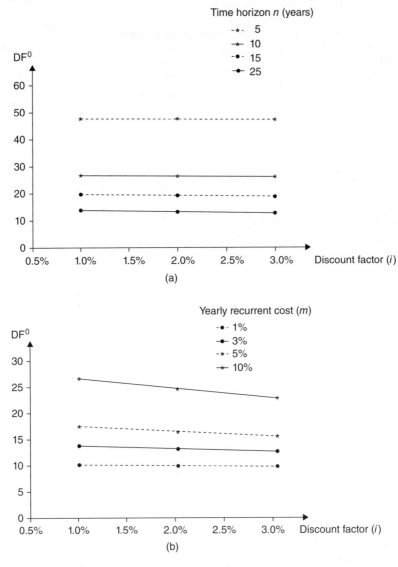

Figure 5.10 Relationship between DF^0 [the disproportion factor (DF) that makes the net present value equal to zero] and the discount factor: (a) for different time horizons; (b) for different yearly recurrent costs.

higher the DF^0 should be to make the investment desirable, due to a shorter time period for the return on investment.

Still considering safety investment number 2, the same accident scenario analyzed previously, and using the technical and financial parameters as in Table 5.12, Figure 5.10 illustrates the relationship between the discount factor used to calculate the NPV and the DF^0. Figure 5.10(a) shows that the DF^0 remains stable for different values of the discount rate given a fixed time horizon. Moreover, the longer the time horizon over which the investment is

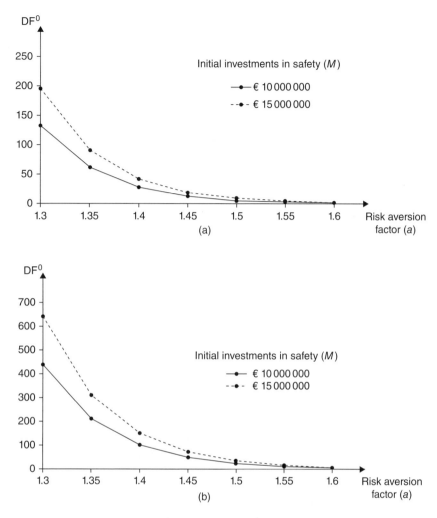

Figure 5.11 Relationship between DF^0 [the disproportion factor (DF) that makes the net present value equal to zero] and the risk factor for different accident scenarios. (a) Potential benefits equal to €2 million; (b) potential benefits equal to €5 million.

analyzed, the lower the DF^0. As shown in Figure 5.10(b), the DF^0 is not significantly affected by a variation in the discount factor as long as it remains under a certain level. However, the DF^0 can vary depending on the maintenance cost that is considered. In addition, when the maintenance costs are high, an increase in the discount factor can trigger a decrease in the DF^0.

Impact of the Risk Aversion Factor on the DF^0

Finally, the relationship between a decision-maker's risk attitude and the DF^0 can be assessed. Figure 5.11 represents the evolution of the DF^0 when the risk aversion factor assumes different values for two different safety investments. Moreover, the lower the potential benefit, the lower the DF^0 that would be used by the decision-maker to evaluate the investment (see

Figure 5.11a,b). As expected, the lower the investment required for a safety measure, the lower the DF^0 that is needed to make the investment proportional to the potential benefit. When the value of the risk aversion factor increases, it means that the decision-maker is more risk-averse and will therefore be more inclined toward safety investments. As both the discount factor and the risk aversion factor can be used as a bias in favor of safety, when the value of a increases there is a reduced need to use the DF^0 as an incentive for investments in safety.

5.7.5   Recommendations with Regard to Using the DF^0

A structured approach to carry out a quantitative CBA for safety investments aimed at HILP (type II) risks has been presented. A DF has been used to make safety costs comparable with the hypothetical benefits that such investments might trigger due to accident prevention. A decision model is presented showing a possible approach to assessing and evaluating mid- to long-term safety decisions considering technical, financial and risk management aspects. The goal of the methodology is to define the level of DF that makes the investment profitable from a financial point of view. The approach can be used to classify and rank alternative investments, rather than define ideal levels of DF that make the safety investment profitable. These latter levels may depend on several factors such as the type of industry, the type of accident, the reputation of the organization, the legal framework, and so on. However, to support the decision process and the selection of the ideal safety investment for a certain accident scenario, one should consider that reasonable values of DF might lie in the range 1–30 [30]. In light of this, safety investments with DF^0 (i.e., the DF at which the NPV becomes zero) outside of these ranges can be discarded. For this reason the methodology presented here can be used to shortlist and rank alternative options and provide quantitative rationales for decision-makers to justify safety investments.

Similar to other financial indicators and ratios (e.g., NPV, IRR), in which recommendations are provided to support decision-making, here it is suggested that, assuming all safety investments require the same initial investment, the alternative with the lowest DF^0 falling in an interval that is considered reasonable would be the best option from a financial point of view.

A simulation has been performed, analyzing the relationships between the main financial and technical parameters and the DF. This information can be used by decision-makers to compare and analyze alternative safety investments that can be effectively adopted to prevent and mitigate specific accident scenarios. To conclude, the quantitative assessment of safety investments using the DF and the NPV is a complex topic. Despite the sensitivity simulation approach proposed in this section being quite flexible and easily adjustable by the decision-maker, it represents a first attempt to introduce this topic in the literature (see also Section 8.12).

5.8   Advantages and Disadvantages of Analyses Based on Costs and Benefits

It should be remembered that the lack of accuracy associated with CBAs can give rise to significantly different outcomes in assessments of the same issues by different people. In addition, it is often much easier to assess all kinds of costs than to identify and evaluate the (hypothetical) benefits and their probabilities.

Cost-benefit analysis can best be used for type I risks, where a sufficient amount of information and data are available to be able to draw sufficiently precise and reliable conclusions. If it is used for type II risks, it risks creating an image of accuracy and precision that is unrealistic. In that case, it is important to take moral factors into account in the calculation, besides the rational information based on the group risk curve. This can be realized by using a so-called adjusted DF (see Section 4.11.2.3) in the case of CBA for type II risks. Furthermore, investigating the DF with respect to several parameters to verify when the NPV becomes zero, and making decisions based on this sensitivity assessment knowledge are also recommended (see also Section 5.7).

In summary, it is often difficult to incorporate realistic calculations of the NPV of future costs and benefits into the analyses. This is a very difficult prediction exercise with a lot of uncertainty involved. Nonetheless, techniques based on costs and benefits that are used to make prevention decisions have one major undeniable strength: if used correctly, they allow limited financial resources to be allocated efficiently and adequately for carrying out safety investments.

5.9 Conclusions

Cost-benefit analyses in case of safety investments constitute much more than calculating the costs of actual accidents, or determining the costs of prevention. Hypothetical benefits, the benefits gained from accidents that have not occurred, should be considered, and type I as well as type II risks should be taken into account when dealing with prevention investment choices. Decisions concerning safety investments make up for a complex decision problem where opportunity costs, perception and human psychology, budget allocation strategies, the choice of cost-benefit ratios, and so on, all play an important role.

The NPV can be defined as the net value on a given date of a payment or series of payments made at other times. If the payments are made in the future, they are discounted to reflect the time value of money and other factors such as investment risk. NPV calculations are widely used in business and economics to provide a means to compare cash flows at different times on a meaningful "like for like" basis. Discounted values reflect the reality that a sum of money is worth more today than the same sum of money at some time in the future. Therefore, in CBAs, prevention costs incurred today should be compared with hypothetical benefits obtained at some time in the future, but equated to today's values.

References

[1] Jackson, D. (2012) *Healthcare Economics Made Easy*. Scion Publishing Ltd, Banbury.
[2] Boardman, A.E., Greenberg, D.H., Vining, A.R., Weimer, D.L. (2011). *Cost-Benefit Analysis. Concepts and Practice*, 4th edn. Pearson Education, Inc. , Upper Saddle River, NJ.
[3] Begg, D., Fischer, S., Dornbusch, R. (1991). *Economics*. Maidenhead, McGraw-Hill.
[4] Fuller, C.W., Vassie, L.H. (2004). *Health and Safety Management. Principles and Best Practice*. Prentice Hall, Essex.
[5] Campbell, H.F., Brown, R.P.C. (2003). *Benefit-Cost Analysis. Financial and Economic Appraisal Using Spreadsheets*. Cambridge University Press, New York.
[6] Meyer, T., Reniers G. (2013). *Engineering Risk Management*. De Gruyter, Berlin.
[7] Gavious, A., Mizrahi, S., Shani, Y., Minchuk, Y. (2009). The cost of industrial accidents for the organization: developing methods and tools for evaluation and cost-benefit analysis of investment in safety. *Journal of Loss Prevention in the Process Industries*, **22**(4), 434–438.

[8]   Brijs, T. (2013) Cost-benefit analysis and cost-effectiveness analysis tool to evaluate investments in safety related to major accidents. Master's thesis, University of Antwerp.

[9]   Reniers, G.L.L., Audenaert, A. (2009). Chemical plant innovative safety investments decision-support methodology, *Journal of Safety Research*, **40**, 411–419.

[10] Heinrich, H.W. (1959) *Industrial Accident Prevention. A Scientific Approach*. McGraw-Hill, London.

[11] Simmonds, R.H. (1951) Estimating Industrial Accident Costs. HBR 29 (January, no. 1), pp. 107–118.

[12] Sun, L., Paez, O., Lee, D., Salem, S., Daraiseh, N. (2006). Estimating the uninsured costs of work-related accidents. *Theoretical Issues in Ergonomic Science*, **7**(3), 227–245.

[13] Buncefield Major Incident Investigation Board (2008) *The Buncefield Incident 11 December 2005: The Final Report of the Major Incident Investigation Board*. Buncefield Major Incident Investigation Board, London.

[14] Brijs, T. (January 7, 2013) Legal aspects of major accidents (interview of G. Moeyersoms). Cost-benefit analysis and cost-effectiveness analysis tool to evaluate investments in safety related to major accidents. Masters' thesis. University of Antwerp, Antwerp, Belgium.

[15] AFP (September 24, 2012) Total Subsidiary Ex-boss Jailed for Deadly French Blast. Retrievde March 10, 2013, from AFP – L'Agence France Presse.

[16] Chakravarty, P. (June 21, 2010) India to Seek Extradition of US Bhopal Boss. Retrieved March 10, 2013, from AFP – L'Agence France Presse.

[17] BP Incident Investigation Report (2010) Deepwater Horizon Accident Investigation Report. BP, London.

[18] Vierendeels, G., Reniers, G., Ale, B. (2011) Modeling the major accident prevention legislation change process within Europe. *Safety Science* **49**, 513–521.

[19] De Standaard (July 31, 2004) Leeswijzer. Trieste Precedenten. De Standaard, Antwerp, the Netherlands.

[20] Reuters (June 19, 1995) Dertien doden bij explosie benzinepomp. De Volkskrant, Rotterdam, the Netherlands.

[21] Alain, G. and Conraads, D. (June 19, 1995) Explosion du restoroute a eynatten: 16 morts. *Le Soir*, Paris

[22] Depré, E. (2012) *Insurance: Global Aspects*. Dilbeek.

[23] EDConsulting (2013). *Insurances for the Petrochemical Industry*. Dilbeek.

[24] Depré, E. (January 4, 2013) Insurance Aspects of Major Accidents (T. Brijs, interviewer). Dilbeek.

[25] Miller, T. (1997) Estimating the cost of injury to U.S. employers. *Journal of Safety Research*, **28**, 1–13.

[26] LaBelle, J.E. (2000). What do accidents truly cost. *Professional Safety* **45**, 38–43.

[27] Jallon, R., Imbeau, D., De Marcellis-Warin, N. (2011). Development of an indirect cost-calculation model suitable for workplace use. *Journal of Safety Research*, **42**, 149–164.

[28] Head, L., Harcourt, M. (1998). The direct and indirect costs of work injuries and diseases in New Zealand. *Asia Pacific Journal of Human Resources*, **36**, 46–58.

[29] OGP (2000) Fire System Integrity Assurance. Report no. 6.85/304, International Association of Oil and Gas Producers (OGP), London.

[30] Goose, M.H. (2006) *Gross Disproportion, Step by Step – A Possible Approach to Evaluating Additional Measures at COMAH Sites*. Institution of Chemical Engineers Symposium Series, vol. **151**, p. 952. Institution of Chemical Engineers.

[31] Rushton, A. (April 4, 2006) CBA, ALARP and Industrial Safety in the United Kingdom.

[32] Thomas, P., Jones, R. (2010). Extending the j-value framework for safety analysis to include the environmental costs of a large accident. *Process Safety and Environmental Protection*, **88**(5), 297–317.

[33] Center for Chemical Process Safety (2008). *Guidelines for Chemical Transportation Safety, Security and Risk Management*. John Wiley & Sons, Inc., Hoboken, NJ.

[34] Talarico, L., Reniers, G., Sörensen, K., Springael, J. (2015). Mistral: a game-theoretical model to allocate security measures in a multi-modal chemical transportation network with adaptive adversaries. *Reliability Engineering and System Safety*, **138**, 105–114.

[35] Janssens, J., Talarico, L., Reniers, G., Sörensen, K. (2015). A decision model to allocate protective safety barriers and mitigate domino effects. *Reliability Engineering and System Safety*, **143**, 44–52.

6

Cost-effectiveness Analysis

6.1   An Introduction to Cost-effectiveness Analysis

Although the previous chapter on cost-benefit analyses stated that if the so-called net present value (NPV) is positive, a company should implement the new safety investment, it is not always realistic to assume that companies are able to implement all the safety investments with positive NPV, as they will face budget limitations. Safety managers looking for efficient safety investments but facing constraints that prevent them from implementing all recommendations stemming from cost-benefit analyses may find cost-effectiveness analyses very useful. Essentially, the approach gives an idea of whether an investment is "affordable" or not.

But the question arises as to what "affordable" means with respect to safety. If the budget available for safety were infinite, it would be possible to invest so much in safety that almost all occupational and transport incidents and accidents would be eliminated. There would be no tension at all between productivity and safety, and organizations would follow the high-reliability principles at all time. Industry could be much more automated with respect to the "risky" jobs, further avoiding and reducing incidents and accidents. However, regretfully, reality tells a different story: the budgets are not infinite, not for governments and not for private organizations. A lot of safety measures are very expensive and cannot be purchased by many companies. Or, at least, many companies are very careful about spending and allocating the available safety resources. The challenge, then, is to work out how best to spend the money, so that the quantity and quality of safety can be maximized. This is the objective of a cost-effectiveness analysis.

Boardman *et al.* [1], however, indicate that cost-effectiveness analyses may circumvent three well-known problems associated with cost-benefit analyses. The first boils down to the problems of monetization of certain costs and benefits. For instance, people may be willing to predict the number of lives saved by a safety investment, but they may be unwilling to put a value on human life. The second problem stems from incompleteness of costs and benefits, and the resulting fact that not all impacts of a safety investment are monetized. Third, analysts may be dealing with intermediate impacts whose linkage to certain end-goals or preferences is not fully clear.

Operational Safety Economics: A Practical Approach Focused on the Chemical and Process Industries, First Edition. Genserik L.L. Reniers and H.R. Noël Van Erp.
© 2016 John Wiley & Sons, Ltd. Published 2016 by John Wiley & Sons, Ltd.

This book takes the viewpoint that cost-effectiveness analyses involve computing cost-effectiveness ratios (CERs) and using these ratios to select safety investments or policies that are most effective. This can be done without explicit constraints, but usually in organizations, carrying out a cost-effectiveness analysis is most useful when considering a certain budget constraint (see Section 6.3).

Although monetized benefits are clearly the easiest to use (and to process in further calculations) and to interpret, it should be clear that a cost-effectiveness analysis does not strictly require the monetization of benefits. Nonetheless, a cost-effectiveness exercise always involves: (i) costs measured in monetary units (e.g., euros); and (ii) effectiveness (corresponding to the benefits associated with the costs), which may be measured in units such as lives saved, amount of chemical spill avoided, employees' competences improved, and the like. The ratio of the two measures can then be determined, which can be employed as a basis for ranking alternative investments or policies.

6.2   Cost-effectiveness Ratio

Consider two options, i and j. Note that one of these options may be the current situation (i.e., doing nothing, no investment). The CER of investment i relative to investment j, CER_{ij}, is given by the following formula:

$$CER_{ij} = \frac{Costs_i - Costs_j}{Effectiveness_i - Effectiveness_j} = \frac{(C_i - C_j)}{(Eff_i - Eff_j)}$$

where C_i represents the costs accompanying option i and E_i represents the number of effectiveness units produced by option i.

The simplest application of this formula occurs when a single safety investment is being assessed as an addition to the status quo. The CER then simply becomes "costs divided by effectiveness." For instance, a company may wish to know whether an extra safety investment (on top of the existing health and safety investments) of €1 000 000 leading to two avoided fatalities over a period of 5 years should be made. The CER (calculated per year) would then be (€1 000 000/5 years)/2 avoided fatalities, or €100 000 per year per avoided fatality. By itself, this CER does not indicate whether this safety investment is efficient. It does, however, indicate that other safety investments costing more than €100 000 per year per avoided fatality are less efficient than this investment.

The application of the CER becomes more complicated when choosing among multiple investment options. To illustrate this, consider several safety investment options (SIOs) for improving lost time injury figures, as shown in Table 6.1.

If management is interested in the incremental CER where the comparison option is "no investment" (thus with zero safety costs and zero effectiveness), column 4 of Table 6.1 can be used. This ratio is also referred to as the "average CER." The smallest average ratio is SIO_1. If safety management seeks to compare different SIOs with each other, SIO_1 can be used as the starting point for comparisons. For calculating the incremental CER, as has been done in the last column of Table 6.1, figures are calculated relative to the SIO, each time providing the smallest incremental increase (see Table 6.1).

Assume now a SIO that costs €2400 per employee, and delivers an effectiveness of 6.5. In this case, SIO_2 and this new SIO have the same costs, but SIO_2 offers a larger gain

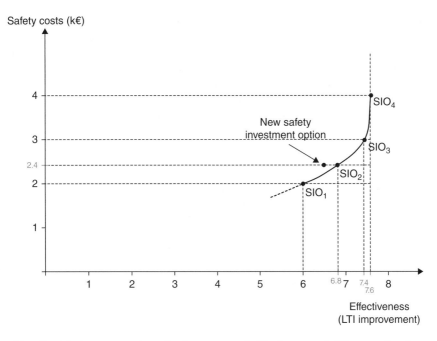

Figure 6.1 Graphical representation of safety costs and effectiveness of different safety investment options. SIO, safety investment option; LTI, lost time injury.

(effectiveness). Hence, SIO_2 strictly dominates this new SIO, and it would never make sense to choose this new option. In fact, this dominance of options can be illustrated graphically, as shown in Figure 6.1.

The dominance of SIO_2 over the new SIO is shown in Figure 6.1, plotting the cost of each investment option on the vertical axis and its effectiveness on the horizontal axis. Holding cost (or effectiveness) constant, one should obviously always prefer an option that has a larger

Table 6.1 Cost-effectiveness ratios – an illustrative example

	Safety costs (€ per employee)	Effective-ness (estimated LTRFR improvement)	CER (relative to no safety investment)	SIO_j (basis for comparison)	Δ(costs) (relative to SIO_j)	Δ(effectiveness) (relative to SIO_j)	Δ(costs)/Δ(effectiveness) (incremental cost-effectiveness ratio)
SIO_1	2000	6.0	333	—	—	—	—
SIO_2	2400	6.8	353	SIO_1	400	0.8	500
SIO_3	3000	7.4	405	SIO_2	600	0.6	1000
SIO_4	4000	7.6	526	SIO_3	1000	0.2	5000

SIO, safety investment option ; LTI, lost time injury.

effectiveness (or lower cost). In fact, the figure shows that the segments connecting the different options, which in theory have slopes equal to their CERs, map out a frontier of the best possible outcomes. The more toward the bottom right of the figure, the better (i.e., the more optimized). The initial step in comparing different options using CERs should therefore always be to remove all options that are dominated from further consideration. In absence of more information about the decision maker's preferences, there is no option preferred over another one if one considers all options situated on the frontier. Hence, new information such as budget constraint or cost limitation needs to be introduced into the problem formulation.

6.3   Cost-effectiveness Analysis Using Constraints

Several approaches are feasible for imposing constraints on the cost-effectiveness problem formulation. Two practices are typically available to carry out a cost-effectiveness analysis with constraints: (i) a minimum acceptable level of effectiveness (Eff_{min}) and (ii) a maximum acceptable cost or the use of a so-called safety budget (Bu_{tot}).

If a minimum level of effectiveness is imposed, either the costs or the CER may be minimized. This translates into the following possible problem formulations:

$$Min(C_i)$$

$$s.t. \ \text{Eff}_i \geq \text{Eff}_{min}$$

or:

$$Min(\text{CER}_i)$$

$$s.t. \ \text{Eff}_i \geq \text{Eff}_{min}$$

If only the costs are minimized, such as in the first problem formulation, the decision-maker does not value additional units of effectiveness. In the case of the second problem formulation, the most cost-effective option is determined that also satisfies the effectiveness constraint. This latter case will usually lead to higher costs than the former case.

If the constraint of a maximum safety budget is used, then one of two options may be selected: that leading to the largest number of units of effectiveness, or the one yielding the lowest CER. The two problem formulations are as follows:

$$Max(\text{Eff}_i)$$

$$s.t. \ C_i \leq Bu_{tot},$$

or:

$$Min(\text{CER}_i)$$

$$s.t. \ C_i \leq Bu_{tot}.$$

In the first problem formulation of this limited budget approach, cost savings beyond the budget are not valued, which for some applications may be a disadvantage. In the second problem formulation, the imposed budget constraint is met in the most cost-effective way.

In the remainder of this chapter, the choice of a combination/portfolio of safety investments with maximizing effectiveness and under safety budget constraint is further elaborated. In principle, it would also be possible to work out the other problem formulations with constraints in an analogous way, i.e., using the same theoretical stepwise approach. Clearly, different input information from that described in the next section will be required in the case of the other problem formulations.

6.4 User-friendly Approach for Cost-effectiveness Analysis under Budget Constraint

6.4.1 Input Information

In industrial practice, companies are faced with budget limitations. Consistent with the previous section, the available yearly budget for prevention related to safety is called Bu_{tot}. When possible prevention investments exceed this budget, they cannot be carried out. Therefore, only the preventative measures having a cost within Bu_{tot} will be considered in the approach explained hereafter (see also Reniers and Sörensen [2]).

Based on the risk matrix approach (see, e.g., Meyer and Reniers [2]), a risk matrix that is divided into four consequence grades and five likelihood grades can be used. Consequence grades can be expressed in financial terms, while likelihood grades are expressed in the number of times per year that a risk leads to an accident in an organization (e.g., based on generic historical information). Financial losses linked to accident consequences, may be estimated through direct and indirect costs that might occur due to the risk, such as those described in Chapter 5.

Table 6.2 illustrates the risk matrix used in the remainder of this section. Every cell of the risk matrix corresponds to a risk class. For each cell (and thus each risk class), the financial consequence value is multiplied by the likelihood value, and the total expected yearly costs per risk class are determined and shown.

The numbers shown in Table 6.2 are illustrative in a sense that every company may develop its own risk matrix with its preferred numbers, applicable to the company. As an example, to guarantee the usefulness of the matrix for the method explained for type I risks, a cut-off consequence class with a maximum financial impact of €2 500 000 is used. The matrix can be designed for type II risks in an analogous way, but the remainder of this section is focused on type I risks. Every type I risk that is part of the cost-effectiveness safety investment study is assumed to be subdivided into one of the risk classes shown in Table 6.2. Cox [4] indicates that a certain risk in each of the cells of any risk matrix is not equally large (or small) due to the classification into risk classes. Hence, a risk cell may contain different varieties of risks.

Table 6.2 Risk matrix used in this section for the purpose of this illustrative approach

Likelihood (per year)	Cell assignments (in €/year)			
< 1	7 500	75 000	750 000	2 500 000
< 10^{-1}	750	7 500	75 000	250 000
< 10^{-2}	75	750	7 500	25 000
< 10^{-3}	7.5	75	750	2 500
< 10^{-4}	0.75	7.5	75	250
Consequence classes/financial impact (€) →	< 7 500	< 75 000	< 750 000	< 2 500 000

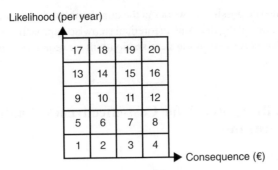

Figure 6.2 Discretization of the risk matrix (for explaining the approach of the cost-effectiveness analysis with budget constraint).

This does not create a problem with the approach, as the aim is to compare groups or bundles of risks, not individual risks.

A discretization of the risk matrix into n cells is illustrated in Figure 6.2. Every risk cell is numbered from 1 to n (in the example, $n = 20$).

Although not strictly needed for carrying out the cost-effectiveness analysis, the discretization of the risk matrix is presented to provide a better understanding of the method that is elaborated, and the numbering will be used to explain the approach. The risk matrix can be refined by relating the risk classes (thus the risk cells) to a cost-benefit analysis. In this way, a decision support instrument can be developed which can be used to determine, taking a certain safety budget into account, the risk reduction measures or precaution measures leading to the most optimal and cost-effective result within an organization. To be able to carry out the approach, certain actions need to be fulfilled by the user and certain input information is also needed. All risks should have been classified into one of the risk cells of the risk matrix.

Every cell i corresponds to a potential expected cell cost C_i, determined by, (Eq. 2.1):

$$C_i = l_i \times c_i,$$

where:

      C_i = expected costs resulting from an accident related to a risk from risk cell i;
      l_i = likelihood corresponding to risk cell i;
      c_i = financial impact (consequences) corresponding to risk cell i.

Table 6.2 illustrates the "expected accident cost" figures (expressed in €/year) for the risk matrix (which is used for illustrative purposes). Other risk matrix configurations are, of course, possible in real-life industrial practice. Evidently, it is also possible for organizations to use actual (real) accident cost figures (to calculate such figures, the reader is referred to Chapter 5), which may lead to more accurate results using the approach.

When precautionary investments are made to decrease risks situated within cell i toward cell j (note that j is characterized by lower consequences and/or likelihood), the potential cell costs become C_j. The expected hypothetical benefits in that case can be calculated as $C_i - C_j$.

The required information for application of the approach is shown in Table 6.3.

When the data from Table 6.3 are known, it is possible to use the approach (discussed in the next section) to determine the most cost-effective prevention measures bundle, where the constraint of a maximum budget is imposed, and the costs and effectiveness (expressed in monetary units) of an option are compared with the situation of no investment. Notice, as already mentioned, that the other problem formulations are only variants of the problem formulation used in this book, and that an analogous approach for a different cost-effectiveness analysis (i.e., a cost-effectiveness analysis with a different problem formulation) can be easily worked out by an organization.

It should be noted that in this illustrative example, the information needed is at a risk matrix cell level, and as such, no individual risk information is needed, only aggregated risk information. Hence, the scope of the suggested approach is on an aggregated scale and requires the user to estimate, in aggregated and composite terms, the prevention costs to go from one risk matrix cell to any other (lower) risk matrix cell (irrespective of the different kinds of risk that may be situated within the cells). This aggregated level is an approximation of real circumstances, and it is possible that in industrial practice more accurate and detailed information will be available which can be used to carry out the cost-effectiveness study. Using more detailed information will lead to better results and more optimal decision-making.

6.4.2 Approach Cost-effectiveness Working Procedure and Illustrative Example

The first step in the first part of the approach is the categorization of risks into the risk classes of the risk matrix. Assume that N_c risk cells (out of the n risk cells in total) contain one or more risks. Safety costs to go from risk cell i to risk cell j (note that $j < i$) are written as CoP_{ij}. If the safety costs are higher than the yearly prevention budget, Bu_{tot}, no investment will be made in these safety measures, and hence these safety costs are excluded at the beginning of the approach. The expected hypothetical benefits corresponding to a decrease in risk cell from i to j are calculated by subtracting C_j from C_i (as explained before).

Following an analysis and preliminary work, a list of possible SIOs will have been drawn up. Using this list, the optimal safety investment portfolio can be determined using a mathematical optimization. In its simplest form, determining the optimal safety investment portfolio is equal to solving a "knapsack problem." The knapsack problem derives its name from the fact that a person having to fill his/her fixed size knapsack with the most valuable items faces a similar problem. The knapsack problem is one of the most fundamental problems in combinatorial optimization and has many applications, e.g., in stock portfolio management, as well as many extensions.

Table 6.3 Required information for application of the approach

n is the total number of cells in the risk matrix
N_c is the number of cells where risks do exist for the organization ($N_c \subset n$)
C_i is the potential expected cell cost (i.e., risk)
Bu_{tot} is the available yearly budget for prevention related to safety
CoP_{ij} is the costs of prevention for going from risk cell i to risk cell j, $\forall i, j$ whereby $i \in N_c$

In the basic version of this problem, a set of decision variables, x_i, is defined where variable x_i (corresponding to measure i) takes on a value 1 if this measure i is chosen as part of the portfolio, and 0 if it is not. A mathematical formulation of the knapsack problem is as follows:

$$\max \sum_{i=1}^{n} B_i x_i$$

$$s.t. \sum_{i=1}^{n} C_i x_i \leq Bu_{\text{tot}}$$

$$x_i \in \{0, 1\}.$$

The first equation expresses the total benefit from the selected portfolio (B), which should be maximized. The second equation expresses the fact that the total cost of the selected measures should not exceed the budget. The third constraint implies that a measure is either fully taken or not taken at all.

A number of assumptions are thus implicitly made in this formulation:

- a measure is either taken or not (it cannot be partially taken);
- the total benefit of all measures taken is the sum of the individual benefits of the chosen measures;
- the total cost of all measures taken is the sum of the costs of the individual measures;
- measures can be independently implemented, without consequences for the other measures.

Some of these assumptions are not completely realistic. In the following section, this observation will be discussed.

Although the knapsack problem is non-deterministic polynomial-time hard (NP hard), [1] it can be solved efficiently even for very large instances [5]. The advantage of using the knapsack formulation is that it can be solved by standard off-the-shelf commercial software for mixed-integer programming. Moreover, even spreadsheet software such as Excel (very popular in many organizations) include a solver that can be used to optimize the safety measures portfolio using the approach described.

6.4.3 Illustrative Example of the Cost-effectiveness Analysis with Safety Budget Constraint

Consider the following example to illustrate the approach. Notice that the annual budgeting is assumed to be designed in such a way that no reservations can be made for capital-intensive items. Based on the risk matrix displayed in Table 6.1 and the basic information of Tables 6.4 and 6.5, the execution of the approach is explained.

An optimal allocation of safety measures with a maximum of one prevention measure from each of the risk cells which can be assigned needs to be determined. As explained before, to solve this problem, four conditions have to be met: (i) the total benefit of measures taken needs

---

[1] An optimization problem is NP hard if the running time of the fastest known algorithm to solve it increases exponentially with the problem size. For example, if the problem size (the number of possible items in the knapsack) is n, the computing time required by the fastest known knapsack algorithm can be written as $a \times e^n$, where a is a constant.

to be maximized; (ii) the available budget constraint needs to be respected; (iii) a maximum of one decrease per risk cell is allowed; and (iv) a measure can be taken, or not. These conditions translate into the following mathematical expressions:

(i) $\text{Max}_{i,j} \sum B_{ij} x_{ij}$

(ii) $\sum_{i,j} \text{CoP}_{ij} \leq Bu_{\text{tot}}$

(iii) $\sum_j x_{ij} \leq 1$

(iv) $x_{ij} \in \{0, 1\}$.

Solving these equations for the unknown x_{ij} yields the optimal solution of the illustrative example represented by Tables 6.4 and 6.5. In this solution the measures taken are shown in Table 6.6 with a total cost of €49 987 and a total hypothetical benefit of €97 499.25. The total hypothetical profit for the illustrative example is thus equal to €47 512.25.

It should be stressed that this illustrative example only serves to explain how to use the knapsack software to determine an optimal allocation of safety resources. It is possible, and recommended, that exact figures be used to determine the hypothetical benefits, simply by calculating all the real expected costs per year of the risk cells for the feasible scenarios in the organization, and thus creating a "real matrix" with real cell assignment figures (instead of general figures).

Furthermore, it is possible to further include more advanced conditions for real-case problems and situations with the method. The next section elaborates on what refinements might be carried out.

6.4.4 Refinements of the Cost-effectiveness Approach

In general, the portfolio of safety investments/measures chosen by a company is subject to a number of extra constraints, expressing relationships between these investments/measures. Fortunately, these relationships are generally easily added to the knapsack approach, usually by introducing additional constraints. This section discusses some of these relationships and shows how they can be expressed in the approach using additional constraints.

6.4.4.1 Binary Relationships

If risk cell r is decreased, risk cell t also has to be decreased and vice versa. This situation occurs when measures are mutually dependent on each other and taking one measure without

Table 6.4 Information of our illustrative example, for application of the approach

$n = 20$

$N_c = 6$; risk cells 3, 7, 10, 12, 13, and 15 in Figure 6.2

C_i as in risk matrix in Figure 6.1

$Bu_{\text{tot}} = $ €50 000

CoP_{ij} as in see Table 6.5

Table 6.5 Costs of prevention CoP$_{ij}$ and hypothetical benefits for the illustrative case

Prevention measure ij	(Illustrative) costs of prevention for going from i to j (CoP$_{ij}$) (€)	Hypothetical benefits for going from i to j (€)
Start = Risk cell 3		
3 2	35	67.5
3 1	42	74.25
Start = Risk cell 7		
7 6	325	675
7 5	460	742.5
7 3	295	675
7 2	420	742.5
7 1	590	749.25
Start = Risk cell 10		
10 9	330	675
10 6	350	675
10 5	390	742.5
10 2	400	742.5
10 1	880	749.25
Start = Risk cell 12		
12 11	13 500	17 500
12 10	13 750	24 250
12 9	14 800	24 925
12 8	13 000	22 500
12 7	15 000	24 250
12 6	16 500	24 925
12 5	26 000	24 992.5
12 4	13 900	24 750
12 3	17 000	24 925
12 2	27 500	24 992.5
12 1	38 000	24 999.25
Start = Risk cell 13		
13 9	410	675
13 5	550	742.5
13 1	700	749.25
Start = Risk cell 15		
15 14	31 000	67 500
15 13	36 650	74 250
15 11	29 880	67 500
15 10	38 000	74 250
15 9	52 000	74 925
15 7	41 440	74 250
15 6	48 990	74 925
15 5	64 450	74 992.5
15 3	50 000	74 925
15 2	62 250	74 992.5
15 1	88 000	74 999.25

Table 6.6 Solution of the illustrative example from Tables 6.4 and 6.5

			Chosen	Cost	Benefit
Start = Risk cell 3			1		
3 2	35	67.5	0	0	0
3 1	42	74.25	1	42	74.25
Start = Risk cell 7			1		
7 6	325	675	0	0	0
7 5	460	742.5	0	0	0
7 3	295	675	1	295	675
7 2	420	742.5	0	0	0
7 1	590	749.25	0	0	0
Start = Risk cell 10			0		
10 9	330	675	0	0	0
10 6	350	675	0	0	0
10 5	390	742.5	0	0	0
10 2	400	742.5	0	0	0
10 1	880	749.25	0	0	0
Start = Risk cell 12			1		
12 11	13 500	17 500	0	0	0
12 10	13 750	24 250	0	0	0
12 9	14 800	24 925	0	0	0
12 8	13 000	22 500	1	13 000	22 500
12 7	15 000	24 250	0	0	0
12 6	16 500	24 925	0	0	0
12 5	26 000	24 992.5	0	0	0
12 4	13 900	24 750	0	0	0
12 3	17 000	24 925	0	0	0
12 2	27 500	24 992.5	0	0	0
12 1	38 000	24 999.25	0	0	0
Start = Risk cell 13			0		
13 9	410	675	0	0	0
13 5	550	742.5	0	0	0
13 1	700	749.25	0	0	0
Start = Risk cell 15			1		
15 14	31 000	67 500	0	0	0
15 13	36 650	74 250	1	36 650	74 250
15 11	29 880	67 500	0	0	0
15 10	38 000	74 250	0	0	0
15 9	52 000	74 925	0	0	0
15 7	41 440	74 250	0	0	0
15 6	48 990	74 925	0	0	0
15 5	64 450	74 992.5	0	0	0
15 3	50 000	74 925	0	0	0
15 2	62 250	74 992.5	0	0	0
15 1	88 000	74 999.25	0	0	0
				49 987	97 499.25

the other makes no sense. An example is when the use of a new device that enhances safety requires training. It does not make sense to install the device without the training, and it does not make sense to give the training without installing the device.

This relationship between risk cell decreases from r to s and from t to u can be expressed in the approach by the extra constraint

$$x_{r\to s} = x_{t\to u}.$$

Another (less flexible) way to include this relationship in the approach is to combine risk cell decreases $r \to s$ and $t \to u$ in a single risk cell decrease.

Suppose, for example, that measures $15 \to 13$ and $13 \to 9$ either need to be taken together or not at all in our illustrative example. We add constraint $x_{15\to 13} = x_{13\to 9}$ to the knapsack problem and get solution $\{3 \to 1,\ 7 \to 1,\ 10 \to 1,\ 12 \to 9,\ 13 \to 1,\ 15 \to 14\}$ with total cost €48 012 and total hypothetical benefit €94 747. The total hypothetical profit in this case would be €44 735.

Another situation that might occur is the following: *if risk cell r is decreased, risk cell t also has to be decreased, but the reverse is not true.* As an example, to prevent fire from spreading between departments, a company is considering installing a fire-resistant door. The length of time the door resists a fire can be increased by adding an extra layer of fireproof coating. Clearly, applying the coating without installing the door makes no sense, but the reverse does.

The relationship between risk cell decrease $r \to s$ (installing the door) and risk cell decrease $t \to u$ (installing the fireproof coating) can be expressed as:

$$x_{r\to s} \leq x_{t\to u}.$$

Suppose that when measure $15 \to 13$ is taken, measure $3 \to 2$ also needs to be taken. The constraint $x_{15\to 13} \leq x_{3\to 2}$ is added, the problem is resolved and the optimal solution is $\{3 \to 2,\ 7 \to 3,\ 12 \to 8,\ 15 \to 13\}$ with a total cost of €49 980 and a total hypothetical benefit and profit, respectively , or €97 492.5 and €47 512.5.

Yet another possible situation is: *either risk cell r or risk cell t needs to be decreased, but not both risk cells at the same time.* This situation can occur if two measures duplicate each other's effects and the company judges it superfluous to invest in both measures simultaneously. For example, a machine can be protected by a concrete casing or a steel casing, but not by both.

This can be mathematically expressed as follows:

$$x_{r\to s} = 1 - x_{t\to u}.$$

Another possibility is: *either risk cell r or risk cell t, or both, needs to be decreased.* Such a situation can occur, for example, if the company management has decided that it will install either smoke detectors or fire doors, but may also decide to install both. Translated into a mathematical constraint, this situation can be included in the safety measures allocation problem as follows:

$$x_{r\to s} + x_{t\to u} \geq 1.$$

Yet another feasible situation is: *if risk cell t is decreased, risk cell r cannot be decreased, and vice versa.* But the possibility exists that both measures are not taken. This situation occurs, for example, when management has decided that smoke detectors might be installed, but two types are available and only one type will be selected at most.

This can be expressed as follows:

$$x_{r \to s} \leq 1 - x_{t \to u}.$$

6.4.4.2 Other Relationships

In principle, all relationships between measures can be expressed as constraints in the knapsack problem. Essentially, the decision of whether to decrease risk cell i can be seen as literal in a propositional logic system, in which logical relationships are expressed by the operators NOT (risk cell i is not decreased), AND (risk cell i and risk cell j are both decreased), OR (risk cell i or risk cell j is decreased), and IMPLICATION (if risk cell i is decreased, then risk cell j is decreased). These operators can be used to create arbitrarily complex relationships to express the most complex logical relationships between safety measures.

For instance, if both the automatic fire door and the alarm system are installed, and the electricity system is not upgraded, then either a backup generator should be installed or a link to an additional power system should be purchased. Each of these relationships can be converted to constraints of the knapsack problem. This section is restricted to one example. For a more elaborate discussion, including details on how to transform a logical relationship to constraints, refer to Martello et al. [5]., Cavalier et al. [6], Mendelsohn [7] and Raman and Grossmann [8].

Consider, for example, the following measures:

M_1 – an automatic fire door is installed (e.g., $x_{4 \to 2}$);
M_2 – an alarm system is installed (e.g., $x_{6 \to 2}$);
M_3 – the electricity system is upgraded (e.g., $x_{3 \to 1}$);
M_4 – a backup generator is installed (e.g., $x_{7 \to 3}$);
M_5 – a link to an additional electricity system is installed (e.g., $x_{5 \to 2}$).

The condition that if both the automatic fire door and the alarm system are installed, and the electricity system is not upgraded, then either a backup generator should be installed or a link to an additional power system should be purchased is logically equivalent to: [2]

$$(M_1 \wedge M_2) \wedge \neg M_3 \Rightarrow M_4 \vee M_5.$$

This can be converted into its "conjunctive normal form":

$$(\neg M_1 \vee \neg M_2 \vee M_4 \vee M_5) \wedge (M_3 \vee M_4 \vee M_5),$$

which translates to the following two constraints:

$$\begin{cases} x_{3 \to 1} + x_{7 \to 3} + x_{5 \to 2} \geq 1, \\ \left(1 - x_{4 \to 2}\right) + \left(1 - x_{6 \to 2}\right) + x_{7 \to 3} + x_{5 \to 2} \geq 1. \end{cases}$$

---

[2] The symbols in this equation are the following: $\wedge$ is AND, $\vee$ is OR, $\neg$ is NOT, and $\Rightarrow$ is IMPLIES

6.4.4.3 Non-additivity

For some situations, the benefits or costs of measures are not simply additive. Suppose, for instance, that two fire doors can be installed in series to prevent fire from spreading to the next room. Clearly, the effect of installing one door instead of none will be larger than the effect of installing two doors instead of one. In other words, there will be a diminishing rate of return on the second door.

This can be easily handled by identifying such situations and creating "virtual" measures in the cost-benefit table to represent the action of taking both measures. To ensure that each measure is only taken once, some additional constraints are also necessary.

As an example, suppose that the effect of combining risk cell decreases $3 \rightarrow 1$ and $7 \rightarrow 3$ in the example does not yield a benefit of $74.25 + 675 = 749.25$, but rather yields only 640. Suppose further that the cost of implementing both $3 \rightarrow 1$ and $7 \rightarrow 3$ is not $42 + 295 = 337$, but that a discount of €30 is given.

This can be handled by adding an extra risk cell decrease with cost 307 and hypothetical benefit 640. Additionally, constraints are necessary to ensure that this extra measure is not taken if either $3 \rightarrow 1$ or $7 \rightarrow 3$ are taken. The additional constraints translate mathematically into:

$$x_{\text{extra risk cell decrease}} \leq 1 - x_{3 \rightarrow 1},$$

$$x_{\text{extra risk cell decrease}} \leq 1 - x_{7 \rightarrow 3}.$$

Additionally, we need to ensure that measures $3 \rightarrow 1$ and $7 \rightarrow 3$ are not both chosen at the same time:

$$x_{3 \rightarrow 1} + x_{7 \rightarrow 3} \leq 1.$$

In this case, resolving our illustrative example leads to an optimal solution, where the following risk reductions are carried out: $3 \rightarrow 2$, $12 \rightarrow 8$, and $15 \rightarrow 13$. The solution displays a total cost equal to €49 992 and a total hypothetical benefit of €97 817.5 and a total hypothetical profit of €47 825.5.

6.5 Cost-effectiveness Calculation Often Used in Industry

A popular method used in industry to determine the cost-effectiveness of a safety investment is to calculate the expected hypothetical benefit (calculated via definition (ii) in Section 4.2.4), which is usually called the "reduced risk" in industry, and dividing it by the cost of the safety investment for achieving the reduction in risk. Using this approach, it is possible in a very simple way to calculate investments that are more cost-effective than other investments. A well-known example of this approach is that of Kinney and Wiruth [9], where an effectiveness factor is defined and, depending on the value, investment is recommended or not. However, this method does not allow one to determine optimal budget allocations at all, and does not utilize constraints for the problem formulation. In this sense, it should instead be considered as an industrial practice of common sense, but it is too simplistic to be considered a true cost-effectiveness analysis. It should be considered more as a kind of cost-benefit analysis (see also Section 5.5.1).

6.6 Cost–Utility Analysis

Although the terms are sometimes used interchangeably, economists distinguish between cost-effectiveness analyses and cost–utility analyses. A cost–utility analysis can be considered a specific type of cost-effectiveness analysis in which utility measures are used in the analysis.

In operational safety economics, the purpose of a cost–utility analysis can be seen as an estimate of the ratio between the cost of a safety investment and the (theoretical) benefit it produces expressed in utils (for type I risks) or in quality-adjusted accident probabilities (QAAPs) (for type II risks) (see also Chapter 3). The outcome of a cost–utility analysis is a cost per util or QAAP gained (this is also called 'the incremental cost-effectiveness ratio (or ICER) in health economics).

> As an illustrative example, assume that safety investment A (€30 000) for dealing with type II events leads to 1.8 QAAPs, while investment B (€20 000) delivers 1.5 QAAPs. Then, the outcome of the cost–utility analysis would be €10 000/0.3 QAAP, which corresponds to €33 333 per QAAP gained.

An advantage of the cost–utility analysis is that it allows comparison across different safety investments and policies, which are otherwise difficult to compare, by using a common unit of measure for both types of prevention investments, the "util" for type I prevention, and the "QAAP" for type II prevention. However, the downside is that measuring utils or QAAPs is very difficult due to their being largely subjective, and this may prove to be even more difficult than expressing all benefits in monetary values and carrying out a "monetary" cost-effectiveness analysis. Hence, if the utils or the QAAPs are not collected or determined properly by an organization, then the whole cost–utility analysis can be open to question. Nevertheless, a cost–utility analysis, when conducted well, can be helpful for safety decision-makers.

6.7 Conclusions

Optimizing prevention investments and making investment decisions in a cost-efficient way is essential for corporations. To this end, a user-friendly knapsack-based approach to take cost-efficient prevention decisions has been described in this chapter. The approach employs some essential data that can be easily determined by any organization and displayed using a risk matrix. Prevention costs are weighed against hypothetical benefits following the preventive measures taken, and the most cost-efficient preventive measures are determined following the knapsack algorithm, given a certain prevention budget available.

References

[1] Boardman, A.E., Greenberg, D.H., Vining, A.R., Weimer, D.L. (2011). *Cost-Benefit Analysis. Concepts and Practice*, 4th edn. Pearson Education Inc., Upper Saddle River, NJ.

[2] Reniers, G., Sörensen, K. (2013). An approach for optimal allocation of safety resources: using the knapsack problem to take aggregated cost-efficient preventive measures. *Risk Analysis*, **33**(11), 2056–2067.

[3] Meyer, T., Reniers G. (2013). *Engineering Risk Management*. De Gruyter, Berlin.

[4] Cox, L.A. (2008). What's wrong with risk matrices. *Risk Analysis*, **28**(2), 497–512.

[5] Martello, S., Plisinger, D., Toth, P. (2000). New trends in exact algorithms for the 0-1 knapsack problem. *European Journal of Operational Research*, **123**(2). 325–332.

[6] Cavalier, T.M., Pardalos, P.M., Soyster, A.L. (1990). Approaching and integer programming techniques applied to propositional calculus. *Computers and Operations Research*, **17**(6) 561–570.

[7] Mendelsohn, E. (1997). *Introduction to Mathematical Logic*, 4th edn. Chapman & Hall, London.

[8] Raman, R., Grossmann, I.E. (1991). Relation between milp modelling and logic inference for chemical process synthesis. *Computers and Chemical Engineering*, **15**(2), 73–84.

[9] Kinney, G. and Wiruth, A. (1976) Practical Risk Analysis for Safety Management. NWC Technical Publication 5865. Naval Weapons Center, China Lake, CA.

7

Beyond the State-of the Art of Operational Safety Economics: Bayesian Decision Theory

7.1 Introduction

The Bayesian decision theory, which is presented in this book as a novel contribution to the decision theoretical field, is a neo-Bernoullian utility theory which also aims to improve on expected utility theory, as did game theory and prospect theory before it. But in its approach it takes the middle road, just as Daniel Bernoulli himself did when he wrote his St. Petersburg paper (see also Chapter 4), in that it recognizes both the desirability of mathematical first principles as well as the necessity for any mathematical theory of human rationality to be able to stand to the benchmark of "common sense."

The structure of this chapter on the Bayesian decision is as follows. First a theoretical discussion of the Bayesian decision theory is given. Thereafter, the alternative criterion of choice which is proposed in the Bayesian decision theory is used to discuss and explain why type II events are perceived to be more risky than type I events.

In this section, a theoretical discussion of the Bayesian decision theory is given. This is done by relating the Bayesian decision theory to the expected outcome and expected utility theories that came before it. Expected outcome theory has been around since the seventeenth century, when the rich merchants of Amsterdam bought and sold expectations as if they were tangible goods. As already expounded in this book, the algorithmic steps of expected outcome theory are very simple:

1. For each possible decision: construct an outcome probability distribution; i.e., for each possible decision, assign to every conceivable contingency both an estimated net monetary consequence and a probability.
2. Choose that decision which maximizes the expectation values (i.e., means) of the outcome probability distributions.

Operational Safety Economics: A Practical Approach Focused on the Chemical and Process Industries,
First Edition. Genserik L.L. Reniers and H.R. Noël Van Erp.
© 2016 John Wiley & Sons, Ltd. Published 2016 by John Wiley & Sons, Ltd.

In the eighteenth century, Bernoulli provided a fundamental contribution to expected outcome theory in that he proposed that it was not the actual gains and losses that move us, but rather that the utility of these gains and losses. Moreover, Bernoulli offered up a specific function by which to translate these gains and losses to their corresponding utilities:

$$u = q \log \frac{m+x}{m}, \quad q > 0, \tag{7.1}$$

where q is some scaling constant that falls away in the decision theoretical (in)equalities, m is the initial wealth of the decision-maker, and x is either a gain or a loss.

In the utility function (Eq. 7.1), the initial wealth functions as a reference point in the following sense. For increments x which are small relative to the initial wealth m, the utility function (Eq. 7.1) becomes linear, as losses are weighted the same as corresponding gains, whereas for increments x which are large relative to the initial wealth m, the utility function (Eq. 7.1) becomes non-linear, as losses are weighted more heavily than corresponding gains. So, Bernoulli's utility function predicts that the psychological phenomenon of loss aversion will hold for large consequences such as, say, the burning down of a house, but not for small consequences such as the breaking of an egg. This is commensurate with intuition. Bernoulli, having provided both the concept and the quantification of utilities, proposed his expected utility theory as a straightforward generalization of the expected outcome theory. The algorithmic steps of expected utility theory are as follows [1]:

1. For each possible decision: construct an outcome probability distribution; i.e, for each possible decision, assign to every conceivable contingency both an estimated net monetary consequence and a probability.
2. Transform outcome probability distributions to their corresponding utility probability distributions; i.e., convert the outcomes of the outcome probability distributions to their corresponding utilities, using Bernoulli's utility function.
3. Choose that decision which maximizes the expectation values (i.e., means) of the utility probability distributions.

Bernoulli's utility concept remained uncontested in the centuries that followed his 1738 paper, but the same cannot be said for his utility function (Eq. 7.1). Nonetheless, Bernoulli's utility function has been demonstrated also to hold for sensory stimuli perception [2], and not only for monetary stimulus perception. Moreover, Bernoulli's utility function may be derived from sound first principles [1, 3].

In von Neumann and Morgenstern's [4] reintroduction of Bernoulli's expected utility theory, the specific form of the utility function was left unspecified. This added degree of freedom in the expected utility theory opened the way for alternative utility functions such as, for example, the following utility function:

$$u = (-x)^a, \quad \text{for } a > 0 \text{ and } x \le 0. \tag{7.2}$$

This alternative function (Eq. 7.2), however, does not, like Bernoulli's utility function (Eq. 7.1), have an explicit reference point by which to modulate the strength of the loss aversion effect as a function of both the current asset position and the increment in that asset position. Moreover, Eq. (7.2) lacks the general validity that Eq. (7.1) enjoys as the psycho-physical Fechner–Weber law that guides our human sense perception [2]. Finally, Bernoulli's utility function admits a consistency derivation which shows that the only consistent utility function is either the utility function (Eq. 7.1) or some transformation thereof [3], and it may be shown that Eq. (7.2) does not belong to this class of consistent utility functions. Note that a power function like Eq. (7.2) is found in Eq. (2.2). But there it is not so much an utility transformation, as it is a correction that will ensure a higher modeled risk for type II risks (see also Section 7.5).

7.2   Bayesian Decision Theory

The Bayesian decision theory is "neo-Bernoullian" in that it proposes that the utility function Eq. (7.1) is the most appropriate function by which to translate, for a given initial wealth, gains and losses to their corresponding utilities. But it deviates from both the expected outcome theory and the expected utility theory in that it questions the appropriateness of the criterion of choice where one has to choose the decision that maximizes the expectation values or, equivalently, the means of the outcome probability distributions under the different decisions.

7.2.1   The Criterion of Choice as a Degree of Freedom

Let D_1 and D_2 be two actions we have to choose from. Let x_i, for $i = 1, \dots, n$, and x_j, for $j = 1, \dots, m$, be the monetary outcomes associated with, respectively, actions D_1 and D_2. Then in the Bayesian decision theory – as in both expected outcome theory and utility theory – the two outcome distributions that correspond with these decisions are constructed:

$$p(x_i|D_1), \quad \text{and} \quad p(x_j|D_2). \tag{7.3}$$

One then proceeds – as in expected utility theory – to map utilities to the monetary outcomes in Eq. (7.4), by way of the Bernoulli utility function (Eq. 7.1). This leaves us with the utility probability distributions:

$$p(u_i|A_1), \quad \text{and} \quad p(u_j|A_2). \tag{7.4}$$

Now, the most primitive intuition regarding the utility probability distributions (Eq. 7.4) is that the action which corresponds with the utility probability distribution lying more to the right (i.e., providing the highest utility of both actions) will also be the action that promises to be the most advantageous. So, when making a decision, we ought to compare the positions of the utility probability distributions on the utility axis and then choose that action which maximizes the position of these utility probability distributions. This all sounds intuitive enough. But how is the position of a probability distribution defined?

Ideally there would be some formal (consistency) derivation of what constitutes a position measure of a probability distribution. But in the absence of such a derivation, ad hoc common-sense considerations need to be used. Stated differently, the criterion of choice in the decision theory constitutes a degree of freedom.

7.2.1.1   The Probabilistic "Worst Case," "Most Likely Case," and "Best Case" Scenarios

From the introduction of expected outcome theory in the seventeenth century and expected utility theory in the eighteenth century, the implicit assumption has been that the expectation value of a given probability distribution is a position of its measure [1, 5]. The expectation value is a measure for the location of the center of mass of a given probability distribution; as such it may give one a probabilistic indication of the most likely scenario:

$$E(X) = \sum_{i=1}^{n} p_i x_i. \tag{7.5}$$

The qualifier "probabilistic indication" is used here to point to the fact that the expectation value or, equivalently, the mean need not give a value that one would necessarily expect.

In the Value at Risk (VaR) methodology used in the financial industry (see also Section 3.4.2) the probabilistic worst-case scenarios are taken as a criterion of choice, rather than the most likely scenarios (in the probabilistic sense). In the VaR methodology the probabilistic worst-case scenario is operationalized as the 1 (or up to 5%) percentile. But instead of percentiles, one may also use the confidence lower bound to operationalize a probabilistic worst-case scenario.

The absolute worst-case scenario is:

$$a = \min(x_1, \ldots, x_n). \tag{7.6}$$

The criterion of choice in Eq. (7.6) is also known as the minimax criterion of choice [6]. The k-sigma lower bound of a given probability distribution is given as

$$lb(k) = E(X) - k \, std(X), \tag{7.7}$$

where k is the sigma level of the lower bound and where

$$std(X) = \sqrt{\sum_{i=1}^{n} p_i x_i^2 - [E(X)]^2} \tag{7.8}$$

is the standard deviation. The probabilistic worst case scenario then may be quantified as an undershoot-corrected lower bound (Eqs. 7.6 and 7.7):

$$LB(k) = \begin{cases} lb(k), & lb(k) \geq a, \\ a, & lb(k) < a. \end{cases} \tag{7.9}$$

Note that the probabilistic worst-case scenario (Eq. 7.9) holds the minimax criterion of choice (Eq. 7.6), as a special case for large k in Eq. (7.7). For $k = 1$, the criterion of choice (Eq. 7.9) constitutes a still likely worst-case scenario (in the probabilistic sense).

One may also imagine – in principle – a decision problem in which one is only interested in the probabilistic best-case scenarios. The absolute best-case scenario is:

$$b = \max(x_1, \ldots, x_n). \tag{7.10}$$

The criterion of choice in Eq. (7.10) is also known as the maximax criterion of choice. The k-sigma upper bound of a given probability distribution is a given as:

$$ub(k) = E(X) + k \, std(X), \tag{7.11}$$

where k is the sigma level of the upper bound. The probabilistic best-case scenario may then be quantified as an overshoot-corrected upper bound (Eqs. 7.10 and 7.11):

$$UB(k) = \begin{cases} b, & ub(k) > b, \\ ub(k), & ub(k) \leq b. \end{cases} \tag{7.12}$$

Note that the probabilistic best-case scenario (Eq. 7.12) holds the maximax criterion of choice, Eq. (7.10), as a special case for large k in Eq. (7.11). For $k = 1$, the criterion of choice Eq. (7.12) constitutes a still likely best-case scenario (in the probabilistic sense).

If a criterion of choice as in Eq. (7.5) is considered, then one neglects what may happen in the worst and the best of possible scenarios. If a criterion of choice as in Eq. (7.9) is taken, then one neglects what may happen in the most likely and the best of possible scenarios. If one takes as a criterion of choice Eq. (7.12), then one neglects what might happen in the worst and the most likely of possible scenarios. An exclusive commitment to any of the criteria of choice (Eqs. 7.5, 7.9, and 7.12), will thus necessarily leave out some pertinent information in one's decision theoretical considerations. So, how does one untie this Gordian knot?

7.2.1.2   A Probabilistic Hurwitz Criterion of Choice

In Hurwitz's criterion of choice, the absolute worst- and best-case scenarios are both taken into account; for a balanced pessimism coefficient of $\alpha = 1/2$ we have that (Eqs. 7.6 and 7.10):

$$\text{Hurwitz's criterion of choice} = \frac{a+b}{2}. \tag{7.13}$$

Now, the absolute worst- and best-case scenarios are replaced in Eq. (7.13) with their corresponding probabilistic undershoot- and overshoot-corrected counterparts (Eqs. 7.9 and 7.12):

$$\text{Probabilistic Hurwitz criterion of choice} = \frac{LB(k) + UB(k)}{2}. \tag{7.14}$$

Under the criterion of choice in Eq. (7.14), undecidedness between D_1 and D_2 translates to the decision theoretical equality:

$$\frac{LB(k|D_1) + UB(k|D_1)}{2} = \frac{LB(k|D_2) + UB(k|D_2)}{2}, \tag{7.15}$$

or, equivalently,

$$LB(k|D_1) - LB(k|D_2) = UB(k|D_2) - UB(k|D_1), \tag{7.16}$$

a trade-off between the losses/gains in the probabilistic worst-case scenarios (Eq. 7.9), and the corresponding gains/losses in the probabilistic best-case scenarios (Eq. 7.10). It follows, seeing that Eq. (7.13) is a limit case of Eq. (7.14), that for a balanced pessimism coefficient of $\alpha = 1/2$ Hurwitz's criterion of choice provides a balanced trade-off between the differences in the absolute worst case scenarios and the differences in the absolute best case scenarios.

The probabilistic Hurwitz criterion of choice (Eq. 7.14) translates to Eqs. (7.9 and 7.12):

$$\frac{LB(k) + UB(k)}{2} = \begin{cases} E(X), & lb(k) \geq a, \ \ ub(k) \leq b, \\ \dfrac{a + E(X) + k\,\text{std}(X)}{2}, & lb(k) < a, \ \ ub(k) \leq b, \\ \dfrac{E(X) - k\,\text{std}(X) + b}{2}, & lb(k) \geq a, \ \ ub(k) > b, \\ \dfrac{a+b}{2}, & lb(k) < a, \ \ ub(k) > b. \end{cases} \tag{7.17}$$

It follows that the alternative criterion of choice (Eq. 7.14), which takes into account what may happen in the worst and the best of possible scenarios, holds both the traditional expected value criterion of choice (Eq. 7.5) as a special case as well as Hurwitz's criterion of choice with a balanced pessimism factor (Eq. 7.13). However, it may be found that the criterion of choice in Eq. (7.14) – and by implication also the Hurwitz criterion of choice (Eq. 7.13) – is vulnerable to a simple counter-example.

Imagine two utility probability distributions having equal lower and upper bounds LB(k) and UB(k), but one distribution being right-skewed and the other being left-skewed. Then the criterion of choice in Eq. (7.14) will leave its user undecided between the two decisions, whereas intuition would give preference to the decision corresponding with the left-skewed outcome probability distribution, as the bulk of the probability distribution of the left-skewed distribution will be more to the right than that of the right-skewed distribution.

7.2.2   The Proposed Criterion of Choice

The probabilistic Hurwitz criterion of choice (Eq. 7.17) is an alternative to the expectation criterion of choice (Eq. 7.5) , which also takes into account the standard deviation of a given probability distribution, by way of the positions of the under- and overshoot-corrected lower and upper bounds (Eqs. 7.9 and 7.12). But the universality of this proposal is compromised by way of the simple counter-example of a right-skewed and a left-skewed distribution that have the same lower and upper bounds. It follows that a criterion of choice, in order to be universal, should not only take into account the trade-off between the probabilistic worst- and best-case scenarios, as is done in Eq. (7.17), but also the location of the probabilistic bulk of the probability distribution.

The following position measure for a probability distribution accommodates the intuitive preference for the left-skewed distribution of the counter-example, while taking into account the probabilistic worst and best cases:

$$\frac{LB(k) + E(X) + UB(k)}{3} = \begin{cases} E(X), & lb(k) \geq a, \ ub(k) \leq b, \\ \dfrac{a + 2E(X) + k\,\text{std}(X)}{3}, & lb(k) < a, \ ub(k) \leq b, \\ \dfrac{2E(X) - k\,\text{std}(X) + b}{3}, & lb(k) \geq a, \ ub(k) > b, \\ \dfrac{a + E(X) + b}{3}, & lb(k) < a, \ ub(k) > b. \end{cases} \quad (7.18)$$

Note that the alternative criterion of choice (Eq. 7.18), which takes into account what may happen in the worst, the most likely, and the best of possible scenarios, holds the traditional expected value criterion of choice (Eq. 7.5), as a special case.

In any problem of choice, one will endeavor to choose that action which has a corresponding utility probability distribution that is lying most to the right on the utility axis; i.e., one will choose to maximize the utility probability distributions. In this there is little freedom. But one is, of course, free to choose the measures of the positions of the utility probability distributions any way one sees fit. Nonetheless, it is held to be self-evident that it is always a good policy to take into account all the pertinent information at hand.

First, if only the expectation values of the utility probability distributions are maximized, then, by definition, the information that the standard deviations of the utility probability distributions bring to bear on the problem of choice at hand will be neglected.

Second, if only one of the confidence bounds of the utility probability distributions is maximized, while neglecting the other, then a probabilistic minimax or maximax analysis will be performed, and, consequently, the possibility of either the (catastrophic) losses in the lower bound or the (astronomical) gains in the upper bound will be neglected.

Third, if only the sum of the lower and upper bounds are maximized, or a scalar multiple thereof, then a trade-off between the probabilistic worst- and best-case scenarios is made. But in the process, one will, for unimodal distributions, be neglecting the location of the bulk of the probability distributions.

In light of the these three considerations, the scalar multiple, the sum of the undershoot-corrected lower bound, expectation value, and overshoot-corrected upper bound, i.e., Eq. (7.18), is currently believed to be the most all-round position measure for a given probability distribution, as it takes into account the position of the probabilistic worst- and best-case scenarios (Eqs. 7.9 and 7.12), as well as the position of the probabilistic most likely scenario (Eq. 7.5).

7.2.3   The Algorithmic Steps of the Bayesian Decision Theory

The algorithmic steps of the Bayesian decision theory are as follows (van Erp *et al.* [3]):

1. For each possible decision, construct an outcome probability distribution; i.e. for each possible decision, assign to every conceivable contingency both an estimated net monetary consequence and a probability.
2. Transform outcome probability distributions to their corresponding utility probability distributions; i.e., convert the outcomes of the outcome probability distributions to their corresponding utilities, using Bernoulli's utility function.
3. Maximize a scalar multiple of the sum of the lower bound, the expectation value, and the upper bound of the utility probability distributions; i.e., the criterion of choice in Eq. (7.18).

Note that the Bayesian decision theory is just Bernoulli's expected utility theory, except for the alternative criterion of choice (Eq. 7.18), which is to be maximized. But in the case that the k-sigma confidence lower bound (Eq. 7.7) does not undershoot the absolute minimum (Eq. 7.6), and the confidence upper bound (Eq. 7.11) does not overshoot the absolute maximum (Eq. 7.10), then the criterion of choice (Eq. 7.18) collapses to the expectation value (Eq. 7.5) and, as a consequence, the Bayesian decision theory becomes equivalent to Bernoulli's expected utility theory [1].

7.3   The Allais Paradox

Allais constructed his famous paradox in order to demonstrate the very profound psychological reality of the preference for security in the neighborhood of certainty [7]. For example, if one has to choose between the following two options:

1A. 10% chance of winning €100 million and 90% chance of winning nothing;
1B. 9% chance of winning €500 million and 91% chance of winning nothing,

then most of us – in correspondence with expected utility theory – will prefer option 1B, which has the greater expected utility. However, if one has to choose between the two options:

2A. absolute certainty of winning €100 million;
2B. 90% chance of winning €500 million and 10% chance of winning nothing,

then most of us would prefer the secure option 2A, even though the uncertain option 2B has the greater expected utility, providing one is not a multi-billionaire, as we opt for security in the neighborhood of certainty.

Allais offered up his paradox because he felt that the exclusive focus of the expected utility theory on the means (i.e., expectation values) of the utility probability distributions neglects the spread of the probabilities of the utility values around their mean, a spread which – according to Allais – represented the fundamental psychological element of the theory of risk [7].

7.3.1   The Choosing of Option 1B

The outcome probability distributions of the options of the first bet are:

$$p(O_i|D_{1A}) = \begin{cases} 0.10, & O_1 = 100 \times 10^6 \\ 0.90, & O_2 = 0 \end{cases} \tag{7.19}$$

and

$$p(O_j|D_{1B}) = \begin{cases} 0.09, & O_1 = 500 \times 10^6 \\ 0.91, & O_2 = 0 \end{cases} \tag{7.20}$$

The corresponding utility probability distributions, for a monthly initial asset position of, say, $m = 3000$ for groceries and the like, are (Eq. 7.1):

$$p(U_i|D_{1A}) = \begin{cases} 0.10, & U_1 = q\log\dfrac{3000 + 100 \times 10^6}{3000} \\ 0.90, & U_2 = q\log\dfrac{3000}{3000} = 0 \end{cases} \tag{7.21}$$

and

$$p(U_j|D_{1A}) = \begin{cases} 0.09, & U_1 = q\log\dfrac{3000 + 500 \times 10^6}{3000} \\ 0.91, & U_2 = 0 \end{cases} \tag{7.22}$$

The expected utilities of Eqs. (7.21) and (7.22) are, respectively (Eq. 7.5):

$$E(U|D_{1A}) = 1.041q, \text{ and } E(U|D_{1B}) = 1.082q \tag{7.23}$$

It can be seen in Eq. (7.23) that the expected utility solution of the first bet – for any $q > 0$ – points to option 1B as the predicted preferred choice.

The standard deviations of Eqs. (7.21) and (7.22) are, respectively (Eq. 7.8):

$$\text{std}(U|D_{1A}) = 3.124q, \text{ and } \text{std}(U|D_{1B}) = 3.441q \tag{7.24}$$

It follows from Eqs. (7.23) and (7.24) that the $k = 1$ sigma confidence intervals (Eqs. 7.7 and 7.11), of the utility probability distributions (Eqs. 7.20 and 7.21) are, respectively:

$$lb(k = 1|D_{1A}) = -2.083q, \quad ub(k = 1|D_{1A}) = 4.166q \tag{7.25}$$

and

$$lb(k = 1|D_{1B}), = -2.359q, \quad ub(k = 1|D_{1B}) = 4.523q. \tag{7.26}$$

The absolute minimal and maximal values (Eqs. 7.6 and 7.10) of the utility probability distributions (Eqs. 7.20 and 7.21) are, respectively:

$$a(D_{1A}) = 0, \quad b(D_{1A}) = q \log \frac{3000 + 100 \times 10^6}{3000} = 10.414q \tag{7.27}$$

and

$$a(D_{1B}) = 0, \quad b(D_{1B}) = q \log \frac{3000 + 500 \times 10^6}{3000} = 12.024q. \tag{7.28}$$

Comparing Eqs. (7.25)–(7.27), one sees that for the utility probability distribution Eq. (7.21) there occurs a lower bound undershoot, and the same can be seen for the utility probability distribution Eq. (7.22), when one compares Eqs. (7.26)–(7.28). It follows that for both options of the first bet, the appropriate criterion of choice is the second row of Eq. (7.18):

$$R(U|D_i) = \frac{a(D_i) + 2E(U|D_i) + std(U|D_i)}{3}, \tag{7.29}$$

where $k = 1$. Substituting Eqs. (7.23), (7.24), (7.27), and (7.28) into Eq. (7.29), we obtain the alternative criteria of choice:

$$R(U|D_{1A}) = 1.736q, \quad \text{and} \quad R(U|D_{1B}) = 1.868q. \tag{7.30}$$

It can be seen in Eq. (7.30), as $q > 0$, Eq. (7.1), that the Bayesian decision theory solution of the first bet points to option 1B as the predicted preferred choice.

7.3.2   The Choosing of Option 2A

The outcome probability distributions of the options of the second bet are:

$$p(O|D_{2A}) = \{1.0, \quad O = 100 \times 10^6, \tag{7.31}$$

and

$$p(O_j|D_{2B}) = \begin{cases} 0.9, & O_1 = 500 \times 10^6, \\ 0.1, & O_2 = 0. \end{cases} \tag{7.32}$$

The corresponding utility probability distributions, for a monthly initial asset position of, say, $m = 3000$ for groceries and the like, are (Eq. 7.1):

$$p(U|D_{2A}) = \left\{ 1.0, \quad U = q \log \frac{3000 + 100 \times 10^6}{3000}, \right. \tag{7.33}$$

and

$$p(U_j|D_{2B}) = \begin{cases} 0.9, & U_1 = q\log\dfrac{3000 + 500 \times 10^6}{3000}, \\ 0.1, & U_2 = 0. \end{cases} \tag{7.34}$$

The expected utilities of Eqs. (7.33) and (7.34) are, respectively (Eq. 7.5):

$$E(U|D_{2A}) = 10.414q, \text{ and } E(U|D_{2B}) = 10.821q. \tag{7.35}$$

It can be seen in Eq. (7.35), as $q > 0$, Eq. (7.1), that the expected utility solution of the second bet points to option 2B as the predicted preferred choice, which contradicts our basic intuition.

The standard deviation of Eqs. (7.33) and (7.34) are, respectively (Eq. 7.8):

$$\text{std}(U|D_{2A}) = 0, \text{ and } \text{std}(U|D_{2B}) = 3.607q. \tag{7.36}$$

It follows from Eqs. (7.35) and (7.36) that the $k = 1$ sigma confidence intervals (Eqs. 7.7 and 7.11) of the utility probability distributions Eqs. (7.33) and (7.34) are, respectively:

$$\text{lb}(k = 1|D_{2A}) = 10.414q, \quad \text{ub}(k = 1|D_{2A}) = 10.414q \tag{7.37}$$

and

$$\text{lb}(k = 1|D_{2B}), = 7.214q, \quad \text{ub}(k = 1|D_{2B}), = 14.429q. \tag{7.38}$$

The absolute minimal and maximal values (Eqs. 7.6 and 7.10) of the utility probability distributions Eqs. (7.33) and (7.34) are, respectively:

$$a(D_{2A}) = b(D_{2A}) = q\log\frac{3000 + 100 \times 10^6}{3000} = 10.414q \tag{7.39}$$

and

$$a(D_{2B}) = 0, b(D_{2B}) = q\log\frac{3000 + 500 \times 10^6}{3000} = 12.024q. \tag{7.40}$$

Comparing Eqs. (7.37)–(7.39), one sees that for the utility probability distribution Eq. (7.33) there is neither a lower bound undershoot nor an upper bound overshoot. It follows that for the first option of the second bet, the appropriate criterion of choice is the first row of Eq. (7.18):

$$R(U|D_{2A}) = E(U|D_{2A}). \tag{7.41}$$

Comparing Eqs. (7.38)–(7.40), one sees that for the utility probability distribution Eq. (7.34), an upper bound overshoot occurs. It follows that for the second option of the second bet, the appropriate criterion of choice is the third row of Eq. (7.18):

$$R(U|D_{2B}) = \frac{2E(U|D_{2B}) - \text{std}(U|D_{2B}) + b(D_{2B})}{3}, \tag{7.42}$$

where $k = 1$. Substituting Eqs. (7.35), (7.36), (7.39), (7.40) into Eqs. (7.41) and (7.42), we obtain the alternative criteria of choice:

$$R(U|D_{2A}) = 10.414q \text{ and } R(U|D_{2B}) = 10.020q. \tag{7.43}$$

It can be seen in Eq. (7.43), as $q > 0$, Eq. (7.1), that the Bayesian decision theory solution of the second bet points to option 2A as the predicted preferred choice, which – unlike the expected utility solution Eq. (7.35) – corresponds to our basic intuition.

7.3.3   How to Resolve an Allais Paradox

The Bayesian decision theory solution aligns itself with the basic intuition to "take the money and run," i.e., the choosing of option 2A in the second bet, by taking into account the probabilistic worst- and best-case scenarios, as well as the most likely scenario. The probabilistic worst-case scenario (Eq. 7.9) under both options are (Eqs. 7.37–7.40):

$$\text{LB}(k = 1|D_{2A}) = 10.414q, \quad \text{and} \quad \text{LB}(k = 1|D_{2B}), = 7.214q. \tag{7.44}$$

The probabilistic most likely scenario (Eq. 7.5) under both options are (Eq. 7.35):

$$E(U|D_{2A}) = 10.414q, \quad \text{and} \quad E(U|D_{2B}) = 10.821q. \tag{7.45}$$

The probabilistic best-case scenario (Eq. 7.12) under both options are (Eqs. 7.37–7.40):

$$\text{UB}(k = 1|D_{2A}) = 10.414q, \quad \text{and} \quad \text{UB}(k = 1|D_{2B}), = 12.024q. \tag{7.46}$$

If we take the secure option 2A as our reference point, then this choice represents a probabilistic gain in the lower bound (Eq. 7.44):

$$\Delta_{\text{LB}} = (10.414 - 7.214)q = 3.200q, \tag{7.47}$$

a probabilistic loss in the expectation value (Eq. 7.45):

$$\Delta_{E(U)} = (10.414 - 10.821)q = -0.407q, \tag{7.48}$$

and a probabilistic loss in the upper bound (Eq. 7.46):

$$\Delta_{\text{UB}} = (10.414 - 12.024)q = -1.610q. \tag{7.49}$$

It follows that the secure option 2A represents a net probabilistic gain, as the gain in the lower bound compensates for the losses in both the expectation value and the upper bound (Eqs. 7.47–7.49):

$$\Delta_{\text{LB}} + \Delta_{E(U)} + \Delta_{\text{UB}} = 1.183\,q, \tag{7.50}$$

for any $q > 0$.

According to Allais the fundamental psychological element of the theory of risk is the spread of the probabilities of the utility values around their mean [7]. The alternative criterion of choice takes this spread around the mean into account by way of the probabilistic worst- and best-case scenarios (Eqs. 7.44 and 7.46).

7.4   The Ellsberg Paradox

Ellsberg [8] presents two experiments, the first consisting of two urns with red and black balls, and the second a single urn with combinations of red, black, yellow balls. The first of these experiments is described here; the second experiment essentially shows the same results. Ellsberg's first urn experiment involves the two urns:

Urn 1 – 100 balls, 50 red, 50 black;
Urn 2 – 100 balls, red and black with proportions not specified,

with payoffs defined as:

    I. "Payoff on red1" – draw from urn 1, receive $1 if red, $0 if black;
   II. "Payoff on black1" – draw from urn 1, receive $0 if red, $1 if black;
  III. "Payoff on red2" – draw from urn 2, receive $1 if red, $0 if black;
  IV. "Payoff on black2" – draw from urn 2, receive $0 if red, $1 if black.

The gambles posed are:

1. Which do you prefer, "payoff on red1" or "payoff on black1"?
2. Which do you prefer, "payoff on red2" or "payoff on black2"?
3. Which do you prefer, "payoff on red1" or "payoff on red2"?
4. Which do you prefer, "payoff on black1" or "payoff on black2"?

Results according to Ellsberg, from introspection and non-experimental surveying of colleagues, are:

(a) Majority will be indifferent in gambles (1) and (2), which indicates that subjective probabilities of red versus black are 50/50 for both urn 1 and urn 2,
(b) Majority prefers red1 in (3) and black1 in (4) – in other words, most people prefer urn 1 (known 50/50) over urn 2 (unknown split between red and black).

The Ellsberg observations (a) and (b) become paradoxical only if one interprets the preference for urn 1 over urn 2 in the gambles (3) and (4) to mean that the probabilities for red1 and black1 are greater than the corresponding probabilities, red2 and black2. As such, an interpretation of observation (b) would be in contradiction with observation (a), which states that the probabilities of red1, red2, black1, and black2 are all 1/2. However, this paradox may be trivially resolved if one realizes that:

> *All things being equal probability wise*, one is perfectly free to prefer one Ellsberg urn over the other without being inconsistent.

As one may witness a clear preference for the urn with known proportions over the urn with unknown proportions – i.e., observation (b) – we may conclude that people prefer gambles with crisp and clear probabilities to gambles with "fuzzy" probabilities, even if these probabilities are ultimately equivalent – i.e., observation (a). This preference is called "ambiguity aversion," [8].

In closing, the Ellsberg observations (a) and (b) would only be paradoxical for, say, the alternative urns:

Urn 1 – 100 balls, 25 red, 75 black;
Urn 2 – 100 balls, red and black with proportions not specified,

as simple introspection suggests that for these alternative urns, one would prefer red2 over red1 and black1 over black2. This concludes our discussion of the Ellsberg paradox.

7.5   The Difference in Riskiness Between Type I and Type II Events

Due to many consistency problems with, on the one hand, predictive risk analysis and expected utility theory and, on the other hand, common-sense considerations of risk acceptability in some well-constructed counter-examples, it has been difficult to maintain the case for the feasibility of a predictive theory of choice – even to such an extent that one may read in a textbook like Reith [9] that the usefulness of the notion of "risk" lies not in its ability to correctly predict future outcomes, but rather in its ability to provide a basis for decision-making.

In van Erp *et al.* [3] it is postulated that the expectation value as an index of risk should be used with a critical mind, and that this value:

$$E(X) = \sum_{i=1}^{n} p_i x_i, \tag{7.51}$$

may lie at the heart of problems of expected utility theory to model choice preferences. For it is found that the alternative risk index (Eq. 7.18):

$$\frac{LB(k) + E(X) + UB(k)}{3} = \begin{cases} E(X), & \mathrm{lb}(k) \geq a, \ \mathrm{ub}(k) \leq b, \\ \dfrac{a + 2E(X) + k \, \mathrm{std}(X)}{3}, & \mathrm{lb}(k) < a, \ \mathrm{ub}(k) \leq b, \\ \dfrac{2E(X) - k \, \mathrm{std}(X) + b}{3}, & \mathrm{lb}(k) \geq a, \ \mathrm{ub}(k) > b, \\ \dfrac{a + E(X) + b}{3}, & \mathrm{lb}(k) < a, \ \mathrm{ub}(k) > b. \end{cases} \tag{7.52}$$

accommodates not only the Allais paradox, as previously discussed, but also the prospect theoretical counter-examples to expected utility theory; i.e., the reflection effect and the inverted *S*-shape in certainty bets (see van Erp *et al.* [3]). Moreover, this alternative risk index may also shed some light on common-sense observations such as the one made in Section 2.2 that most people judge a type II event as more undesirable than a type I event, even if the expected consequences of the two events are exactly the same.

In this section it will further be demonstrated that the alternative risk index (Eq. 7.52) will assign higher risks to type II events than to type I events, even if the expectation values of both events are so constructed that they are equal for both types of events.

7.5.1   *Outcome Probability Distributions with Equal Expectation Values*

Imagine that one has two dichotomous events, one of which is a type I event, T_1, and the other a type II event, T_2. If (i) the negative consequence of the type II event occurring is n times the negative consequence of the type I event occurring, (ii) the consequence of the non-occurrence of either event is zero, and (iii) the probability of the type II event is an nth fraction of the probability of the type I event, i.e.:

$$p(O_i | T_1) = \begin{cases} \theta, & O_1 = x, \\ 1 - \theta, & O_2 = 0, \end{cases} \tag{7.53}$$

and

$$p(O_i|T_2) = \begin{cases} \dfrac{\theta}{n}, & O_1 = nx, \\ 1 - \dfrac{\theta}{n}, & O_2 = 0. \end{cases} \tag{7.54}$$

then both outcome probability distributions will admit the same expectation values, i.e.:

$$E(O|T_1) = x\theta = nx\frac{\theta}{n} = E(O|T_2). \tag{7.55}$$

Note that in the outcome probability distributions (7.53) and (7.54) the negative consequences are represented by positive numbers, rather than negative numbers. As a consequence, the risk index Eq. (7.52) is to be minimized, rather than maximized.

7.5.2 The Risk of the Type I Event

Suppose that the type I event has a probability of occurring of:

$$\theta = \frac{1}{2}. \tag{7.56}$$

Then the expectation value and standard deviation of the corresponding outcome probability distribution (7.55) are given as:

$$E(O|T_1) = \frac{x}{2} \tag{7.57}$$

and

$$std(O|T_1) = x\sqrt{\frac{1}{2}\left(1 - \frac{1}{2}\right)} = \frac{x}{2}. \tag{7.58}$$

So the k-sigma confidence bounds of Eq. (7.53) under Eq. (7.56) are given as Eqs. (7.57) and (7.58):

$$lb(k|T_1) = E(O|T_1) - k\,std(O|T_1) = \frac{1-k}{2}x \tag{7.59}$$

and

$$ub(k|T_1) = E(O|T_1) + k\,std(O|T_1) = \frac{1+k}{2}x \tag{7.60}$$

If one corrects for lower bound undershoot and upper bound overshoot, then the corrected confidence bounds are (Eqs. 7.53 and 7.59):

$$LB(k|T_1) = \begin{cases} 0, & lb(k|T_1) < 0, \\ lb(k|T_1), & lb(k|T_1) \geq 0, \end{cases}$$

$$= \begin{cases} 0, & k \geq 1, \\ \dfrac{1-k}{2}x, & k < 1, \end{cases} \tag{7.61}$$

and (Eqs. 7.54 and 7.60):

$$\mathrm{UB}(k|T_1) = \begin{cases} \mathrm{ub}(k|T_1), & \mathrm{ub}(k|T_1) \le x, \\ x, & \mathrm{ub}(k|T_1) > x, \end{cases}$$

$$= \begin{cases} \dfrac{1+k}{2}x, & k < 1, \\ x, & k \ge 1. \end{cases} \tag{7.62}$$

Substituting Eqs. (7.57), (7.61), and (7.62) into the left-hand side of Eq. (7.52) one obtains a computed risk for the type I event of:

$$R(O|k, T_1) = \frac{\mathrm{LB}(k|T_1) + E(O|T_1) + \mathrm{UB}(k|T_1)}{3} = \frac{x}{2}, \tag{7.63}$$

for any sigma level k. So, for the particular case of Eqs. (7.53) and (7.56), the alternative criterion of choice (Eq. 7.52) collapses to the traditional expectation value (Eq. 7.57) as the criterion of choice.

7.5.3 The Risk of the Type II Event

Assume that the type II event has an n times smaller probability of occurring than the type I event, and the consequence of the type II event is n times larger than the consequence of the type I event (Eqs. 7.53 and 7.54). The expectation value and standard deviation of the type II outcome probability distribution is given as (Eq. 7.55):

$$E(O|T_2) = \frac{x}{2} \tag{7.64}$$

and

$$\mathrm{std}(O|T_2) = nx\sqrt{\frac{1}{2n}\left(1 - \frac{1}{2n}\right)} = \frac{x\sqrt{2n-1}}{2}. \tag{7.65}$$

So the k-sigma confidence bounds of Eq. 7.54 under Eq. (7.56) are given as (Eqs. 7.64 and 7.65):

$$\mathrm{lb}(k|T_2) = E(O|T_2) - k\,\mathrm{std}(O|T_2) = \frac{1 - k\sqrt{2n-1}}{2}x \tag{7.66}$$

and

$$\mathrm{ub}(k|T_2) = E(O|T_2) + k\,\mathrm{std}(O|T_2) = \frac{1 + k\sqrt{2n-1}}{2}x. \tag{7.67}$$

If one corrects for lower bound undershoot and upper bound overshoot, then the corrected confidence bounds are (Eqs. 7.54 and 7.66):

$$\mathrm{LB}(k|T_2) = \begin{cases} 0, & \mathrm{lb}(k|T_2) < 0, \\ \mathrm{lb}(k|T_2), & \mathrm{lb}(k|T_2) \ge 0, \end{cases}$$

$$= \begin{cases} 0, & k \ge \dfrac{1}{\sqrt{2n-1}}, \\ \dfrac{1 - k\sqrt{2n-1}}{2}x, & k < \dfrac{1}{\sqrt{2n-1}}. \end{cases} \tag{7.68}$$

and (Eqs. 7.62 and 7.67):

$$
\mathrm{UB}(k|T_2) = \begin{cases} \mathrm{ub}(k|T_2), & \mathrm{ub}(k|T_2) \leq nx, \\ nx, & \mathrm{ub}(k|T_2) > nx, \end{cases}
$$

$$
= \begin{cases} \dfrac{1 + k\sqrt{2n-1}}{2} x, & k < \sqrt{2n-1}, \\ nx, & k \geq \sqrt{2n-1}. \end{cases} \tag{7.69}
$$

Substituting Eqs. (7.64), (7.68), and (7.69) into the left-hand side of Eq. 7.52, one may obtain for the high–impact, low-probability (HILP) event the following risk index as a function of the chosen k-sigma level:

$$
\frac{\mathrm{LB}(k|T_2) + E(O|T_2) + \mathrm{UB}(k|T_2)}{3} = \begin{cases} \dfrac{x}{2}, & k < \dfrac{1}{\sqrt{2n-1}}, \\ \dfrac{x}{6}(2 + k\sqrt{2n-1}), & \dfrac{1}{\sqrt{2n-1}} \leq k < \sqrt{2n-1}, \\ \dfrac{x}{6}(1 + 2n), & k \geq \sqrt{2n-1}. \end{cases} \tag{7.70}
$$

For both large n, i.e., pronounced type II events, and k-sigma levels in the reasonable range of, say, 1–6, or, equivalently, $1/\sqrt{2n-1} \leq k < \sqrt{2n-1}$, the risk (Eq. 7.52), of the type II event translates to:

$$
R(O|k, T_2) = \frac{\mathrm{LB}(k|T_2) + E(O|T_2) + \mathrm{UB}(k|T_2)}{3} = \frac{x}{6}(2 + k\sqrt{2n-1}). \tag{7.71}
$$

So, for the particular case of Eqs. (7.54) and 7.56, and a variable n much greater than 1, i.e., $n \gg 1$, the alternative criterion of choice (Eq. 7.52), goes to a value differing markedly from the traditional expectation value (Eq. 7.57).

7.5.4   Comparing the Risks of the Type I and Type II Events

The type I event, T_1, is operationalized in this section as an event having a probability of $1/2$ and a loss of x (Eqs. 7.53 and 7.56). The type II event, T_2, is operationalized relative to the type I event as the event that has a probability n times smaller and a loss n times x greater (Eqs. 7.54 and 7.56), i.e., the more severe the type II event, the less likely its occurrence.

The expectation values of the type I and type II events are equal (Eq. 7.55). Nonetheless, for a given "severity" variable $n > 1$, the alternative risk index (Eq. 7.52) will assign a higher risk to the type II event than to the type I event by a factor of (Eqs. 7.63 and 7.71):

$$
\frac{\mathrm{Risk}_{\text{type II event}}}{\mathrm{Risk}_{\text{type I event}}} = \frac{R(O|k, T_2)}{R(O|k, T_1)} = \frac{1}{3}(2 + k\sqrt{2n-1}). \tag{7.72}
$$

If for $k = 1, 2, 3, 6$ sigma levels the risk ratio Eq. 7.72 is plotted as a function of the severity variable n, Figure 7.1 is obtained.

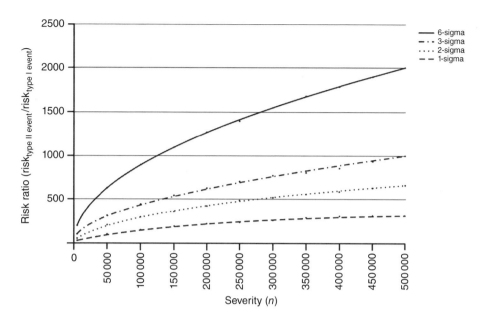

Figure 7.1   Risk ratios as a function of the severity variable n.

In Figure 7.1 it can be seen that for a severity variable $n = 500.000$, or, equivalently, a probability of 10^{-6} for the type II event and losses that are an order of magnitude of 500.000 more severe than those under a type I event having a probability of 0.5, and sigma levels of $k = 1, 2, 3, 6$, one observes risk ratios of about 333, 667, 1000, and 2000, respectively, as may be seen from Eq. 7.72:

$$\frac{\text{Risk}_{\text{type II event}}}{\text{Risk}_{\text{type I event}}} = \frac{2 + k\sqrt{10^6 - 1}}{3} \approx 333k, \tag{7.73}$$

So, the adjusted criterion of choice (Eq. 7.52) – which is the location measure taking into account the probabilistic worst-case scenario, expected scenario, and best-case scenario (Eq. 7.18) – corroborates and quantifies the common-sense observation that most people judge a type II event as more undesirable than a type I event, even if the expected consequences of the two events are exactly the same.

7.6   Discussion

Although it seems barely conceivable today, the historical record shows clearly and repeatedly that the notion of "expectation of profit" to the first researchers in probability theory was perhaps even more intuitive than the notion of probability itself [5]. Moreover, it was so obvious to many that a person acting in pure self-interest should always behave in such a way as to maximize the expected profit that the prosperous merchants in seventeenth-century Amsterdam bought and sold mathematical expectations as if they were tangible goods.

A new insight into risk science came in 1738 with the St. Petersburg paper [1]. This paper is truly important on the subject of risk as well as on human behavior, as it introduced the pivotal idea that the true value to a person of receiving a certain amount of money is not measured simply by the amount received; it also depends upon how much the person already possesses: "Utility resulting from any small increase in wealth will be inversely proportionate to the quantity of goods previously possessed." This idea is as simple as it is intuitive. For it stands to reason that a beggar will assign a much higher utility to the gain of a thousand euros than will a billionaire, and a modern economist is expressing the same idea when he speaks of the "diminishing marginal utility of wealth." So, where decision theory, by way of the maximization of the expectation of profit, hitherto had only taken profits and the probabilities of these profits materializing into account, Daniel Bernoulli identified the initial wealth of the decision-maker as a third decision theoretical factor of consequence.

Bernoulli's expected utility theory laid the intellectual groundwork for many aspects of microeconomic theory and decision theory. Furthermore, by way of its utility function – which measures that which cannot be counted – it also managed to elevate psychology in 1860 from a mere metaphysical pastime to an exact science [2], as it was found by Fechner that Bernoulli's utility function could model our perception of sensory stimuli for a given background stimulus intensity [10]. The decibel scale, for example, is an instance where Bernoulli's utility function converts objective loudness stimuli to a corresponding subjective scale. Moreover, with the introduction of operational risk management in the mid-twentieth century, in the aftermath of the Second World War and at the beginning of the atomic era, Bernoulli's expected utility theory has remained as relevant as ever. Many improvements by many authors have been proposed to Bernoulli's initial 1738 proposal, the most notable of these being game theory and its offshoots [4, 11] and prospect theory [12].

von Neumann and Morgenstern [4] formulated a lofty predictive mathematical theory on their axiomatic scaffolding and proclaimed that their theory had superseded Bernoulli's expected utility theory as a more general theory. But Tversky and Kahneman were then quick to point out that both von Neumann and Morgenstern's game theory and Bernoulli's expected utility theory could be demonstrated to violate common-sense rationality in certain instances and, as a reaction, counter-postulated that human rationality can never be captured by simple mathematical maximization principles [13]. Tversky and Kahneman then proceeded to propose a mathematical descriptive decision theory that was rooted in empirical observation of psychological betting experiments, rather than abstract mathematical first principles [12].

So persuasive was the case made by Tversky and Kahneman that the descriptive paradigm of behavioral economics came to dominate the decision theoretical field. But just as prospect theory, in its initial inception, was mainly a reaction to von Morgenstern and Neumann's game theory, which put too high a premium on mathematics, while failing to appeal to our common sense, so the Bayesian decision theory [3] is a reaction to Tversky's and Kahneman prospect theory, which puts too high a premium on bare bones empiricism of hypothetical betting experiments, while failing to appeal to our mathematical need for compelling first principles.

A Bayesian decision theoretical research program for the future is one where all the reported inconsistencies of expected utility theory are reassessed and where it is studied if these inconsistencies – if legitimate inconsistencies – may resolve themselves by applying this neo-Bernoullian utility theory. Moreover, it should be gauged how much of prospect theory, which, like the Bayesian decision theory, purports to resolve the inconsistencies that have plagued expected utility theory, remains relevant after: (i) the utility function of the utility theory is set to Bernoulli's original 1738 proposal (see Section 7.1); and (ii) the expectation value is replaced by the alternative position measure, which not only takes into account the most likely scenario – in the probabilistic sense – but also the worst- and best-case scenarios (see Section 7.2.2). The potential benefit of such a research program would be that it might establish the Bayesian decision theory as the normative theory of risk which accurately predicts the way people choose when faced with uncertain outcomes. Such a normative theory of risk could be a boon for operational risk management, as it would put at their disposal a simple but effective mathematical tool to guide them in their decision-making.

7.7 Conclusions

We have introduced in this chapter the Bayesian decision theory as an improvement on Bernoulli's utility theory and as an alternative to prospect theory. It has been demonstrated that this neo-Bernoullian utility theory accommodates, non-trivially, the Allais paradox and the common-sense observation that type II events are more risky than type I events, as well as – very trivially, as the Ellsberg paradox is arguably not that paradoxical – the Ellsberg paradox. In Section 8.12 the Bayesian decision theory will be applied to a simple non-trivial decision problem. It will be demonstrated in this decision theoretical toy problem that the alternative criterion of choice (Eq. 7.18), relative to the traditional criterion of choice (Eq. 7.5), leads to a more realistic estimation of hypothetical benefits of type II risk barriers.

References

[1] Bernoulli, D. (1738). *Exposition of a New Theory on the Measurement of Risk*. Translated from Latin into English by Dr Louise Sommer from 'Specimen Theoriae Novae de Mensura Sortis', Commentarii Academiae Scientiarum Imperialis Petropolitanas, Tomus **V**, pp. 175–192.

[2] Fancher, R.E. (1990). *Pioneers of Psychology*. W.W. Norton and Company, London.

[3] van Erp, H.R.N., Linger, R.O., Van Gelder, P.H.A.J.M. (2015). Fact Sheet Research on Bayesian Decision Theory. 1409.8269v4 arXiv. Online: http://arxiv.org/abs/1409.8269 (accessed March 2016).

[4] von Neumann, J., Morgenstern, O. (1944). *Theory of Games and Economic Behavior*, Princeton, NJ, Princeton University Press.

[5] Jaynes, E.T. (2003). *Probability Theory; the Logic of Science*. Cambridge University Press, Cambridge.

[6] Lindgren, B.W. (1993). *Statistical Theory*. Chapman & Hall, Inc., New York.

[7] Allais, M. (1988). An Outline of My Main Contributions to Economic Science. Nobel Lecture, December 9.

[8] Ellsberg D. (1961). Risk, ambiguity, and the savage axioms. *Quarterly Journal of Economics,* **75**, 643–699.

[9] Reith, G. (2009). Uncertain times: the notion of 'risk' and the development of modernity, In: *The Earthscan Reader on Risk* (eds Lofstedt, R. & Boholm, A.). Earthscan, London.

[10] Masin S.C., Zudini V., Antonelli M. (2009). Early alternative derivations of Fechner's law. *Journal of Behavioural Sciences*, **45**, 56–65.

[11] Savage, L.J. (1954). *The Foundations of Statistics*. John Wiley & Sons, Inc., New York.

[12] Tversky A., Kahneman D. (1992). Advances in prospect theory: cumulative representation of uncertainty. *Journal of Risk and Uncertainty*, **5**, 297–323.

[13] Bernstein, P.L. (1998). *Against the Gods. The Remarkable Story of Risk*. John Wiley & Sons, Inc., New York.

8

Making State-of-the-Art Economic Thinking Part of Safety Decision-making

8.1 The Decision-making Process for an Economic Analysis

The aim of economic analyses applied in an operational safety context is to support safety investment decision-making, and not to produce numbers, figures, or tables. However, even taking opportunity costs and benefits into consideration, in most decision-making situations with respect to operational safety, there is no simple scale of preference and the outcomes are not easily observable. Moreover, in the case of operational safety, using outcomes as a basis for evaluating the "goodness" of a safety investment decision is problematic. If accidents occur despite the safety investment, does this mean that the investment was a wrong decision, or, on the contrary, that it avoided even more accidents? Or, if no incidents or accidents occur, does this mean that the safety investment was a good decision, or is the absence of accidents the result of pure coincidence, or is there yet another safety measure that prevented all incidents? Hence, the validation of the goodness of a safety investment is not easy.

Yet, as Aven [1] also indicates, outcome-centered thinking is important and helps to give a clear view on objectives and preferences. The problem is, however, that such thinking usually leads to making decisions without contemplating and calculating a variety of alternatives. In economic thinking, the concept/principle of "opportunity costs and benefits" is nonetheless a central one, and this should be fully recognized by safety managers. Using economic analyses to make good safety investment decisions is very much related to determining and calculating different options, and knowing their accuracy and the uncertainties involved in the outcomes.

The basic structure of the decision-making process for an economic analysis is given in Figure 8.1.

The starting point of the decision-making process is "reality" in an organization, and within society at large. The perception of reality (with respect to operational safety and risks) is then formed by processing available information and data, such as accident scenarios, probabilities, consequences, liabilities, utilities, costs of prevention, and costs of accident scenarios.

Operational Safety Economics: A Practical Approach Focused on the Chemical and Process Industries,
First Edition. Genserik L.L. Reniers and H.R. Noël Van Erp.
© 2016 John Wiley & Sons, Ltd. Published 2016 by John Wiley & Sons, Ltd.

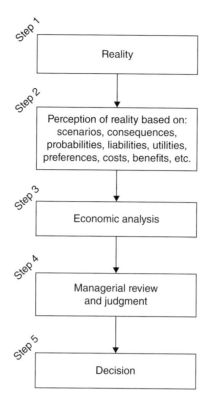

Step 1 — Reality

Step 2 — Perception of reality based on: scenarios, consequences, probabilities, liabilities, utilities, preferences, costs, benefits, etc.

Step 3 — Economic analysis

Step 4 — Managerial review and judgment

Step 5 — Decision

Figure 8.1 Basic structure of the decision-making process for an economic analysis. (Source: Based on Aven [1].)

Depending on the amount of information and data available regarding a certain event or risk, and on its variability, perception of reality is of high or low quality (and resembles reality itself to a greater or lesser extent) and the decision-maker knows which domain of Figure 2.5 he or she is situated in (A, B, C, or D). Information can subsequently be processed via economic analysis models (as described in this book), and, depending on the domain (A, B, C, or D), a different analysis approach and/or method can be employed and the resulting economic information will be of higher or lower accuracy. Based on these economic analyses, a further managerial review can be carried out, and a decision on operational safety investments can finally be made. In this approach, it is clear that an economic analysis should provide input to the decision-maker, but it does not provide the decision itself.

As a very simple example, assume that two options are available for dealing with a certain risk scenario, option A and option B. Option A leads to a hypothetical benefit of €220 000 (with a probability of 75% and a variability of 40%), and option B to a hypothetical benefit of €90 000 (with a probability of 80% and a variability of 28%). The rest-risk probability of a fatality for option A is equal to 0.2%, whereas in the case of option B it is equal to 0.1%. The organization chooses a value of statistical life (VoSL) of €8 000 000 for these probabilities. This information is available at the end of step 2. The safety manager of the organization then uses this information to assess that the risk scenario is situated in domain C (of Figure 2.5). He therefore chooses a straightforward cost-benefit analysis model available for type I events, and calculates the cost-benefit value in both cases. For option A, the gain boils down to $[0.75 \times 220\,000 - (2/1000) \times 8\,000\,000] = €149\,000$. For option B, the expected gain is equal to $[0.8 \times 90\,000 - (1/1000) \times 8\,000\,000] = €64\,000$. Hence, in this example, according to this economic analysis, option A is preferable to option B. This is the end of step 3. This information is passed to management, who then make their decision, based on all the information available (also, the information that the likelihood of fatalities in the case of option A is twice that in option B). The final step (step 5) is management's decision.

Besides the well-known cost-benefit (see Chapter 5), cost-effectiveness, and cost–utility analyses (see Chapter 6), there are other modeling and calculation approaches to help decision-makers assess the possible options or decision alternatives. The following sections discuss cost-benefit analyses in practice, as well as some of the other available techniques.

8.2   Application of Cost-Benefit Analysis to Type I Risks

In order to illustrate the application of a cost-benefit analysis and parameters such as the payback period (PBP) and the internal rate of return, an extensive example is discussed in this section.

Assume that a firm specialized in the production of preassembled wooden houses is evaluating two different investment options to improve the safety conditions inside the production area. Currently the production is semi-automated with the cutting operations carried out manually, with machines being used to support the movement of the raw materials (mainly logs and beams) from the warehouse to the cutting area. In the cutting area, the personnel have to manually orientate the logs and define the type of cut, depending on the features of the logs, using a plumbline and a pencil. The guidelines and reference points to be followed during the cutting are manually traced directly on the log. Once the log is manually secured by fixing it on some tracks, the cutting operations can start. During the preliminary operations before the final cut, the manual operations significantly expose the workers to many potential work-related accidents, such as falls, cuts, diseases to the joints due to the handling of heavy weights. At the same time, the quality of the final products can potentially be compromised by human error.

The firm has decided to improve the working conditions in the cutting area, defining a safety budget of €500 000. Two investment options are analyzed, aimed at improving the safety levels within the company, increasing the risk awareness and spreading good safe practices in the working environment. The first case refers to the purchase of a new machine to improve the safety conditions in the production area, while the second refers to new equipment to reduce the dust and improve the overall working conditions. The features of these alternative investment options are presented in the following sections.

8.2.1   Safety Investment Option 1

The first option involves installing a computerized machine to automate the log cutting at high speed and to ensure enhanced levels of safety for operators working in the production area. With the new machine, the level of dust in the air, as well as the risk of repetitive strain injury, cuts, crushing injuries of fingers or toes, excessive noise, excessive mental problems, and lower back pain would be significantly reduced. The duration of the investment is assumed to be 10 years.

Using the new machine, the cutting process, usually performed by a team of three carpenters and a supervisor, would be performed by three operators: one technical operator would be responsible for the cutting machine, another would be responsible for the maintenance activities, and the third would be in charge of the final product design. With this new working configuration, the same workload would be performed in 60% less time. As a result of the cutting optimization, the number of defects would be reduced and the quality of the final products would also be significantly enhanced. The beneficial effect of the improved quality of the product has also been quantified in terms of additional profits of €25 000 per year. Moreover, a team responsible for the risk assessment has estimated a yearly saving in the production operations of €135 000 due to waste reduction and a reduction in raw materials. On the investment side, the initial purchase price of the machine is €280 000. Additional costs due to training, redesign of the layout in the production area, feasibility studies are to be initially sustained. The company has conducted an economic analysis associated with the investment, and quantified the main benefits and costs, as shown in Tables 8.1 and 8.2. The initial investment and yearly benefits and costs associated with safety investment option 1 are shown in Table 8.3.

The costs and the hypothetical benefits summarized in Tables 8.1 and 8.2 can be further distinguished in terms of initial costs/benefits and yearly recurring costs/benefits. Initial costs/benefits are supposed to be sustained in year 0 (in which the evaluation is made). Yearly recurring costs/benefits are sustained during the time horizon in which the investment is evaluated. Figure 8.2 shows a graphical representation of the evolution of the net yearly cash flow. Yearly cash flows are given by the difference between costs (negative cash flows) and

Table 8.1   List of costs associated with safety investment option 1

Categories of costs	Subcategories of costs	Value
Initial costs (€)	Investigation and preliminary study	15 400
	Machine purchase costs	280 000
	Initial training	25 000
	Changing layouts and production operations	110 500
Installation costs (€)	Machine configuration and testing	5 500
	Equipment costs	15 400
	Installation team costs	25 000
Operating costs (€/year)	Energy costs	38 500
Maintenance costs (€/year)	Material costs	15 000
	Maintenance team costs	7750
Inspection costs (€/year)	Inspection team costs	2500
Other safety costs (€/year)	Other safety costs	2500

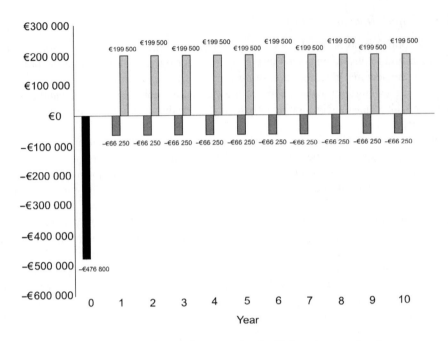

Figure 8.2   Yearly cash flows associated with investment option 1.

Table 8.2   List of hypothetical benefits associated with safety investment option 1

Type of benefits	Subcategory	Value
Supply chain benefits (€/year)	Production savings	135 000
	Expected additional profits due to increased sales	25 000
Damage benefits (€/year)	Damage to own material/property	2500
Legal benefits (€/year)	Fines	10 000
Insurance benefits (€/year)	Insurance premium	20 000
Human and environmental benefits (€/year)	Yearly reduction of days of illness	2500
Other benefits (€/year)	Cleaning	4500

Table 8.3   Initial investment and yearly benefits and costs associated with safety investment option 1

Description	Value (€)
Initial costs (€)	−476 800
Yearly costs (€/year)	−66 250
Yearly benefits (€/year)	199 500

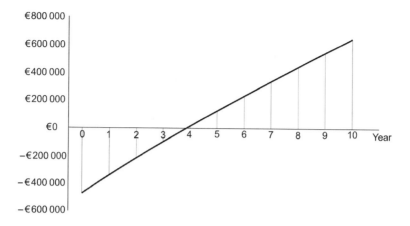

Figure 8.3 Payback period for investment option 1.

benefits (positive cash flows) from year 0 to year 10. Analyzing the nature of costs and benefits it can be seen that the total initial investment required by the company is €0.477 million.

The well-known formula in Eq. (8.1) (see also Chapter 5, Eq. (5.1)) has been used to compute the net present value (NPV) associated with the investment, assuming for each year t the previously mentioned negative cash flows (C_t) and positive cash flows (B_t) and the discount factor $i = 3\%$:

$$\text{NPV} = \sum_{t=0}^{5} \frac{(B_t - C_t)}{(1 + i)^t}. \tag{8.1}$$

The total NPV associated with the investment in this case is equal to €702 501. As this value is greater than zero, the investment is profitable for the company. The PBP is equal to 3.33. This means that after 3 years and 4 months the investment will cover the costs and it will start producing profits for the firm as shown in Figure 8.3.

Another indication that is often used by decision-makers to assess the profitability of an investment is the internal rate of return (IRR), as also mentioned in Chapter 5. This measure represents the discount factor that makes the NPV equal to zero, or, in other words, the value of the interest rate at which an investment reaches a breaks-even point. In this case, the IRR is equal to 24.93% as shown in Figure 8.4.

8.2.2 Safety Investment Option 2

The second investment option is to drastically improve the ventilation system and thus enhance the working conditions in the production area. Due to the cutting operation, the level of dust in the production area is high. The exposure of the personnel to dust is very close to the maximum value allowed by law. In addition, dust and wood particles also reduce the visibility conditions during the cutting operations, exposing the personnel to a risk of significant work-related injury. Safety conditions are legally respected even if the wellness of the personnel is currently not being fully addressed, with high levels of illness due to employees suffering the effects of elevated levels of wood dust. In the past, to improve the quality of the air, the company introduced longer breaks to allow personnel to rest in areas with a lower amount of dust.

The company is currently considering buying a dedicated aspiration and ventilation system, which uses cyclotron technology; it will significantly reduce the level of dust and thereby

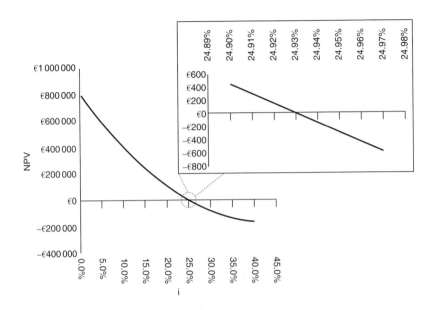

Figure 8.4   Internal rate of return associated with investment option 1.

improve the working conditions in the production area. As a side-effect, the company's productivity could be increased by 25% and the quality of the final products slightly improved. The health of the personnel would be significantly improved and the turnover reduced. Moreover, as the dust and the wood particles are automatically expelled by the new system and compacted into special bags, the waste could be sold to a specialized company for recycling. The equipment inside the production area, due to a lower level of dust, would require less maintenance and cleaning activities. Moreover, the downtime would be reduced by 30%, with a significant impact on the production costs. After a thorough assessment, a team of experts has estimated the benefits and costs associated with the investment; these are summarized in Tables 8.4 and 8.5, respectively.

The costs and benefits can be further analyzed to distinguish between the initial investments and the yearly recurring costs/benefits; these are shown in Table 8.6.

As done for safety investment option 1, the NPV is computed considering a time horizon of 10 years. As the time horizon is the same as the one used to evaluate the first investment option, the comparison between the alternative investments will be simplified, as it will use the same criteria. In addition, the amount of money required for the initial investment (€0.479 million) is comparable to that in the previous investment option.

In this case, the NPV is equal to €490 884, which is lower than the €702 501 associated with the previous investment option. In addition, the PBP shown in Figure 8.5 is slightly longer than 3.3 years. Therefore, this investment option seems to be inferior to investment option 1. Given the company's limited safety budget, safety investment option 1 is to be preferred.

If additional resources were found, investment option 2 should also be considered and implemented as its profitability in the medium to long term is guaranteed by a positive NPV and a relatively limited PBP.

The IRR associated with the second safety investment option is computed as shown in Figure 8.6. The IRR in this case is equal to 18.89%. Therefore, in this case too, investment option number 2 is clearly worse than the new cutting machine.

Table 8.4 List of costs associated with safety investment option 2

Categories of costs	Subcategories of costs	Value
Initial costs (€)	Investigation costs	8300
	Selection and design costs	10 200
	Material costs	85 500
	Training costs	4 500
	Changing guidelines and informing costs	6500
	Purchase costs	195 500
Installation costs (€)	Start-up costs	23 500
	Equipment costs	58 500
	Installation costs	86 500
Operating costs (€/year)	Energy consumption costs	10 000
Maintenance costs (€/year)	Material costs	3500
	Maintenance team costs	1500
Inspection costs (€/year)	Inspection team costs	1500
Other safety costs (€/year)	Other safety costs	1400

Table 8.5 List of hypothetical benefits associated with safety investment option 2

Type of benefits	Subcategory	Value
Supply chain benefits (€/year)	Production loss	8700
	Waste recycling	5000
	Expected additional sales due to better product quality	5500
	Saving in personal protective equipment	4000
Damage benefits (€/year)	Damage to own material/property	41 000
Legal benefits (€/year)	Fines	15 000
Insurance benefits (€/year)	Insurance premium	15 000
Human and environmental benefits (€/year)	Injured employees	5000
	Environmental damage	5000
Other benefits (€/year)	Manager working time	2400
	Cleaning	25 000

Table 8.6 Initial investment and yearly benefits and costs associated with safety investment option 2

Description	Value (€)
Initial costs (€)	−479 000
Yearly costs (€/year)	−17 900
Yearly benefits (€/year)	131 600

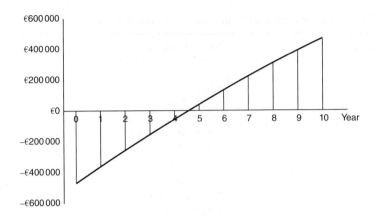

Figure 8.5   Payback period associated with investment option number 2.

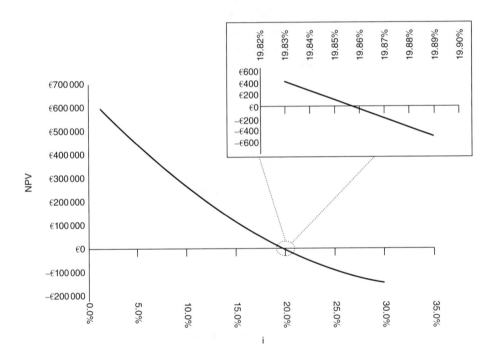

Figure 8.6   Internal rate of return associated with investment option number 2.

8.3   Decision Analysis Tree Approach

In the case of both type I and type II risks, it is possible to draft and use a decision analysis tree if a number of alternative outcomes are possible, as was theoretically explained in Section 3.4.1. The probabilities of safety measures in the case of certain events need to be taken into account, and the expected values need to be calculated. As already explained, expected value theory stems from a concept where the value of an improbable outcome can be weighed against the likelihood that the event will occur (see Chapters 4 and 7).

8.3.1 Scenario Thinking Approach

One way to develop a decision analysis tree is by using scenario thinking. For example, possible scenarios are identified using specific risk assessment techniques, such as event tree analysis. Using the event tree analysis approach (for more information, see, e.g., Greenberg & Cramer [2], the Centre for Chemical Process Safety [3] and Meyer & Reniers [4]), a graph such as that in Figure 8.7 is obtained. The illustrative example of a decision analysis tree for a runaway reaction event is given. All costs and probabilities are assumed to be yearly.

Based on the event tree of Figure 8.7, an analysis of safety investment decisions can be made. The costs of the safety measures can be collected and displayed in the figure (e.g., in the case of this illustrative runaway reaction event, a total prevention cost of €250 000 is obtained), as well as the expected total costs of the event. The expected total costs (which can be seen as the expected hypothetical benefits) can then be compared with the prevention costs. In this illustrative example, the expected total costs are equal to $P_1[P_2 n + (1 - P_2)C_F]$, and should be compared with €250 000 when formulating a decision recommendation. In this way, a decision can be made regarding this safety investment portfolio composed of three barriers (i.e., the cooling system, the operator manual intervention, and the automatic inhibition intervention).

8.3.2 Cost Variable Approach

It is possible to draft a tree where the cost of each decision is presented as a variable named "cost" (see also Muennig [5]). The cost is a running total of the costs of each event in a given event pathway. At each box in the tree, the running total is increased by the cost assigned to the box. Figure 8.8 presents an illustrative example.

> For instance, if companies decide to proactively collaborate with respect to prevention of domino effects, e.g., by developing and maintaining a piece of software for warning and helping neighbors, it would cost them each €10 000. If nothing were to happen, the total cost for the company would be €10 000 (for the collaboration) + €0 (nothing happens) = €10 000. If a large-scale fire were to break out inside the company with severe cross-border potential despite collaboration efforts, the company would also incur the initial costs of the large-scale fire (€12 000 to deal with the fire). If the fire accident did not affect other companies in a severe way, or if it remained internal, no extra costs would be incurred for the company, and the final value of the cost variable at the end of the bottom pathway in Figure 8.8 would be €22 000. An analogous reasoning can be made for all other pathways, resulting in Figure 8.9.
>
> From Figure 8.9, it follows that, for the probabilities and costs assumed, the overall cost of domino effect collaboration would be €10 705, and with no collaboration the cost would be €13 200. Hence, in this illustrative example, "no collaboration" would be some €2500 more expensive in expected costs overall than "collaboration".

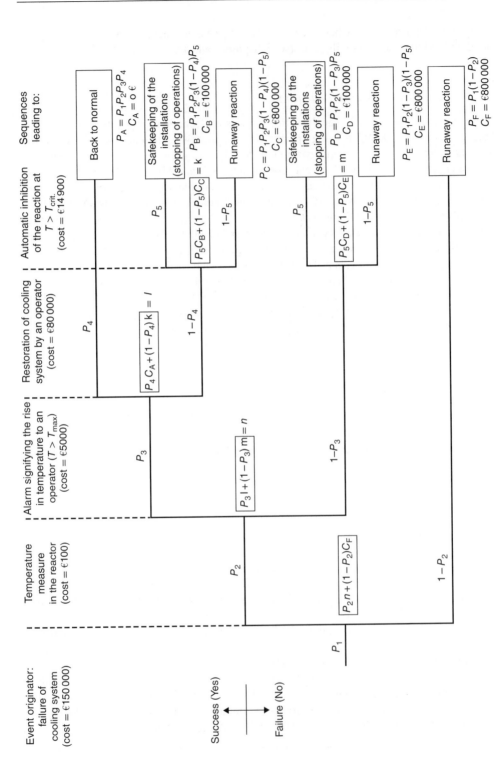

Figure 8.7 Illustrative decision analysis tree for runaway reaction event.

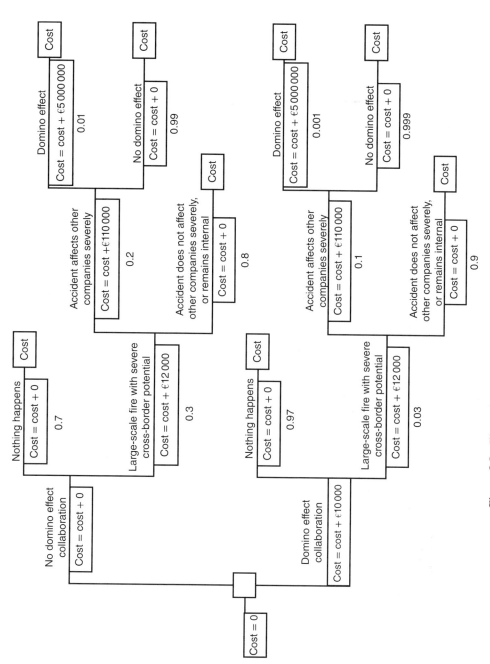

Figure 8.8 Illustrative event pathway for domino effect prevention.

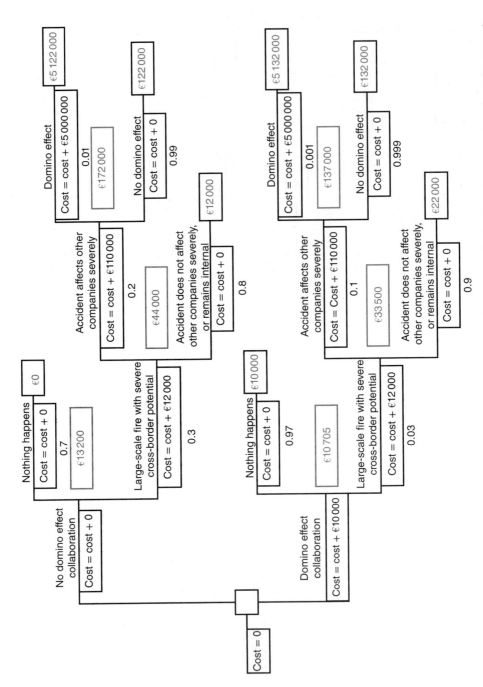

Figure 8.9 Decision analysis tree rolled back to reveal the total cost of each strategy in the domino effect prevention illustrative example.

8.4   Safety Value Function Approach

Recall from Section 3.3 that a value function is a real-valued mathematical function defined over an evaluation criterion that represents an option's measure of "goodness" over the levels of the criterion. A measure of goodness reflects a decision-maker's judged value in the performance of an option (in this case, a safety measures portfolio) across the levels of the criterion. Value functions can be used to determine safety budget allocations for different alternatives of safety measures portfolios. For this, the impact of portfolios are assessed by decision-makers on different criteria. To this end, value functions can be designed as a way to quantitatively express the consequences of a safety measures portfolio's impact on operational safety.

The criteria against which operational safety are evaluated are derived from, for instance, The Egg Aggregated Model (TEAM) of safety culture (see also Chapter 2). Five criteria could then be used: observable technology, observable procedures, observable human behavior, safety climate (i.e., the aggregated perception of workers on safety in the organization), and personal psychological factors of safety.

Figure 8.10 presents illustrative safety value functions for expressing the impact on an organization's safety culture dimensions. Figure 8.10(a) can, for instance, be interpreted as a decision-maker (e.g., the safety manager or the safety management team of an organization) and his or her monotonically increasing preferences for safety measure/investment portfolios that score increasingly higher levels along the observable technology safety culture dimension impact scale (see also Garvey [6]). Thus, the higher a safety measure portfolio score along the value function of Figure 8.10(a), the greater its observable technology impact on operational safety.

The ordinal scale level from 1 to 5 in Figure 8.10(a) could, for example, be drafted/decided upon by safety management as in Table 8.7.

Two different types of value function, shown in Figures 8.10(d) and (e), were, for illustrative purposes, also designed by organizational safety management. These value functions captured a safety measure portfolio's impacts on an organization's safety climate (Figure 8.10d) and on the personal psychological factors of safety (Figure 8.10e) within the organization. These impacts are shown as single dimensional, monotonically increasing exponential value functions. The illustrative value functions vary continuously across their levels/scores. From the curves, it follows that the formulas for these value functions are given by the following equations:

Safety climate – function:

$$S_{(d)}(x) = 1.096(1 - e^{-x/8.2}) \tag{8.2}$$

Personal psychological factors of safety – function:

$$S_{(e)}(x) = 1.018(1 - e^{-x/4.44}) \tag{8.3}$$

It is also possible, however, to represent these safety culture dimension value functions in an ordinal context. An example for safety climate is given in Table 8.8.

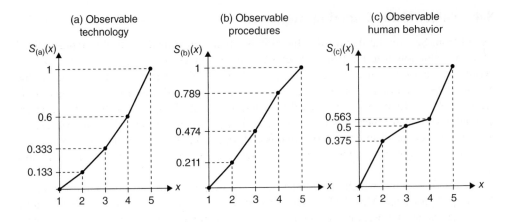

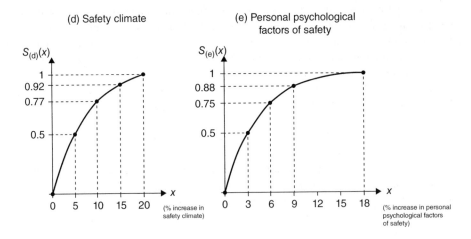

Figure 8.10 Illustrative safety value functions for impacts on safety culture dimensions.

There are many ways to combine the single dimensional value functions of Figure 8.10 into an overall measure of impact. The additive value function is further used as a means to determine a safety measure portfolio's overall impact score on operational safety. For this, it is assumed that an additive safety value function S_{OS}(safety measure portfolio) comprises n single dimensional safety value functions $S_{(a)}(x_1), S_{(b)}(x_2), \ldots, S_{(n)}(x_n)$ satisfying:

$$S_{OS}(x_1, x_2, \ldots, x_n) = w_1 S_{(a)}(x_1) + w_2 S_{(b)}(x_2) + \ldots + w_n S_{(n)}(x_n)$$

where w_i for $i = 1, \ldots, n$ are non-negative weights (representing the importance of the safety dimensions), conditioned on $w_1 + w_2 + \ldots + w_n = 1$.

Table 8.7   A constructed scale for observable technology (illustrative)

Ordinal scale level	A constructed scale for observable technology
1	A portfolio of safety measures that impacts observable safety-related technology in a way that results in a negligible effect on overall operational safety
2	A portfolio of safety measures that impacts observable safety-related technology to the extent that operational safety results fall far below stated objectives but above minimum acceptable levels
3	A portfolio of safety measures that impacts observable safety-related technology to the extent that operational safety results fall below stated objectives but far above minimum acceptable levels
4	A portfolio of safety measures that impacts observable safety-related technology to the extent that operational safety results fall on or above stated objectives and far above minimum acceptable levels
5	A portfolio of safety measures that impacts observable safety-related technology to the extent that operational safety results fall far above stated objectives and far above minimum acceptable levels

Table 8.8   An ordinal scale representation for safety climate (illustrative)

Ordinal scale level	Definition/context: safety climate
1	A portfolio of safety measures that will cause a $< 2\%$ increase in operational safety
2	A portfolio of safety measures that will cause a $> 2\%$ increase, but $\leq 5\%$ increase in operational safety
3	A portfolio of safety measures that will cause a $> 5\%$ increase, but $\leq 10\%$ increase in operational safety
4	A portfolio of safety measures that will cause a $> 10\%$ increase, but $\leq 15\%$ increase in operational safety
5	A portfolio of safety measures that will cause a $> 15\%$ increase, but $\leq 20\%$ increase in operational safety

If decision-makers wish to compare two safety measure portfolio options, and also take into account economic issues, such as budget allocation among safety dimensions, they can use the safety value function approach as follows.

Assume that the weights represent the safety budget allocation among the different safety dimensions, as decided upon by an organization's safety management. Assume that safety management wishes to invest 30% in observable technology, 10% in observable procedures, 15% in observable behavior, 20% in safety climate, and 25% in personal psychological factors of safety. Then, the following weights can be derived: $w_1 = 0.3; w_2 = 0.1; w_3 = 0.15; w_4 = 0.2; w_5 = 0.25$.

Two portfolios A and B are then compared by computing the two additive value functions of the portfolios using the following function:

$$S_{OS}(i) = 0.3S_{(a)}(x_1) + 0.1S_{(b)}(x_2) + 0.15S_{(c)}(x_3) + 0.2S_{(d)}(x_4) + 0.25S_{(e)}(x_5)$$

with $i = $ A, B.

Assume then the following safety management assessments for the single dimensional value functions for both portfolios:

- *Portfolio A*:
  - Observable technology: level 4 → from Figure 8.10(a): $S_{(a)}(4) = 0.6$
  - Observable procedures: level 4 → from Figure 8.10(b): $S_{(b)}(4) = 0.79$
  - Observable behavior: level 3 → from Figure 8.10(c): $S_{(c)}(3) = 0.5$
  - Safety climate: 12% increase → from Eq. (8.2): $S_{(d)}(12) = 1.096(1 - e^{-12/8.2}) = 0.84$
  - Personal psychological factors of safety: 4% increase → from Eq. (8.3): $S_{(e)}(4) = 1.018(1 - e^{-4/4.44}) = 0.60$
- *Portfolio B*:
  - Observable technology: level 3 → from Figure 8.10(a): $S_{(a)}(3) = 0.33$
  - Observable procedures: level 4 → from Figure 8.10(b): $S_{(b)}(4) = 0.79$
  - Observable behavior: level 5 → from Figure 8.10(c): $S_{(c)}(5) = 1$
  - Safety climate: 8% increase → from Eq. (8.2): $S_{(d)}(8) = 1.096 \cdot (1 - e^{-8/8.2}) = 0.68$
  - Personal psychological factors of safety: 10% increase → from Eq. (8.3): $S_{(e)}(10) = 1.018(1 - e^{-10/4.44}) = 0.91$

The additive value functions of the two alternative safety investment/measure portfolios A and B can then be calculated: $S_{OS}(A) = 0.65$ and $S_{OS}(B) = 0.69$. Hence, in this illustrative example, the recommendation based on a safety value function approach is to invest in safety measure portfolio B.

8.5 Multi-attribute Utility Approach

For illustrative purposes, a simple example will be offered to give the reader an idea of how a multi-attribute approach can work in industrial practice. The assumptions made in the examples may differ greatly from company to company, but the method for using the approach remains the same.

Consider the problem of safety management having to decide whether or not to invest in a certain safety investment strategy. The safety investment would cost €400 000, while the hypothetical benefits of that strategy would be the avoidance of accidents with given probabilities: assume there is a probability of 0.016 that €2 000 000 is avoided, a probability of 0.064 that €1 000 000 is avoided, and a probability of $(1 - 0.016 - 0.064) = 0.920$ that no accidents will be avoided at all.

First, the objectives of the multi-attribute approach have to be specified. The company sets two main goals: (i) minimize safety investments (x_1); and (ii) maximize hypothetical benefits (x_2). Both objectives are thus measured in financial terms (i.e., money). Second, a utility function $u(x_1, x_2)$ needs to be drawn up. Suppose that individual utility functions $u(x_1)$ and $u(x_2)$ have been drafted for the attributes x_1 and x_2. Now the organization may decide to employ a weighted average of these individual utility functions in the safety investment case:

$$u(x_1, x_2) = k_1 u(x_1) + k_2 u(x_2)$$

with the weights k_1 and k_2, and where $k_1 + k_2 = 1$. Of course, other multi-attribute utility forms besides this additive form may be used to compute the multi-attribute utility value.

Next, the exercise is carried out for the *safety investment strategy* question.

In step 1, the utility function for the prevention investments is drafted. Assume that the maximum budget is €600 000. This investment is given a utility of 0. The minimum budget is zero, which is given a utility of 1. Between these values, a linear utility curve is used (for instance), as the company behaves risk neutrally and the company's attitude to investments and risk within the interval [0; €600 000] is thus expressed by the expected value. Therefore, the safety investment strategy of €400 000 has a utility value of:

$$u(€400\ 000) = (600\ 000\ -\ 400\ 000)/600\ 000 = 0.333$$

In step 2, the utility function of attribute 2, the hypothetical benefits, needs to be established. In this case, how to assess or to give a value to the utility of avoided accidents with respect to the option of "not investing" can be reasoned as follows. The more accident costs that have been avoided via a safety investment, the more that accidents can cost when this safety investment has not been carried out. Hence, the "lower" the level of avoided accident costs, the "better" when calculating the utility of not investing. Therefore, the highest potential benefit, i.e., €2 000 000, is given a utility value of 0, while the lowest potential benefit, which is €0, is given a utility value of 1. The company decides to be risk-neutral, and therefore a linear utility function is used. Therefore, the utility of €1 000 000 is set to be 0.5.[1]

---

[1] Note that the company could have decided to adopt a risk-averse attitude. In this case, a risk-averse utility curve should be used. Going from zero to €1 000 000 of accident costs is then worse than going from €1 000 000 to €2 000 000 of costs. Therefore, €1 000 000 may have a utility value of, for example (depending on the exact risk-averse utility curve), $u(€1 000 000) = 0.4$.

In step 3, the constants k_1 and k_2 need to be determined. As $u(0; €2\,000\,000) = k_1$ and $u(€600\,000; 0) = k_2$, $k_1/k_2 = 600\,000/2\,000\,000 = 3/10$. Solving the following system of equations:

$$\begin{cases} k_1 + k_2 = 1 \\ \dfrac{k_1}{k_2} = \dfrac{3}{10} \end{cases}$$

gives $k_1 = 0.23$ and $k_2 = 0.77$.

In the following step, step 4, the expected utility value for the two options (i.e., "safety investment" and "no safety investment") can be determined. For the safety investment option, the expected utility value can be calculated as follows:

$$E_{\text{Invest}}(u) = k_1 u(x_1 = 400\,000€) + k_2 u(x_2 = 0) = 0.23 \times 0.33 + 0.77 \times 1 = 0.846$$

For the "no safety investment" option, the expected utility value becomes:

$$E_{\text{Not Invest}}(u)$$
$$= k_1 u(x_1 = 0)$$
$$+ k_2[u(x_2 = €2\,000\,000)0.016 + u(x_2 = €1\,000\,000)0.064 + u(x_2 = €0)0.92]$$
$$= k_1 \times 1 + k_2[0 \times 0.016 + 0.5 \times 0.064 + 1 \times 0.92] = 0.96$$

Hence, if the multi-attribute utility approach is used, it is recommended not to carry out this suggested safety investment strategy. To change this recommendation, the hypothetical benefits should be much higher. For instance, if the hypothetical benefits were €15\,000\,000 instead of €2\,000\,000, then the expected utility value in the case of investment would amount to 0.9742, and that of not investing would be 0.9538, and hence it would be recommended to invest in this safety portfolio/strategy.

In the multi-attribute utility approach, it is possible to employ more attributes according to a company's insights. Other factors may, for example, be the acceptability of risks, the acceptability of the litigation distribution among the victims of an accident, and so on. However, as Aven [1] also indicates, to be able to use the approach in a pragmatic way, these other factors will often not be taken into consideration by companies.

In the case where this approach is used, a utility function for every attribute should be developed by the company, and this may be a cumbersome exercise (see also Chapter 4). However, despite the difficulties associated with utility theory, the approach may be used in practice, as it is possible to standardize the needed utility functions to some extent and thus to reduce the work that has to be done in specific cases.

8.6 The Borda Algorithm Approach

The Borda algorithm is mainly used in voting problems [7, 8]. The Borda rule assigns linearly decreasing points to consecutive positions; for example, for three alternatives the points would be 3 for the first place, 2 for the second place, and 1 for the third place. The Borda algorithm can be found in the literature on group decision-making and social choice theory. Readers interested in applications of the algorithm are referred to [9–11]. The algorithm is employed to develop an ordinal ranking of preferences. The Borda rule can also be employed in a risk management context (e.g., [6, 12]). In the context of operational safety decision-making with respect to economics, the Borda algorithm can be employed to develop an ordinal ranking of safety investment options, thereby using several safety investment criteria [13].

In the operational safety investment context, the algorithm can, for example, work as follows. All safety investment options are ranked by a number of criteria. In the case of type I risks, criteria might be the absolute cost of safety (investment amount), the expected hypothetical benefit of safety (expected avoided accident cost), the cumulative probability of the accident scenarios avoided, the PBP of the safety investment, and the internal rate of investment. In the case of type II risks, criteria could be the cost of the safety investment, the maxmax hypothetical benefit, the variability related to the accident scenarios avoided, the information availability related to the safety investment, the equity principle, and the fairness principle. Let us consider type I risks and their safety investment. If there are n safety investment options to be compared, then the first-place option (e.g., according to the absolute cost of safety) receives $(n-1)$ points, the second-place option receives $(n-2)$ points, and so forth, until the last-place option, which receives 0 points. The same rule is used for assigning points according to the expected hypothetical benefit of safety, the cumulative probability, the PBP and the IRR. All the points obtained for the five criteria are summed for all installations and the option with the most points is ranked first, and so on.

Let us explain this concept for the case of four safety investment (SI) options: SI1, SI2, SI3, and SI4. Suppose that the rank-order positions are as follows:

- Absolute cost of safety (investment amount): SI3 > SI2 = SI1 > SI4
- Expected hypothetical benefit: SI1 > SI3 = SI2 = SI4
- Cumulative probability of accident scenarios avoided: SI4 = SI1 > SI2 = SI3
- Payback period: SI1 > SI3 > SI2 = SI4
- Internal rate of investment: SI2 > SI4 > SI1 > SI3.

When ties occur, for example, in the case of the absolute cost of safety SI2 and SI1 being tied, points allocated to these positions are derived from the average, i.e., SI2 and SI1 each will receive $\frac{[(n-2)+(n-3)]}{2}$. In the case of the expected hypothetical benefit, I3, I2, and I4 will each receive $\frac{[(n-2)+(n-3)+(n-4)]}{3}$.

The resulting point distribution is summarized in Table 8.9.

From Table 8.9, it can be concluded that SI1 has the highest Borda count and, therefore, ranks first and is the best safety investment according to the criteria used. The overall rank order of all four safety investment options employing the five criteria is as follows: SI1 > SI2 = SI3 > SI4.

Table 8.9   Ranking investment options for type I risks using the Borda algorithm for a four-option illustrative example

Criteria	Safety investment options			
	SI1	SI2	SI3	SI4
1. Absolute cost of safety	1.5	1.5	3	0
2. Expected hypothetical benefit	3	1	1	1
3. Cumulative probability of accident scenarios avoided	2.5	0.5	0.5	2.5
4. Payback period	3	0.5	2	0.5
5. Internal rate of return	1	3	0	2
Total Borda index for four safety investment options	11	6.5	6.5	6

SI, safety investment.

The sole concern of the developed approach is the investigation of a safety investment option's position relative to other safety investment options if one looks simultaneously at the five criteria for, in the case of the illustrative example, the type I risks. This ranking information may lead to optimizing the allocation of safety budget resources within an organization.

8.7 Bayesian Networks in Relation to Operational Safety Economics

8.7.1 Constructing a Bayesian Network

A Bayesian network (BN) is a network composed of nodes and arcs, where the nodes represent variables and the arcs represent causal or influential relationships between the variables. Hence, a BN can be viewed as an approach to obtain an overview of relationships between causes and effects of a system. In addition, each node/variable has an associated so-called conditional probability table (CPT). Thus, BNs can be defined as directed acyclic graphs, in which the nodes represent variables, arcs signify direct causal relationships between the linked nodes, and the so-called CPTs assigned to the nodes specify how strongly the linked nodes influence each other. The key feature of BNs is that they enable us to model and reason about uncertainty.

The nodes with no arcs directed into them are called "root nodes," and they are characterized by marginal (or unconditional) prior probabilities. All other nodes are intermediate nodes and each one is assigned a CPT. Among intermediate nodes, the nodes having arcs directed into them are called "child nodes" and the nodes having arcs directed away from them are called "parent nodes." Each child has an associated CPT, given all combinations of the states of its parent nodes. Nodes with no children are called "leaf nodes." There are two possible approaches to constructing a BN. The first approach develops a network topological structure to quantitatively determine the correlation and causal relationships between the nodes. The second approach builds a CPT to quantitatively determine the conditional probability of every node in the network.

The probabilities that have to be assigned to the different states of the variables can be determined in different ways. One way is to use historic data and statistical information (e.g., observed frequencies) to calculate the probabilities. Another way, especially when there is no or insufficient statistical information, is to use expert opinion. Hence, both probabilities based on objective data and subjective probabilities can be used in BNs.

Bayesian networks are based on Bayes' theorem and Bayes' theory to update initial beliefs or prior probabilities of events using data observed from the event studied. Bayes' theorem can be expressed using the well-known Bayes' formula:

$$P(A|B) = \frac{P(B|A)P(A)}{P(B)}.$$

In this equation, $P(A)$ is the prior probability of an event, $P(B|A)$ is the likelihood function of the event, $P(B)$ is the probability of information/data observed (commonly called evidence), and $P(A|B)$ is the posterior probability of the event.

Hence, a prior probability of an event is the starting point of knowledge, but the goal is to find out what the posterior probability of this event is, given the evidence "B." In the case where A is a variable with n states, according to the total probability formula, the following equation holds:

$$P(B) = \sum_{i=1}^{n} P(B|A_i)P(A_i).$$

Hence, Bayes' formula transforms into:

$$P(A_i|B) = \frac{P(B|A_i)P(A_i)}{\sum_{i=1}^{n} P(B|A_i)P(A_i)}.$$

For multivariable, multinode BNs, considering the independence and separation theorems of BNs, the joint probability $P(X_1, X_2, \ldots, X_n)$ can be expressed as the product of edge probabilities of each node:

$$P(X_1, X_2, \ldots, X_n) = \prod_{i=1}^{n} P(X_i|\text{parent}(X_i)),$$

where X_i is the i-th node of the BN, and parent(X_i) is the corresponding i-th parent node.

As Fenton and Neil [14] explain, BNs offer several important benefits as compared with other available techniques. In BNs, causal factors are explicitly modeled. In contrast, in regression models, historical data alone are used to produce equations relating dependent and independent variables. No expert judgment is used when insufficient information is available, and no causal explaining is carried out. Similarly, regression models cannot accommodate the impact of future changes. In short, classical statistics (e.g., regression models) are often good for describing the past, but poor for predicting the future. A BN will update the probability distributions for every unknown variable whenever an observation or piece of evidence is entered into any node. Such a technique of revised probability distributions for the cause nodes as well as the effect nodes is not possible in any other approach. Moreover, predictions are made with incomplete data. If no observation is entered then the model simply assumes the prior distribution. Another advantage is that all types of evidence can be used: objective data as well as subjective beliefs.

This range of benefits, together with the explicit quantification of uncertainty and ability to communicate arguments easily and effectively, makes BNs a powerful solution for all types of risk assessment, as well as assessments involving safety investment decisions.

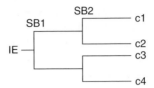

Figure 8.11   Event tree for an accident scenario.

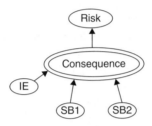

Figure 8.12   Bayesian network for the accident scenario.

8.7.2   Modeling a Bayesian Network to Analyze Safety Investment Decisions

As shown in the event tree of Figure 8.11, an initiating event (IE) could result in four consequences, c1, c2, c3 and c4, based on the function/failure (presence/absence) of two safety barriers, SB1 and SB2. The safety barriers SB1 and SB2 have a cost of €1000 and €4000, respectively. Also, the monetary values of losses, given c1, c2, c3, and c4, are equal to €1000, €10 000, €100 000, and €1 000 000, respectively. The occurrence probability of IE is equal to 0.01 while the failure probabilities of SB1 and SB2 have been estimated as 0.2 and 0.1, respectively. The decision-maker wants to decide: "Given a specific amount of budget and an acceptable (tolerable) level of risk, what would be the best (optimal) allocation of safety barriers? SB1, SB2, both, or none?" The technique of BNs can be employed to answer this question.

The event tree shown in Figure 8.11 can be mapped into a BN, as shown in Figure 8.12. The CPT of the node "consequence" is given in Table 8.10.

Figure 8.13 shows an extended version of the BN in Figure 8.12 in order to account for the effect of safety barriers on both the required budget (node "budget") and the amount of risk. The CPTs for nodes "SB1," "SB2," and "budget" are presented in Tables 8.11–8.13, respectively.

It should be noted that in Table 8.11, if SB1 is installed, it can work or fail (the second column in Table 8.11). However, if SB1 is not installed, the effect on the consequences would be the same as that of the failure of SB1. In other words, it has been assumed that the absence and failure of SB1 would result in the same consequences (the third column in Table 8.11); the same rationale has been used to assign the conditional probabilities for SB2 in Table 8.12.

Figure 8.13 shows the same BN as in Figure 8.12 in which the node "budget" has been given the value of to €5000, implying that both SB1 and SB2 can be afforded. As a result, the cost (required budget) would be €5000, while the value of the risk (according to the node "risk" in

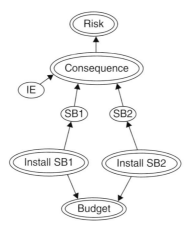

Figure 8.13 An extension of the Bayesian network in Figure 8.12 to account for the effect of safety barriers on the budget and risk.

Figure 8.14) would be:

$$\text{Risk} = 0.0072 \times €1000 + 0.0008 \times €10\,000 + 0.0018 \times €100\,000 + 0.0002$$
$$\times €1\,000\,000 = €395.2$$

Table 8.10 Conditional probability table of the node "consequence" in Figure 8.12

IE	SB1	SB2	Consequence
Yes	Work	Work	c_1
Yes	Work	Fail	c_2
Yes	Fail	Work	c_3
Yes	Fail	Fail	c_4
No	Work	Work	Safe
No	Work	Fail	Safe
No	Fail	Work	Safe
No	Fail	Fail	Safe

IE, initiating event; SB, safety barrier.

Table 8.11 Conditional probability table of node SB1 in Figure 8.13

Install SB1	Yes	No
$P(\text{SB1} = \text{work})$	0.8	0.0
$P(\text{SB1} = \text{fail})$	0.2	1.0

SB, safety barrier.

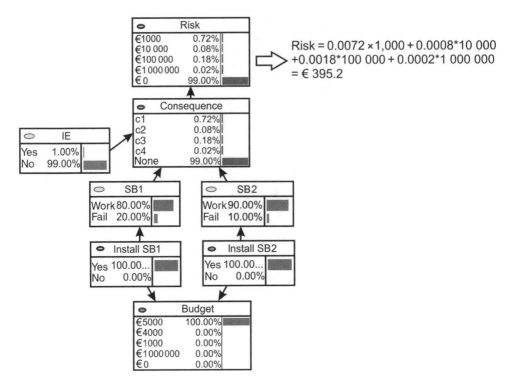

Figure 8.14   The same Bayesian network as in Figure 8.13 where both SB1 and SB2 have been afforded, i.e., "install SB1 = yes" and "install SB2 = yes." SB, safety barrier; IE, initiating event.

Table 8.12   Conditional probability table of node SB2 in Figure 8.13

Install SB2	Yes	No
$P(SB2 = work)$	0.9	0.0
$P(SB2 = fail)$	0.1	1.0

SB, safety barrier.

Table 8.13   Conditional probability table of node budget in Figure 8.13

Install SB1	Yes		No	
Install SB2	Yes	No	Yes	No
$P(budget = 5000)$	1	0	0	0
P(budget = 4000)	0	0	1	0
P(budget = 1000)	0	1	0	0
P(budget = 0)	0	0	0	1

SB, safety barrier.

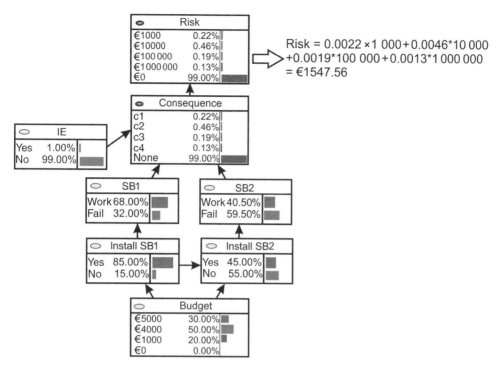

Figure 8.15 An extension of the Bayesian network in Figure 8.12 to account for budget variability. SB, safety barrier; IE, initiating event.

Similarly, the values of cost and risk can be calculated for different combinations of safety barriers, as shown in Table 8.14.

Given the constraints regarding the available budget (to purchase safety barriers) and the level of acceptable risk, multi-criteria decision-making can be employed to find the optimal allocation of safety barriers.

The BN of Figure 8.12 can alternatively be extended as shown in Figure 8.15 to incorporate the uncertainty in the available budget on the selection of safety barriers. Figure 8.15 shows a case in which the probability distribution of budget is $P(€5000, €4000, €1000) = (0.3, 0.5, 0.2)$.

Table 8.14 Cost-risk analysis of safety barriers for the accident scenario shown in Figure 8.11

Safety barrier	Cost (€)	Risk (€)
SB1	1000	2080
SB2	4000	1900
Both	5000	395.2
None	0	10 000

SB, safety barrier.

Also, it has been assumed that in case of a budget deficit, SB1 is given priority over SB2. The CPTs of SB1 and SB2 are given in Tables 8.15 and 8.16, respectively.

According to Figure 8.15, the expected value of the budget can be calculated as $E(B) = 0.3 \times$ €5000 + 0.5 × €4000 + 0.2 × €1000 = €3700. Moreover, the estimated value of the risk (with respect to the node "risk" in Figure 8.14) would be equal to Risk = 0.0022 × €1000 + 0.0019 × €10 000 + 0.0046 × €100 000 + 0.0013 × €1 000 000 = €1781.2. Following the same procedure, the values of risk for other distributions of the budget can be calculated. Similar to the previous approach, a multi-criteria decision-making technique can be used to determine the most cost-effective safety barrier to use.

8.8   Limited Memory Influence Diagram (LIMID) Approach

Limited memory influence diagram (LIMID) is an extension of BN by adding decision nodes (shown as rectangle nodes) and utility nodes (shown as diamond nodes). Other nodes, which represent random variables, are called chance nodes, and these are shown as oval nodes.

A decision node has a finite number of decision alternatives (in this example, install SB1, install SB2, install both, install none) that the decision-maker can take to achieve the desired outcome. The chance nodes whose probability distributions are important to decision-making should be the parents of the decision node, while the chance nodes whose probability distributions are affected by the decision node should be children of the decision node.

A utility node is a random variable whose values reflect the satisfaction the decision-maker gains from each decision alternative. Like other random variables, a utility node holds a table of utility values for all value configurations of its parent nodes. In a LIMID, let $A = \{a_1, a_2, \ldots, a_n\}$ be a set of mutually exclusive decision alternatives, H be the set of influential random variables, and $U(A, H)$ be the utility table of the utility node whose values are determined based on different configurations of A and H. Thus, the expected utility value of a decision

Table 8.15   Conditional probability table of "install SB1" in Figure 8.15

Budget (€)	5000	4000	1000	0
P(install SB1 = yes)	1.0	0.7	1.0	0.0
P(install SB1 = no)	0.0	0.3	0.0	1.0

SB, safety barrier.

Table 8.16   Conditional probability table of "install SB2" in Figure 8.15

Budget (€)	5000		4000		1000		0	
Install SB1	Yes	No	Yes	No	Yes	No	Yes	No
P(install SB2 = yes)	1	1	0	1	0	0	0	0
P(install SB2 = no)	0	0	1	0	1	1	1	1

SB, safety barrier.

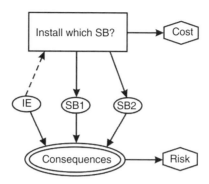

Figure 8.16   Limited memory influence diagram for cost-effective safety barrier selection.

alternative, e.g., a_1, can be calculated as:

$$EU(a_1) = \prod_H P(H|a_1)\, U(a_1, H)$$

As a result, the best decision alternative will be the one with the maximum expected utility. $U(a_1, H)$ are the entries of the utility table of the utility node U. The conditional probability $P(H|a_1)$ can be calculated from the CPT of H given the decision alternative a_1. The BN of Figure 8.12 can be extended as a LIMID by adding a decision node "Install which SB?" and two utility nodes "cost" and "risk," as shown in Figure 8.16.

The decision node includes four decision alternatives: (i) SB1, (ii) SB2, (iii) both, and (iv) none. As SB1 and SB2 are affected by the decision node, they are presented as the child nodes of the decision node. The illustrative CPTs of SB1 and SB2 are presented in Tables 8.17 and 8.18, respectively.

Table 8.17   Conditional probability table of node "SB1" in Figure 8.16

Install which SB?	SB1	SB2	Both	None
$P(SB1 = work)$	0.8	0.0	0.8	0.0
$P(SB1 = fail)$	0.2	1.0	0.2	1.0

SB, safety barrier.

Table 8.18   Conditional probability table of node "SB2" in Figure 8.16

Install which SB?	SB1	SB2	Both	None
$P(SB2 = work)$	0.0	0.9	0.9	0.0
$P(SB2 = fail)$	1.0	0.1	0.1	1.0

SB, safety barrier.

Similar to Tables 8.11 and 8.12, in Tables 8.17 and 8.18 whenever a safety barrier has been present (installed), probabilities have been assigned to its operation (work) and failure (fail). Otherwise, the absence and failure of a safety barrier have been assumed to have similar effects on the consequences. For example, in the second column of Table 8.18, as only SB1 has been installed, the absence of SB2 has been denoted as P (SB2 = work) = 0.0.

The utility tables for the utility nodes "cost" and "risk" in Figure 8.16 are shown in Tables 8.19 and 8.20, respectively.

Table 8.19 Utility table of "cost" in Figure 8.16

Install SB	Cost (€)
SB1	−1000
SB2	−4000
Both	−5000
None	0

SB, safety barrier.

Table 8.20 Utility table of "risk" in Figure 8.16

Consequence	Risk (€)
c1	−1000
c2	−10 000
c3	−100 000
c4	−1 000 000
None	0

c1–c4, consequences.

It should be noted that in Tables 8.19 and 8.20, the direct values of cost and risk (negative values) have been used as utility values in the utility tables of "cost" and "risk." However, in the case of having constraints such as maximum/minimum acceptable values for the cost and risk, appropriate utility functions should be used to convert the direct values to utility values. In this way, satisfaction of the decision-maker from each decision alternative can be better incorporated into decision-making.

For example, assume the maximum available budget and the maximum tolerable risk are €4500 and €200 000, respectively. Then, using the utility function $U_i = 1 - \frac{X_i - X_{max}}{X_{max}}$, the corresponding utility values for the cost of safety barriers and the risk of consequences can be calculated as shown in Tables 8.21 and 8.22.

Table 8.21 Utility values corresponding to the values of cost in Table 8.19

Cost (€)	Utility
5000	0.89
4000	1.11
1000	1.78
0	2.00

Table 8.22 Utility values corresponding to the values of risk in Table 8.20

Risk (€)	Utility
1000	1.99
10 000	1.95
100 000	1.50
1 000 000	−3.00
0	2.00

Using the utility values in Tables 8.21 and 8.22, the expected utility values are calculated for each decision alternative as shown in Table 8.23. As can be seen from Table 8.23, in the case of occurrence of IE, the decision alternative "SB1" has the highest expected utility, 2.74, implying that the optimal decision would be to install SB1 and not SB2. Obviously, in the case of non-occurrence of IE, the optimal decision is not to install any safety barriers, with the highest expected utility as 4.00.

Table 8.23 Expected utility values for each decision alternative

IE	Yes	No
Both	2.67	2.89
SB1	2.74	3.78
SB2	2.16	3.11
None	−1.00	4.00

IE, initiating event; SB, safety barrier.

Nevertheless, it should be noted that the results presented in Table 8.23 are highly dependent on the cost and risk constraints, as well as the type of the utility function used. Thus, the main challenge in the application of LIMID would be to find appropriate utility functions that best reflect the attitude of the decision-maker and also the available resources and limitations.

8.9   Monte Carlo Simulation for Operational Safety Economics

Regarding the application of BN to accident modeling and decision-making, discrete probability values are usually assigned to chance nodes (random variables). However, Monte Carlo simulation can be employed in conjunction with BN to sample chance nodes from continuous probability distributions so that the uncertainty can be more adequately captured in the analysis. For this purpose, the BN of interest can be modeled using a Markov chain (MC) framework such as WinBUGS. Alternatively, the non-parametric BN approach can be employed in which continuous probability distributions are assigned to random variables while dependencies are expressed in the form of rank correlations (instead of CPTs) (e.g., see Hanea [15]).

To illustrate the application of MC simulation, the BN shown in Figure 8.13 has been implemented in WinBUGS. For this purpose, the following assumptions have been made:

1. The occurrence of IE follows an exponential distribution: $P(\text{IE} = \text{yes}) = 1 - \exp(-\lambda t)$ where $t = 1$ year, and λ follows a lognormal distribution as $\lambda \sim$ lognormal (0.01, 0.001), as shown in Table 8.24.
2. Having SB1 installed, i.e., install SB1 = true, the failure probability of SB1 follows a beta distribution as $P(\text{SB1} = \text{fail}) \sim$ Beta (2, 8), as shown in Table 8.25. This beta distribution yields a mean value and variance of 0.2 and 0.015, respectively, for the failure of SB1 (compare with the single value of 0.2 for the failure probability of SB1 in previous sections).
3. Having SB2 installed, i.e., install SB2 = yes, the failure probability of SB2 follows a beta distribution as $P(\text{SB2} = \text{fail}) \sim$ Beta (0.5, 4.5), as shown in Table 8.26. This beta distribution yields a mean value and variance of 0.1 and 0.015, respectively, for the failure of SB2 (compare with the single value of 0.1 for the failure probability of SB2 in previous sections).

Table 8.24   Marginal probability distribution of initiating event (IE) in Figure 8.13

IE	Yes	No
Probability	$1 - \exp(-\lambda)$	$\exp(-\lambda)$

Table 8.25   Conditional probability distribution of SB1 in Figure 8.13

Install SB1	Yes	No
$P(\text{SB1} = \text{work})$	$1 - \text{Beta}(2, 8)$	0
$P(\text{SB1} = \text{fail})$	$\text{Beta}(2, 8)$	1

SB, safety barrier.

Table 8.26 Conditional probability distribution of SB2 in Figure 8.13

Install SB2	Yes	No
$P(SB2 = work)$	$1 - Beta(0.5, 4.5)$	0
$P(SB2 = fail)$	$Beta(0.5, 4.5)$	1

SB, safety barrier.

Figures 8.17–8.19 present the posterior distributions of the chance nodes IE, SB1, SB2, and consequence (risk) for the BN of Figure 8.14 with the above-mentioned probability distributions. In Figure 8.17, the states 1 and 2 refer to non-occurrence and occurrence of IE, respectively. In Figures 8.18(a) and (b), states 1 and 2 refer to the operation and failure of the safety barriers SB1 and SB2.

In Figure 8.19, the states 1, 2, 3, 4, and 5 refer to none, c4, c3, c2 and c1, respectively.

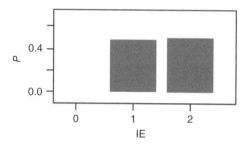

Figure 8.17 Posterior probability distribution of initiating event (IE) in Figure 8.14.

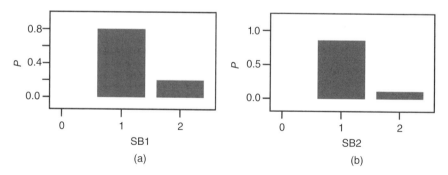

Figure 8.18 Posterior probability distributions of safety barriers (SBs) SB1 (a) and SB2 (b) in Figure 8.14.

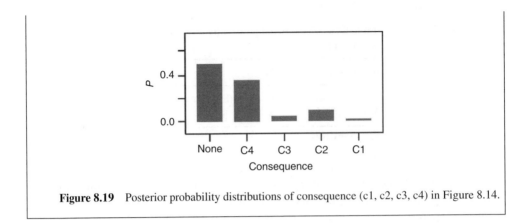

Figure 8.19   Posterior probability distributions of consequence (c1, c2, c3, c4) in Figure 8.14.

8.10   Multi-criteria Analysis (MCA) in Relation to Operational Safety Economics

Sometimes, both quantitative and qualitative indicators need to be evaluated by decision-makers in the safety investment allocation problem. A multi-criteria analysis (MCA) may allow a comparison of different alternatives presenting different scores for the criteria which are considered important by decision-makers. The main benefit of the MCA is to enable decision-makers to consider a number of different criteria, in a consistent way. The MCA can be used by decision makers for many purposes: ranking options, identifying the most suitable investment, or identifying a limited number of options for more detailed appraisal by more extensive economic analyses. The reader is referred to references [16–18] for more details about multi-criteria decision-making.

An MCA enabling the comparison of alternatives considering factors that are not necessarily measured in monetary terms can thus be used as a possible economic assessment approach. The benefit of MCA over other approaches as described in the previous sections is that the multi-criteria technique allows an evaluation of alternative investments based on a wide combination of quantitative and qualitative criteria.

A structured MCA methodology should take into account the following issues:

- A base case (the "do-nothing" scenario) should be used as a reference benchmark.
- All differences between the intervention under analysis and the base case are to be considered and somehow measured. These values should be appropriate, consistent, and possibly expressed in market prices and time but also using appropriate measurement scales.
- The impact of each investment should be assessed over a clearly defined time horizon depending on the duration of the assets or technologies involved.
- A sensitivity analysis should be performed by decision-makers to understand whether certain criteria have a higher impact on the outcomes.

The indicators considered by decision-makers should be somehow significantly correlated to the goals to be achieved, capturing important features of the safety investment that could have

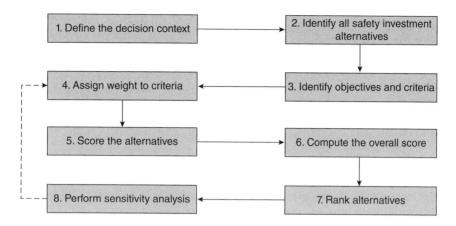

Figure 8.20   Multi-criteria analysis scheme.

an influence on the decision to be made. Therefore, establishing the most appropriate evaluation criteria within an MCA represents the foundation of a more general economic assessment. A structured MCA method should be characterized by the steps shown in Figure 8.20.

The first step involves the definition of the final goals and the decision-makers involved in the decision process. Alternative investments should be identified and related to a "non-investment" base case. Subsequently, the relevant criteria used during the evaluation should be defined. These criteria should reflect the values associated with the impact of every alternative/option. A weight can be associated with every criterion to reflect the relative importance of one criterion over the others according to the decision-makers. Once this framework has been set, the evaluation of the alternatives can be performed by assigning values (scores) to each alternative, for each of these criteria based on a predefined scale. Weights and scores are finally combined to assign a total weighted score for each investment. In so doing, the alternative investment can easily be compared using the same scale. Finally, a sensitivity analysis can be performed to assess whether a change in score or weight, associated with specific criteria, would significantly affect the final rank.

The MCA approach used in the next section is the most common "linear additive model," which has its benefit in its simplicity and hence user-friendliness. An application of this MCA approach to compare alternative investments is shown hereafter with an illustrative example.

To illustrate the MCA approach, we consider a public railway transport organization that has to evaluate complex safety investments aimed at preventing and mitigating railway disasters. The goal is to evaluate alternative investments on train control systems aimed at increasing the overall level of safety in the whole railway network. The level of safety is measured considering the number of trains that are traveling on the railway network, the average number of passengers on each train, the nature of the materials being transported (in the case of freight transport), the environment, and the possibility of the population living in the vicinity of each railway line being affected by a train accident. Given the current railway network and its related level of safety, the firm received a mandate by the government to improve the level of safety by 20% in a time horizon of 5 years, and to do so without exceeding a predefined safety budget.

Some improved safety technologies can be installed both on trains and throughout the network infrastructure to achieve this goal. However, each safety technology presents different features, such as installation costs, maintenance costs, effectiveness (expressed as the ability to improve the current safety level), compatibilities with other technologies and systems, and duration of installation.

The base "no investment" case is represented by the current state of the overall railway network. This latter is composed of railway lines, with their own safety systems, and rolling stock with on-board safety software that is compatible with the safety technology installed on the railway lines on which such trains normally operate. Due to the complexity of the project, a team of experts from different departments has defined eight alternative investments, each one allowing an increase in the overall level of safety using a different approach. In fact, these options present different levels of expenditure and require different changes to be made compared with the current network system. Moreover, each safety investment option has a different timing structure as well as implementation difficulties and criticalities. As mentioned, the assessment of the alternative investment options is carried out based on an MCA. Therefore, each of these potential safety investments has been assessed using the main criteria defined by a team of decision-makers.

The main features of an improved integrated railway network (made by network infrastructure and rolling stock) are essentially related to investment costs, yearly recurrent costs, safety level, compatibility with other ongoing modernization projects, railway capacity requirements, project duration, and implementation risks. As these features represent critical issues that need to be assessed by decision-makers, they can be treated as criteria to be used to evaluate each investment option. Fourteen criteria have been identified that correspond to the set of relevant objectives to be pursued by decision-makers. Therefore, their evaluation will facilitate the comparison of the alternative investment options. The list of criteria that have been selected by decision-makers as well as a description, the evaluation metric, and the weights are summarized in Table 8.27.

As none of these criteria can be properly quantified using monetary or other quantitative measures, each alternative project is compared with the base "non-investment" scenario, and marginal expected differences are quantified using a standard scale based on the judgment of a team of experts. This score is useful to measure relative differences between alternative investment options and to rank each alternative. For each of these criteria, the performances of each alternative investment are scored using a Likert scale (except for criterion 1) with values from 1 to 5. A score of 1 represents a low (generalized) benefit or high (generalized) cost, and a score of 5 represents a high (generalized) benefit or low (generalized) cost, depending on each criterion. Furthermore, a weight is associated with every criterion, reflecting the relative importance of each criterion and making sure that the weights sums to 100%. Criterion number 1 has been excluded as it represents a showstopper that might spoil the whole investment if it is not compliant with transport legislation.

Given the list of weights w_i and metrics (with a min and max score s_i) for each criterion i, summarized in Table 8.27, the total weighted score ($\Sigma_{i=1}^{13} w_i s_i$) associated with a safety investment option should lie in the interval [100; 500]. More specifically, the total weighted score associated with a project is 100 if the option scores a value of 1 for all criteria. At the other extreme, the maximum total weighted score for an investment assumes a value of 500 if the project scores 5 for all criteria.

Table 8.27 Description of the criteria used in the multi-criteria analysis

Id	Criterion	Description	Metric	Weight (%)
1	Transport regulations	The investment option is compliant with all the transport regulations. This criterion represents a clear showstopper	Yes/no	–
2	Sustainability	The investment is feasible and sustainable from a technical, financial, and economical point of view. Moreover, its impact on the environment is limited and within the boundaries defined by law. The higher the score, the more sustainable is the option	Score (1–5)	10
3	Duration	The investment can be completed by the due date (project duration 5 years). A high score for this criterion means that the project can be completed even before the deadline	Score (1–5)	5
4	Initial investment costs	The expected capital investment required by the alternative investment. Assuming that the total investment cost does not exceed the initial safety budget, the higher the score, the lower the economic burden associated with the investment	Score (1–5)	15
5	Yearly recurring costs	The expected recurring costs due, for example, to maintenance costs and energy consumption required by each investment after its competition. The higher the score, the lower the capital required by the investment on a yearly basis	Score (1–5)	10
6	Safety level	The improvement level that can be achieved by the investment due to the installation of enhanced safety technologies on both rolling stock and network infrastructure. Assuming that all alternatives guarantee the minimal safety increase threshold of 20%, the higher the score, the higher the expected safety improvement that can be achieved	Score (1–5)	20
7	Capacity improvement	The total number of trains that can travel on the railway network after the implementation of the investment compared with the current values. The higher the score, the larger the improvement guaranteed by the investment	Score (1–5)	5

Table 8.27   (*continued*)

Id	Criterion	Description	Metric	Weight (%)
8	Compatibility with existing projects	The investment does not present any incompatibility with existing investment projects. The higher the score, the lower the level of incompatibilities with ongoing projects	Score (1–5)	4
9	Availability and reliability	The robustness of the comprehensive railway network after the modernization investment. The higher the score, the better the investment	Score (1–5)	4
10	Risk to delivery	The likelihood that the project will miss the expected due date. The higher the score, the lower the risk of project delays	Score (1–5)	7
11	Implementation difficulties	Level of criticalities and foreseeable problems that might emerge during the implementation phase. The higher the score, the higher the level of confidence for a smooth implementation	Score (1–5)	2
12	Lines impacted	Amount of network lines affected by the investment projects due to modernization and/or installation of new technologies. The higher the score, the lower the extent of network infrastructure affected by the investment	Score (1–5)	6
13	Rolling stock impacted	Number of trains affected by the investment project due to modernization and/or installation of new technologies. The higher the score, the lower the amount of rolling stock affected	Score (1–5)	7
14	External risk	Risk related to the trustworthiness and/or other criticalities that might arise with suppliers, partners, and/or third parties during the implementation of the investment. The higher the score, the lower the expected external risks the investment will face during the implementation	Score (1–5)	5

After having identified the main criteria that generalize the concepts of costs, benefits and risk, every alternative investment is assessed. For each alternative, a team of experts assigned a score based on the impact of a given investment alternative on each of the criteria. In Table 8.28 the score associated with each safety investment is shown. As can be seen, investment option H does not fit the requirement to be compliant with legislation on railway transport and for this reason it has been disregarded. The remaining seven investments (A to G) have been subjected to the next steps of the MCA methodology.

The outcome of the MCA is also shown in Table 8.28 where, based on the total weighted score, all safety investments are ranked. Based on the total weighted score, alternative investment C appears to be the best alternative. It scores 441 out of a maximum of 500. In the light of this result, the recommendation that the decision-makers can provide is that investment C deserves further investigation, as it provides significant additional benefits compared with the other investment options. Investment B ranks second. Compared with investment B, alternative C presents slightly worse initial investments (although lower than the maximum safety budget) but lower maintenance costs and a higher safety level. Moreover, alternative C is fully compatible with other projects. On the other hand, alternative B impacts a lower number of lines and rolling stock and can be concluded a bit earlier than alternative C.

After having ranked the alternative safety investments, decision-makers can shortlist a restricted number of investments with the highest score (e.g., options B, C, and D) and carry out more detailed analysis involving other decision-makers or experts. As also highlighted in

Table 8.28 Evaluation of the alternative investments

Criteria	Weight (%)	Alternative investment options							
		A	B	C	D	E	F	G	H
Transport regulations	—	Yes	Yes	Yes	Yes	Yes	Yes	Yes	No
Sustainability	10	1	3	5	4	2	3	2	
Duration	5	2	5	4	3	2	4	2	
Initial investment costs	15	5	5	4	3	2	3	5	
Yearly recurring costs	10	4	4	5	5	3	5	4	
Safety level	20	4	4	5	4	4	3	4	
Capacity improvement	5	2	3	3	4	5	3	2	
Compatibility with existing projects	4	2	3	5	4	5	4	3	
Availability and reliability	4	5	5	4	4	4	2	4	
Risk to delivery	7	1	4	5	5	5	5	5	
Implementation difficulties	2	5	3	4	4	4	3	4	
Lines impacted	6	3	5	4	4	3	1	5	
Rolling stock impacted	7	3	5	4	4	4	5	4	
External risk	5	5	4	3	3	4	4	5	
Total weighted score	100	334	416	441	392	340	346	389	
Rank		5	2	1	3	7	6	4	

Figure 8.20, in step 8, decision-makers can also conduct a sensitivity analysis to assess the validity of the solution with the highest score in the case of changes in the scores/weights. In this case, the total weighted score is updated and the alternative investments are ranked again. Once the alternative investment project is selected, the implementation phase can start.

8.11 Game Theory Considerations in Relation to Operational Safety Economics

8.11.1 An Introduction to Game Theory

People make decisions all the time. In an organizational setting, these decisions typically are interdependent with decisions made by others. Hence, there is a kind of multi-personal interaction. A central feature of multi-personal interaction is the potential for the presence of strategic interdependence, meaning that a person's well-being (e.g., measured as utility or profit, or as non-accident) depends not only on their own actions but also on the actions of other individuals involved into the situation with strategic interdependence. The actions that are best for the person may depend on actions that other individuals have already taken, or expected to be taken at the same time, and even on the future actions other individuals may take, or decide not to take, as a result of the current actions.

The mathematical tool that is used for analyzing interactions with strategic interdependence is non-cooperative game theory. The term "game" highlights the theory's central feature: the persons (also called "players," "decision-makers," or "agents," as persons can also represent entire organizations or nations, for example) under study are concerned with strategy and winning (in the general sense of expected utility maximization). The agent will have some control over the situation, as their choice of strategy will influence it in some way. However, the outcome of the game is not determined by their choice alone, but also depends on the choices of all other players. This is where conflict and cooperation come into the play. It should be emphasized that the term "non-cooperative game theory" does not mean that non-cooperative theory is unable to explain cooperation within groups of individuals. Rather, it focuses on how cooperation may emerge from rational objectives of the players, given that rationality is a common knowledge, and in the absence of any possibility of making a binding agreement.

It is useful to notice that in the decision-making theory and, consequently, in game theory, individual rationality should not be associated with dispassionate reasoning. It is assumed that while making an optimal decision, an individual will be able to observe the courses of action that are available and determine the outcomes he or she can receive as a result of the interaction with others. An individual should be able to rank those outcomes in terms of preferences, so that the system of preference is internally consistent and can form a basis for choice. Thus a rational player is defined as one who has consistent preferences concerning outcomes and will attempt to achieve preferred outcomes. Rational play will involve complicated individual decisions about how to choose a strategy that will produce a favorable outcome, knowing that other players are trying to choose strategies that will also produce an outcome favorable to them. It will also involve social decisions about how and with whom a player should try to cooperate.

Game theory is a logical analysis of situations of conflict and cooperation. More specifically, a game is defined to be any situation in which:

- There are at least two players. A player may be an individual, but may also be a more general entity, such as a company, an institution, a country, and so on.
- Each player has a number of possible strategies, courses of action that the player may choose to follow.
- The strategies chosen by each player determine the outcome of the game.
- Associated with each possible outcome of the game is a collection of numerical payoffs, one to each player. These payoffs present the value of the outcome to the different players.

Game theory can generally be divided into non-cooperative game theory and cooperative game theory. Furthermore, a distinction can be made between "static games" and "sequential move games." A static game is one in which a single decision is made by each player, and each player has no knowledge of the decision made by the other players before making their own decision. Sometimes such games are referred to as "simultaneous decision games," because any actual order in which the decisions are taken is irrelevant. A sequential game is a game where one player chooses their action before the others choose theirs. Importantly, the later players must have some information of the first player's choice, otherwise the difference in time would have no strategic effect. Another distinction that can be made between the game-theoretic models concerns the information available to the players: games of perfect information should be solved differently from games of imperfect information. In many economically important situations, the game may begin with some player disposing of private information about a relevant issue with respect to their decision-making. These are called "games of incomplete information" or "Bayesian games." Incomplete information should not to be confused with imperfect information in which players do not perfectly observe the actions of other players. Although any given player does not know the private information of an opponent, he or she will have some beliefs about what the opponent knows, and the assumption that these beliefs are common knowledge can be made. Traditional applications of game theory attempt to find equilibria in these games. In an equilibrium, each player of the game has adopted a strategy that they are unlikely to change. Many equilibrium concepts have been developed (in particular, the Nash equilibrium) in an attempt to capture this idea. These equilibrium concepts are motivated differently depending on the field of application, although they often overlap or coincide.

It should be noted that this methodology is not without criticism, and debates continue concerning the appropriateness of particular equilibrium concepts, the appropriateness of equilibria altogether, and the usefulness of mathematical models more generally. For instance, game theory should not be considered as a theory that will prescribe the best course of actions in any situation of conflict and cooperation. First of all, the real-world situations are performed in a quite simplified form of a game. In real life it may be hard to say who the players are, to delineate all conceivable strategies and to specify all possible outcomes, and it is not easy to assign payoffs. What is typically done is to develop a simple model which incorporates some important features of the real situation. Thus building such a model and its analysis may give insights into the original situation. The second obstacle is that game theory deals with play that is rational. Each player logically analyses the best way to achieve their goals, given that the other players are logically analyzing the best way to achieve their own goals. In this way, rational play assumes rational opponents. It means that players are able to tell which outcomes are preferable to others and that they are able to align them in some kind of preference relationship order. However, experiments show that in some cases rationality may fail. Game theory results should thus always be interpreted with caution. For example, game theory does not give a unique prescription for games with more than two players. What game theory offers is a variety of interesting examples, analysis, suggestions, and partial prescriptions for certain situations. Practitioners using game theory should be aware of this.

In summary, a game is a situation where, for two or more individuals, their choice of action or behavior has an impact on others. The outcome of the game is expressed in terms of the strategy combinations that are most likely to achieve the players' goals given the information available to them. Games are often characterized by the way or order in which players move. Games in which players move at the same time or in which the players' moves are hidden from others until the end of the game, are called "simultaneous-move" games or "static" games. Games in which moves are made in some kind of predetermined order are referred to as "sequential move" games or "dynamic" games.

Furthermore, a distinction is made between zero-sum games and non-zero-sum games. In the former case, the payoffs for each outcome add to zero, and the interests of the players are strictly opposed. In the latter case, there is no strict conflict and the payoffs do not add up to zero.

8.11.2   The Prisoner's Dilemma Game

One classic game theory example, called *Prisoner's Dilemma*, is particularly interesting. The Prisoner's Dilemma game is a two-person non-cooperative non-zero-sum game. The Prisoner's Dilemma, as applied to this book, i.e., economics in combination with operational safety in an organization, can be understood as follows. Assume two workers in an organization, having to decide whether to focus on safety *and* production, or production *over* safety. If they both focused on production over safety, they would experience a type I accident once in a while, and their payoff function would be two payoff units for both. If they were both to decide to place emphasis on safety as well as production, no accidents would occur, there would still be a high production rate, and both their payoffs would end up in 3 units. If one of the workers (worker 1) decided to focus on production over safety, and the other worker (worker 2) decided to focus on both safety and production, there would still be an occasional (type I) accident in the organization (due to worker 1, not necessarily affecting worker 1). However, due to the fact that production figures are, for example, deemed more important than safety figures in this particular organization (many organizations still stress production over safety), worker 1 would be more rewarded and receive 4 payoff units, compared with worker 2, who would only receive 1 payoff unit.

The game can be described in a tabular form where the possible courses of action open to each worker are: (i) S = "focus on production and safety"; and (ii) P = "focus on production over safety." The payoffs are given in terms of payoff units (e.g., monetary values). The payoffs for the first worker (player 1) are given first in each pair of entries in the table; those for the other worker (player 2) come second.

Game theoretic analysis suggests the following behavior for each worker in the case of Figure 8.21. First, consider player 1. Player 1 knows that, if player 2 decides to play S as a strategy, then player 1 will do better if he plays P, because that leads to 4 units rather than 3 units. On the other hand, if player 2 decides to play P, then player 1 should also play P, because that leads to 2 units rather than 1. So whatever player 2 does, with the payoffs as displayed in Figure 8.21, player 1 is better off playing P and thus putting production first. Similar reasoning can be applied to player 2. So logically both workers should play P rather than S.

Worker 2
(= Player 2)

		S	P
Worker 1 (= Player 1)	S	(+3, +3)	(+1, +4)
	P	(+4, +1)	(+2, +2)

Figure 8.21   Payoff matrix for the Prisoner's Dilemma game – an illustrative example.

The interest in this game arises from the following observation. Both players, by following their individual self-interest, end up worse off than if they had not followed their self-interest, and followed the organizations' interest instead. One could argue that the workers should follow company rules, which place the emphasis on both production and safety, no matter what. However, when each worker has no absolute assurance that the other will follow this rule, the equilibrium outcome would be (P, P).

The conditions for the game Nash equilibria and Pareto optimal solution result in the following conditions for players' payoffs in the case of a Prisoner's Dilemma game (see Figure 8.22):

$$e \geq a \geq g \geq c,$$

$$d \geq b \geq h \geq f.$$

Player 2

		S	P
	S	(a, b)	(c, d)
Player 1	P	(e, f)	(g, h)

Figure 8.22   Payoff matrix for the Prisoner's Dilemma game – general case.

An organization wishing to avoid this rather unpleasant sub-optimal equilibrium outcome has to intervene in the payoffs of the workers (not necessarily expressed in monetary terms; they can also be expressed in utils) in such a way that these conditions are not met.

8.11.3   The Prisoner's Dilemma Game Involving Many Players

The Prisoner's Dilemma can also be played with more than two players simultaneously. This game is known as the "Tragedy of the Commons." With the help of an illustrative example, using the concept of hypothetical benefits (see earlier), the effect on safety when this game is played in an organizational context can be explained. Assume that four workers (e.g., in a shift) in one single organization receive 10 hypothetical benefit units each (i.e., the company looks after their safety by making safety investments), and, of course, that they themselves have the choice between placing emphasis on safety and production and focusing on production over safety. If they focus on safety and production, there is a pot of joint hypothetical benefit which is even enforced thanks to their personal commitment, and the joint pot is multiplied by 1.5 – this hypothetical benefit profit stems from not having any accidents due to the safety focus of the workers. The total hypothetical benefit can then be divided evenly among all the workers, as the workplace safety is uniform across the company. Obviously, if all four workers decided to pool their hypothetical benefit units and all placed the focus on safety and production, the pool would contain 40 units. Multiplied by 1.5 this leads to 60 units being divided among the four workers. Hence, each worker ends up with 15 hypothetical benefit units. Full cooperation and focus on safety as well as production thus lead to a hypothetical benefit profit of 5 units for every worker.

However, there is a catch. Assume that one worker refuses to focus on safety and production at the same time, and places emphasis instead on production over safety and thus does not put any hypothetical benefit unit in the joint pot? In other words, one worker relies only on his 10 company-given hypothetical benefit units without sharing them. If the other three workers still focused on safety as well as production, the joint hypothetical benefit pot would contain 30 units at this point. Hence, 30×1.5 delivers 45 units to be divided among four workers (as the fourth worker also benefits from the uniform workplace safety created by the other three), thus giving rise to 11.25 hypothetical benefit units per worker. In this situation, while three of the workers have made a profit of only 1.25 units, one worker has made a profit of 11.25 units, and is as safe as the others while putting the focus on production over safety. There is even an extra incentive to deviate from cooperation for the non-sharing worker: he not only receives safety from the company and from his co-workers for free, but he himself focuses on production, and therefore if his production figures are better than the others, he will be rewarded for them too. Of course, such a process is difficult to see and/or to quantify or prove, especially as hypothetical benefits are invisible. Nevertheless, from a theoretical perspective, the Tragedy of the Commons is easy to understand and clearly occurs in real industrial settings.

Most people, also workers, are of course initially optimistic. If a hypothetical benefit investment of 10 units is made, this would lead to a profit of 5 units – that is, if every worker does the same. But clearly there might always be one worker who holds back on investing (does not put the focus on safety) and takes advantage of the others' generosity – a so-called "free rider." When the return on investment is less than 15 hypothetical benefit units, if they could calculate such hypothetical benefits, the workers would realize that someone had not invested his or her fair share. Of course, in that case, other workers would be tempted to hold back as well. In summary, there is a strong incentive to let everyone else focus on safety so that one worker, by placing his or her focus on production and gaining by doing so, and being relatively safe due to the company's investments and other workers' safety focus, can reap the benefits of both focus on production and not having many accidents. If everyone were to think this same way, nobody would focus adequately on safety, and nobody would make the extra hypothetical benefits that can be made by such a safety focus. The rational course of this game is thus not to put too much focus on safety in the company. This is the Tragedy of the Commons for industrial safety.

There is some good news, however. Based on game theoretical experiments surrounding climate change, the Tragedy of the Commons game can indeed be influenced in the right direction, i.e., having the workers cooperate and focus on safety and production, instead of acting only out of self-interest [19]. The first ingredient of cooperation is information. Workers need correct information as regards safety aspects of decisions and there needs to be transparency – as much as possible. The second ingredient of cooperation appears to be reputation. The effect of reputation is surprisingly strong. People really do like to be seen doing the right thing. In industrial safety terminology, this boils down to social control and management by walking around having a positive effect.

8.12   Proving the Usefulness of a Disproportion Factor (DF) for Type II Risks: an Illustrative (Toy) Problem

In this section an illustrative (toy) problem is discussed by way of the three decision theories considered in this book: expected outcome theory, expected utility theory, and Bayesian decision theory. First, the analytical solution of the investment willingness, or, equivalently, hypothetical benefits, is determined. Then some numerical values will be inserted into these analytical expressions and the resulting numerical solutions will be discussed.

8.12.1   The Problem of Choice

The Bayesian framework is now applied to a problem of choice in which a decision-maker must decide how much he or she is willing to invest in order to reduce the probability of a type II risk event occurring. The two decisions under consideration in this simple scenario are:

D_1 = keep the status quo,
D_2 = improve barrier for type II event.

The possible outcomes in the risk scenario remain the same under either decision, and therefore are not dependent upon the particular decision taken. These outcomes are:

O_1 = catastrophic type II event occurs,
O_2 = no type II event.

The hypothetical damages associated with these outcomes are:

$$O_1 = -x \text{ euros,}$$

$$O_2 = 0 \text{ euros,} \tag{8.4}$$

and the investment costs associated with the additional improvement of the type II event barriers are expressed by the parameter:

$$I = \text{investment costs.} \tag{8.5}$$

The decision on whether to improve the type II event barriers or not is of influence in the probabilities of the respective outcomes. Under the decision to make no additional investments in the type II event barriers and keep the status quo, D_1, the probabilities of the outcomes will be, say:

$$P(O_1|D_1) = \theta,$$
$$P(O_2|D_1) = 1 - \theta. \tag{8.6}$$

Under the decision to improve the type II event barriers, D_2, the probability of the catastrophic type II event will be decreased by, say:

$$P(O_1|D_2) = \phi,$$
$$P(O_2|D_2) = 1 - \phi, \tag{8.7}$$

where $\phi < \theta$. Stated differently, the proposed barrier improvements will decrease the chances of the catastrophic type II event by a factor of $c = \theta/\phi$.

In what follows, the solution of this problem of choice will be given for expected outcome theory, expected utility theory, and Bayesian decision theory. These solutions will be given in terms of variables x, θ, and ϕ, respectively (Eqs. 8.4, 8.6 and 8.7).

8.12.2   The Expected Outcome Theory Solution

The prosperous merchants in seventeenth-century Amsterdam bought and sold expectations as if they were tangible goods. It seemed obvious to many that a person acting out of pure self-interest should always behave so as to maximize his expected profit [20]:

$$E(X) = \sum_{i=1}^{n} p_i x_i. \tag{8.8}$$

Combining Eqs. (8.4)–(8.7), one may construct the outcome probability distributions under the decisions D_1 and D_2:

$$p(O_i|D_1) = \begin{cases} \theta, & O_1 = -x, \\ 1 - \theta, & O_2 = 0, \end{cases} \tag{8.9}$$

and

$$p(O_j|I, D_2) = \begin{cases} \phi, & O_1 = -x - I, \\ 1 - \phi, & O_2 = -I, \end{cases} \tag{8.10}$$

where in Eq. (8.10) one may explicitly conditionalize on the investment parameter I, which is to be to estimated. The expected outcomes of these probability distributions are, respectively [21]:

$$E(O|D_1) = -\theta x \tag{8.11}$$

and

$$E(O|I, D_2) = -\phi x - I. \tag{8.12}$$

The decision theoretical equality

$$E(O|D_1) = E(O|I, D_2) \tag{8.13}$$

represents the equilibrium situation, where it will be undecided whether to keep the status quo, D_1, or invest in additional barrier improvements, D_2. Now, if one solves for I in Eq. (8.13), by way of Eqs. (8.11) and (8.12):

$$I = (\theta - \phi)x, \tag{8.14}$$

then we find that investment where one will remain undecided.

Stated differently, any investment cost smaller than Eq. (8.14) will turn Eq. (8.13) into an inequality, where D_2 becomes more attractive than D_1. It follows that the equilibrium investment (Eq. 8.14) is also the maximal investment one will be willing to make in order to improve the type II event barriers, or, equivalently, Eq. (8.14) is the hypothetical benefit of the type II event barrier improvement.

8.12.3 The Expected Utility Solution

For a rich man €100 is an insignificant amount of money. So, the prospect of gaining or losing €100 will fail to move the rich man, i.e., for him an increment of €100 has a utility that tends to zero. For the poor man €100 is a significant amount of money. So the prospect of gaining or losing €100 will most likely move the poor man to action. It follows that for him an increment of €100 has a utility significantly greater than zero.

In 1738 Daniel Bernoulli derived the utility function for the subjective value of objective money by way of a variance argument, in which he considered the subjective effect of a given fixed monetary increment, c, for two persons of different initial wealth. Based on this variance argument, he derived the utility function of going from an initial asset position x to the asset position $x + c$:

$$u(x, x + c) = q \log \frac{x + c}{x}, \tag{8.15}$$

where q is some scaling constant greater than zero [22].

In expected utility theory, the expected values of the utility probability distributions are maximized. Assuming that the decision-maker has a total wealth (i.e., an actual income and asset portfolio) of:

$$M = m \quad \text{euros}, \tag{8.16}$$

then, using Eq. (8.15), or, equivalently,

$$U_i = q \log \frac{M + O_i}{M}, \tag{8.17}$$

one may construct from Eqs. (8.9) and (8.10) the corresponding utility probability distributions as:

$$p(U_i | D_1) = \begin{cases} \theta, & U_1 = q \log \dfrac{m - x}{m}, \\ 1 - \theta, & U_2 = 0, \end{cases} \tag{8.18}$$

and

$$p(U_j | I, D_2) = \begin{cases} \phi, & U_1 = q \log \dfrac{m - x - I}{m}, \\ 1 - \phi, & U_2 = q \log \dfrac{m - I}{m}. \end{cases} \tag{8.19}$$

The expected outcomes of the utility probability distributions are, respectively [21]:

$$E(U|D_1) = q\left(\theta \log \frac{m-x}{m}\right) \tag{8.20}$$

and

$$E(U|I,D_2) = q\left(\phi \log \frac{m-x-I}{m-I} + \log \frac{m-I}{m}\right). \tag{8.21}$$

The decision theoretical equality:

$$E(U|D_1) = E(U|I,D_2) \tag{8.22}$$

represents the equilibrium situation, between the decision to keep the status quo D_1 and the decision to invest in additional barriers D_2. Now, if one substitutes Eqs. (8.20) and (8.21) into Eq. (8.22), then one obtains the closed expression for that investment value where one is indifferent between either decision:

$$\log \frac{m-I}{m} = \theta \log \frac{m-x}{m} - \phi \log \frac{m-x-I}{m-I}. \tag{8.23}$$

Any investment cost smaller than the numerical solution of I in Eq. (8.23) will turn Eq. (8.22) into an inequality, where D_2 becomes more attractive than D_1. It follows that the equilibrium investment (Eq. 8.23) is also the maximal investment one will be willing to make to improve the type II event barriers, or, equivalently, Eq. (8.23) is the hypothetical benefit of the type II event barrier improvement.

8.12.4   The Bayesian Decision Theory Solution

In Bayesian decision theory the scaled sum of the confidence bounds and the expectation value of the utility probability distributions is maximized as the risk measure that captures the position of the underlying utility probability distribution (see Section 7.2.2):

$$R(U|k,D_j) = \frac{LB(k|D_j) + E(U|D_j) + UB(k|D_j)}{3}, \tag{8.24}$$

where the lower confidence bound is corrected for undershooting the worst possible outcome a, (Section 7.2.1.1):

$$LB(U|D_j) = \begin{cases} a, & E(U|D_j) - k\,\text{std}(U|D_j) < a, \\ E(U|D_j) - k\,\text{std}(U|D_j), & E(U|D_j) - k\,\text{std}(U|D_j) \geq a, \end{cases} \tag{8.25}$$

and the upper confidence bound is corrected for overshooting the best possible outcome b, (Section 7.2.1.1):

$$UB(U|D_j) = \begin{cases} E(U|D_j) + k\,\text{std}(U|D_j), & E(U|D_j) + k\,\text{std}(U|D_j) \leq b, \\ b, & E(U|D_j) - k\,\text{std}(U|D_j) > b. \end{cases} \tag{8.26}$$

Substituting Eqs. (8.25) and (8.26) into Eq. (8.24), one obtains the risk index:

$$R(U|k, D_j) = \begin{cases} E(U|D_j), & \text{neither overshoot nor undershoot,} \\[2mm] \dfrac{a + 2E(U|D_j) + k\,\text{std}(U|D_j)}{3}, & \text{undershoot and no overshoot,} \\[3mm] \dfrac{2E(U|D_j) - k\,\text{std}(U|D_j) + b}{3}, & \text{overshoot and no undershoot,} \\[3mm] \dfrac{a + E(U|D_j) + b}{3}, & \text{both overshoot and undershoot,} \end{cases} \tag{8.27}$$

where it is noted that the first row of Eq. (8.27) corresponds with the expected utility theory criterion of choice [23], and the fourth row is a kind of adjusted Hurwitz criterion of choice, which may differentiate two probability distributions that have the same minimal and maximal values while at the same time having an opposite skewness.

In the illustrative toy problem under consideration, a simple type II risk scenario is modeled, which is typically a high-impact, low-probability scenario, i.e., both large monetary costs and small probabilities for the high-impact event, or, equivalently, on the impact side (Eq. 8.5), $x \gg 0$ and, on the probability side (Eqs. 8.6 and 8.7), $\theta, \phi \ll 0.5$. Stated differently, the utility probability distributions (Eqs. 8.18 and 8.19) under consideration will both be highly skewed to the left and, as a consequence, will lead to the third condition in Eq. (8.27):

$$R(U|D_j) = \frac{2E(U|D_j) - k\,\text{std}(U|D_j) + b}{3}. \tag{8.28}$$

The best possible outcome under decision D_1 is Eq. (8.18):

$$b = 0, \tag{8.29}$$

and the standard deviation of Eq. (8.20) is [21]:

$$\text{std}(U|D_1) = -q\sqrt{\theta(1-\theta)} \log \frac{m-x}{m}. \tag{8.30}$$

So, the risk index under the decision to keep the status quo is, substituting Eqs. (8.20), (8.29), and (8.30) into Eq. (8.28):

$$R(U|D_1) = \frac{q}{3}\left[2\theta + k\sqrt{\theta(1-\theta)}\right] \log \frac{m-x}{m}. \tag{8.31}$$

The best possible outcome under decision D_2 is (Eq. 8.19):

$$b = q \log \frac{m-I}{m}, \tag{8.32}$$

and the standard deviation of Eq. (8.19) is [21]:

$$\text{std}(U|I, D_2) = -q\sqrt{\phi(1-\phi)} \log \frac{m-x-I}{m-I}. \tag{8.33}$$

So, the risk index under the decision to invest in additional barriers is, substituting Eqs. (8.21), (8.33), and (8.23) into Eq. (8.28):

$$R(U|I, D_2) = \frac{q}{3} \left[2\phi + k \sqrt{\phi(1 - \phi)} \right] \log \frac{m - x - I}{m - I} + q \log \frac{m - I}{m}. \qquad (8.34)$$

The decision theoretical equality:

$$R(U|D_1) = R(U|I, D_2) \qquad (8.35)$$

represents the equilibrium situation, between the decision to keep the status quo, D_1, and the decision to invest in additional risk barriers, D_2. Now, if one substitutes Eqs. (8.31) and (8.34) into Eq. (8.35), then one obtains the closed expression for that investment value which will leave one undecided:

$$\log \frac{m - I}{m} = \frac{1}{3} \left\{ \left[2\theta + k \sqrt{\theta(1 - \theta)} \right] \log \frac{m - x}{m} - \left[2\phi + k \sqrt{\phi(1 - \phi)} \right] \log \frac{m - x - I}{m - I} \right\}. \qquad (8.36)$$

Any investment smaller than the numerical solution of I in Eq. (8.36) will turn Eq. (8.35) into an inequality, where D_2 becomes more attractive than D_1. It follows that the equilibrium investment (Eq. 8.36) is also the maximal investment one will be willing to make to improve the type II event barriers, or, equivalently, Eq. (8.23) is the hypothetical benefit of the type II event barrier improvement.

Note that the "Weber constant" q has fallen away in both the decision theoretical equalities, Eqs. (8.23) and (8.36). This will hold in general, as both the expectation values and standard deviations in Eqs. (8.28) and (8.29) are linear in the unknown constant q. It follows that one may set, without any loss of generality, $q = 1$.

8.12.5   A Numerical Example Comparing Expected Outcome Theory, Expected Utility Theory, and Bayesian Decision Theory

After the great Dutch floods of 1953, the "Oosterschelde Waterkering" was built. This was a movable dyke that allowed for an improved safety of from 1/100 to 1/4000 per year, while keeping the Oosterschelde connected to the North Sea. This open connection to the North Sea was decided upon in order to keep the salt-sea ecological system of the Oosterschelde river intact.

The total costs of the Oosterschelde Waterkering where about €2.5 billion. The bulk of these costs were due to the movable character of this dyke. Had the Dutch government decided to build an unmovable dyke, then the costs would only have been about €175 million.

The total value of the assets at risk at the time were about 1/20th of the then gross domestic product (GDP):

$$x = €3.75 \times 10^9. \qquad (8.37)$$

The wealth of the decision-maker, i.e., the Dutch government, was about 40% of the Dutch GDP at that time, aggregated over a period of 5 years, this being the total construction time of the movable Oosterschelde dyke:

$$m = €1.5 \times 10^{11}. \qquad (8.38)$$

The status quo probability of a catastrophic flood directly after the great flood was estimated to be (Eq. 8.24):

$$\theta = \frac{1}{100},\tag{8.39}$$

whereas the probability of the catastrophic flood with the improved flood defenses were estimated as (Eq. 8.25):

$$\phi = \frac{1}{4000}.\tag{8.40}$$

Substituting the values in Eqs. (8.37)–(8.40) into Eqs. (8.14), (8.23), and (8.36), one obtains the following solutions for the maximal investment willingness, or, equivalently, the hypothetical benefit, I:

- *Expected outcome theory*:
  - Any sigma level: $I = 36.6 \times 10^6$ euros,
- *Expected utility theory*:
  - Any sigma level: $I = 37.0 \times 10^6$ euros,
- *Bayesian decision theory*:
  - 1-sigma level: $I = 129.8 \times 10^6$ euros,
  - 2-sigma level: $I = 234.9 \times 10^6$ euros,
  - 3-sigma level: $I = 340.1 \times 10^6$ euros.

It is noted here that after the great Dutch flood, the discussion was not about whether to build additional flood defenses or not, but rather whether to choose the expensive solution over the "cheap" solution, which would keep the Oosterschelde salt-sea ecosystem intact. Under expected utility theory, the cheap solution of an unmovable dyke would have been too expensive by a factor of 3, whereas under Bayesian decision theory the cheap solution was well within the 2-sigma bounds.

Let us now move to the present time. The current total value of the assets at risk in the Oosterschelde region are about 1/20th of the current GDP (Eq. 8.23):

$$x = 30 \times 10^9 \text{ euros.}\tag{8.41}$$

The wealth of the decision-maker, i.e., the Dutch government, is about 20% of the current Dutch GDP:

$$m = 1.2 \times 10^{11} \text{ euros.}\tag{8.42}$$

If one assumes the current probability of a catastrophic flood to be 1/4000, and if one assumes that in the absence of any maintenance the flood defenses will have deteriorated such that the probability of a catastrophic flood will double to 1/2000 in 5 years' time, then $\sqrt[5]{2}$ is the implied "doubling" 1 year away from the latest maintenance round. Using this doubling factor of $\sqrt[5]{2}$, the probability of a catastrophic flood becomes (Eq. 8.24):

$$\theta = \frac{\sqrt[5]{2}}{4000}.\tag{8.43}$$

If one assumes that the probability of a catastrophic flooding under the flood defence maintenance is our current probability of a catastrophic flooding, (Eq. 8.5):

$$\phi = \frac{1}{4000}.\tag{8.44}$$

Then one has a scenario where one wishes to prevent a current situation, which is very safe (Eq. 8.44), from sliding into a somewhat less safe situation (Eq. 8.43).

Substituting the values from Eqs. (8.42)–(8.44) into Eqs. (8.14), (8.23), and (8.36), one obtains the following solutions for the maximal investment willingness, or, equivalently, the hypothetical benefit, I:

- *Expected outcome theory*:
  - Any sigma level: $I = 1.1 \times 10^6$ euros,
- *Expected utility theory*:
  - Any sigma level: $I = 1.3 \times 10^6$ euros,
- *Bayesian decision theory*:
  - 1-sigma level: $I = 13.9 \times 10^6$ euros,
  - 2-sigma level: $I = 26.9 \times 10^6$ euros,
  - 3-sigma level: $I = 39.8 \times 10^6$ euros.

It is noted here that in order to obtain the very real safety benefit of preventing the probability of a catastrophic flood of $\phi = 1/4000$ from sliding to $\theta = \sqrt[5]{2}/4000$, expected utility theory is not willing to invest more than €1.3 million, whereas Bayesian decision theory with utility transformation, under a 2-sigma safety level, is willing to invest €26.9 million for the safety maintenance of the Oosterschelde Waterkering.

So it would seem that the Bayesian decision theory solution is more commensurate with observed safety management practices, seeing that the Dutch government spends about €20 million a year to maintain the Oosterschelde Waterkering.

8.12.6 Discussion of the Illustrative (Toy) Problem – Link with the Disproportion Factor

In this section, expected outcome theory, expected utility theory, and Bayesian decision theory were compared for a simple illustrative toy problem regarding the investment willingness to avert a high-impact, low-probability event. It was demonstrated that the adjusted criterion of choice, in which scalar multiples of the sum of the lower confidence bound, expectation value, and upper confidence bound of the utility probability distributions are maximized, though mathematically trivial [23], has non-trivial practical implications for the modeled investment willingness, or, equivalently, for the modeled hypothetical benefits.

To summarize this section, based on the results from a numerical example, it might be argued that it is the insufficiency of the expectation value (Eq. 8.8) as an index of risk that forces a cost-benefit analysis to introduce disproportion factors (DFs) as an ad hoc fix-up; i.e., the DFs are needed in cost-benefit analyses because the currently computed hypothetical benefits (Eq. 8.24) may severely underestimate the actual hypothetical benefits (Eq. 8.36) which are computed by way of the more realistic index of risk (Eq. 8.27).

In the case where one restricts oneself to outcome probability distributions, the hypothetical benefits in a cost-benefit analysis are computed as the difference between the expectation value of the outcome under the additional safety barriers and the expectation value of the outcome under the current status quo (Eq. 8.14):

$$I = (\theta - \phi)x. \tag{8.45}$$

But under the alternative index of risk (Eq. 8.27), which not only takes into account the most likely trajectory but also the worst- and best-case scenarios, the corresponding hypothetical benefits may be computed as (cf. Eq. 8.36):

$$I = \frac{1}{3}\left[\left(2\theta + k\sqrt{\theta(1-\theta)}\right) - \left(2\phi + k\sqrt{\phi(1-\phi)}\right)\right] x. \tag{8.46}$$

So, for the outcome probability distributions Eqs. (8.9) and (8.10), the alternative criterion of choice (Eq. 8.27) implies a theoretical DF which is the ratio of Eqs. (8.42) to (8.41):

$$DF = \frac{1}{3}\left[2 + k\frac{\sqrt{\theta(1-\theta)} - \sqrt{\phi(1-\phi)}}{(\theta - \phi)}\right]. \tag{8.47}$$

In Chapter 5 a study on the DF was performed (based on Goose [24] and Rushton [25]) in which, by way of common sense considerations, it is recommended to calculate and employ DFs for which the NPV becomes zero (see also Section 5.7). Alternatively, in Eq. (8.43) a DF is derived for a specific risk scenario, by comparing the hypothetical benefits under the traditional criterion of choice (Eq. 8.8) and the alternative criterion of choice (Eq. 8.27).

8.13   Decision Process for Carrying Out an Economic Analysis with Respect to Operational Safety

Deciding about operational safety investments in an organization is actually a much more sophisticated and complex undertaking than is often understood and acknowledged. First, in an organizational context, operational safety investment decision-making should be envisioned as an integration of two decision-making processes, one for type I risks and the other for type II risks. Second, within each decision-making process, there is a need to choose the correct economic approaches, assumptions, and/or tools to aid the decision-making. However, there are a lot of different economic methodologies, methods, and defensible viewpoints that can be employed to support operational safety decision-making, as is shown in this book. Moreover, there is often not much experience available inside companies regarding these economic matters in relation to operational safety, making the optimization of decision-making even more difficult.

Nonetheless, the goal of this book is to provide safety-related decision-makers with an enterprise-wide understanding of how to deal with the different types of risk, their physical characteristics (such as consequences and probabilities), their moral characteristics (such as fairness and the division of risks and benefits between risk-takers and risk-bearers), and their economic characteristics and constraints (such as safety budgets and using and choosing the value of life – or not). Ultimately, the decision-making process regarding operational safety aims to establish and maintain a holistic view of decisions about risks across an organization, so that budget allocations are optimal, and capabilities and performance objectives are achieved.

Regretfully, in many decision-making situations with respect to risks and operational safety, there is no simple scale of preference, and it is impossible to observe the outcomes of the decisions. If dealing with type I or type II risks, obviously a number of factors are relevant, including physical, moral, and economic issues. Nowadays, decisions are often focused on one of these issues, and the others are only marginally considered. Furthermore, using the outcomes as a basis for judging the goodness of a decision is clearly problematic in the case of safety investments: it is usually very difficult, if not impossible, to prove that a certain safety measure has prevented a number of undesired events from happening, each with a certain probability. Nevertheless, although difficult, this exact reasoning should be at the heart of operational safety decision-making: non-events have happened, and expected hypothetical benefits have been gained by making the "right" safety investment decisions. Hence, outcome-based thinking is largely insufficient in making good decisions about what safety measures to take, and should be replaced, or at least combined, with evidence-driven, uncertainty-based proactive thinking.

As indicated by Aven [1], it is important to see decision-making as a process with formal risk and decision analysis to provide decision support, followed by an informal managerial judgment and review process resulting in a decision. Figure 8.1 provides an idea of how this decision-making process translates into real-world industrial practice. However, Figure 8.1 needs to be made much more concrete and usable for organizational decision-making, including the variety of approaches, methods, methodologies, and concepts expounded in this book.

As previously explained, the decision-making process actually consists of two parts: one for type I risks and one for type II risks. Therefore, the preliminary task that should be carried out is to decide on the risk type for the safety investment exercise. This can be done by assessing the parameters "variability" and "information availability" and using Figure 2.5 to determine the domain in which the safety investment decision problem is situated: A, B, C, or D. Once the domain of the decision problem is successfully fixed, based on the scenario's likelihood and consequences, and using the risk matrix designed by and for the organization, the region in which the decision problem is situated can be determined: negligible, acceptable, tolerable, or unacceptable. Once the region and the type of risk are determined, it is the possible to decide on the economic approaches, methods, and so on, to employ. Figure 8.23 illustrates this heuristic reasoning.

At this point in the book, all information and conceptual thinking are available for decision-makers within organizations to optimize their operational safety investment policies and strategies. The starting point is reality, and the safety investment decision problems are derived from this real world, influenced by organizational constraints, stakeholder values, political reality, geographical characteristics, globalization facts, economic situations, and so on. Operational safety decision problems are always characterized by an amount of information that is available (or not) and a variability of possible outcomes. Based on these data, Figure 8.23 can be used to limit the number and specificity of economic approaches, models, and methods that can be used. Background parameters such as stakeholder values, perception of reality, and goals, criteria and preferences set by decision-makers further influence and/or determine the entire decision-making process. Experts and managers, for instance, have their own background, values and preferences that could significantly influence the selection process of alternatives.

By using the heuristic and the concepts provided and explained in this book, as shown in Figure 8.24, the subjective element of decision-making, i.e., the personal agendas of people and the personal interests of certain company managers, is deliberately kept to a low level. Furthermore, as Aven [1] indicates, the foundation for good decision-making can also

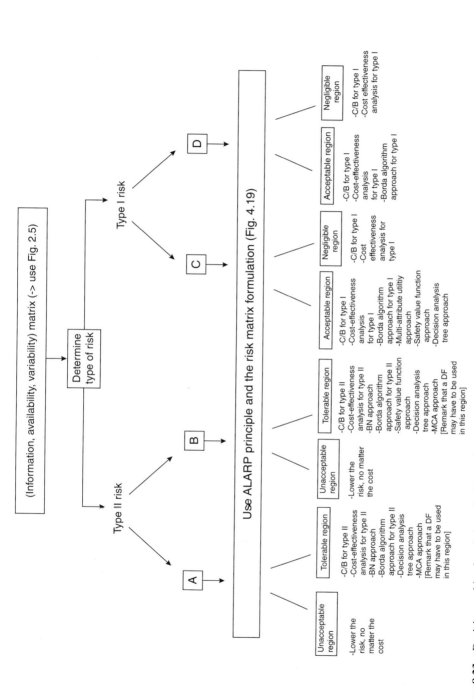

Figure 8.23 Decision-making heuristic to decide on the economic approach/method/procedure to use (illustrative example with suggested approaches; organizations may want to draft their own heuristic). C/B, cost-benefit; BN, Bayesian network; ALARP, as low as reasonably practicable.

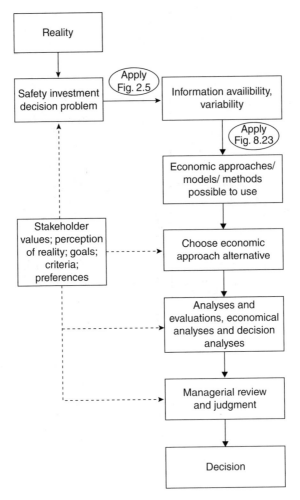

Figure 8.24   Concrete heuristic of the decision-making process for an economic analysis. (Based on Aven [1].)

be laid by ensuring that the decision-making process involves a sufficiently broad group of personnel.

8.14   Conclusions

Several approaches to dealing with operational safety decision-making, and thereby taking economic information and parameters into account, have been outlined and discussed in this chapter. It is obvious that there is a wide range of economic approaches available and that they can be used in many different circumstances. A stepwise approach has also been outlined as an aid to the decision-making process, in order to guide decision-makers in terms of what approach to use in what situation, depending on pre-defined parameters and influenced by certain criteria. This economic decision-making process provides a manner in which to optimize operational safety decision-making within organizations and take financial thinking and safety budget allocation to the next level.

References

[1] Aven, T. (2012). *Foundations of Risk Analysis*, 2nd edn. John Wiley & Sons, Ltd., Chichester.

[2] Greenberg, H.R., Cramer, J.J. (1991). *Risk Assessment and Risk Management for the Chemical Process Industry*. John Wiley & Sons, Ltd., Chichester.

[3] Centre for Chemical Process Safety. (2003). *Guidelines for Investigating Chemical Process Incidents*, 2nd edn. American Institute of Chemical Engineers, New York.

[4] Meyer, T., Reniers, G. (2013). *Engineering Risk Management*. De Gruyter, Berlin.

[5] Muennig, P. (2008). *Cost-Effectiveness Analysis in Health*, 2nd edn. Jossey-Bass, San Francisco, CA.

[6] Garvey, P.R. (2009). *Analytical Methods for Risk Management*. Chapman & Hall/CRC Press, Boca Raton, FL.

[7] Klamler, C. (2005). On the closeness aspect of three voting rules: Borda Copeland maximin. *Group Decisions and Negotiations*, **14**(3), 233–240.

[8] Koch, S., Mitlöhner, J. (2009). Software project effort estimation with voting rules. *Decision Support Systems*, **46**, 895–901.

[9] Martin, W.E., Shields, D.J., Tolwinski, B., Kent, B. (1996). An application of social choice theory to USDA forest service decision making. *Journal of Policy Modeling*, **18**, 603–621.

[10] Zarghami, M. (2011). Soft computing of the Borda count by fuzzy linguistic quantifiers. *Applications in Software Computing*, **11**, 1067–1073.

[11] Sobczak, A., Berry, D.M. (2007). Distributed priority ranking of strategic preliminary requirements for management information systems in economic organisatiuons. *Informations Software Technologies*, **49**, 960–984.

[12] Ni, H., Chen, A., Chen, N. (2010). Some extensions on risk matrix approach. *Safety Science*, **48**, 1269–1278.

[13] Reniers, G.L.L., Audenaert, A. (2014). Preparing for major terrorist attacks against chemical clusters: intelligently planning protection measures wrt domino effects. *Process Safety and Environmental Protection*, **92**, 583–589.

[14] Fenton, N., Neil, M. (2007). *Managing Risk in the Modern World. Applications of Bayesian Networks*. London Mathematical Society, London.

[15] Hanea, A. (2008) Algorithms for non-parametric Bayesian belief nets. PhD thesis. TU Delft, Delft, the Netherlands.

[16] Brans, J.P., Mareschal, B. (2005). *Promethee Methods. Multiple Criteria Decision Analysis: State-of-the-Art Surveys*. Springer-Verlag, New York.

[17] Talarico, L., Sörensen, K., Springael, J. (2014). *A Bi-Objective Decision Model to Increase Security and Reduce Travel Costs in the Cash-in-Transit Sector*. University of Antwerp, Antwerp.

[18] Talarico, L., Reniers, G., Sörensen, K., Springael, J. (2015). Intelligent systems in managerial decision-making. In: *Intelligent Techniques in Engineering Management* (eds Kahraman, C., Onar, S.Ç.), Springer-Verlag, Amsterdam, pp. 377–403.

[19] Nowak, M.A., Highfield, R. (2011). *SuperCooperators: Why We Need Each Other to Succeed*. Simon & Schuster, New York.

[20] Jaynes E.T. (2003). *Probability Theory: The Logic of Science*. Cambridge University Press.

[21] Lindgren B.W. (1993). *Statistical Theory*, Chapman & Hall, Inc., New York.

[22] Bernoulli D. (1738), *Exposition of a New Theory on the Measurement of Risk*. Translated from Latin into English by Dr Louise Sommer from 'Specimen Theoriae Novae de Mensura Sortis' Commentarii Academiae Scientiarum Imperialis Petropolitanas, Tomus V, pp. 175–192.

[23] Van Erp, H.R.N., Linger, R.O., Van Gelder, P.H.A.J.M. (2015) Fact Sheet Research on Bayesian Decision Theory, arXiv: 1409.8269v4 arXiv.

[24] Goose, M.H. (2006). *Gross Disproportion, Step by Step – A Possible Approach to Evaluating Additional Measures at COMAH Sites*. Institution of Chemical Engineers Symposium Series, vol. **151**, p. 952. Institution of Chemical Engineers.

[25] Rushton, A. (2006) CBA, ALARP and Industrial Safety in the United Kingdom.

9

General Conclusions

In this book, theories, models, approaches, and examples within the field of operational safety economics are provided, taking the organizational viewpoint. An overview of useful microeconomic concepts and their relationship with operational safety, are given. Despite the importance that organizations attach to economics, on the one hand, and operational safety, on the other, the combination of both, translated into "operational safety economics," or simply "safety economics," is often disregarded or misunderstood by company management.

In today's organizations, hazards are identified, risks are determined, calculated, and prioritized, and risk management actions are defined and monitored. However, these safety actions and the accompanying investments are often not well understood by company management from a financial and economic perspective. The investments are often based on relatively simple risk assessment methods and they are pursued with some kind of "gut feeling," but without thorough economic assessments and going into depth with regard to the economic aspects of the different types of risk, thereby looking at opportunity costs, expected hypothetical benefits, company management utilities, and so on.

Indeed, a lot has been said and written about economics and economic viewpoints in relation to safety, especially on a macroeconomic level, but to the best of the author's knowledge there has never been published before a comprehensive overview of models and techniques that can easily be used by companies to improve their understanding of operational risks with economic input. Nonetheless, operational safety economics is much more than merely simple cost-benefit analyses or straightforward cost-effectiveness analyses. Despite the obvious impact of safety, or rather un-safety (i.e., accidents), on an organization's profitability, up to now economic considerations within the domain of operational safety have been treated with too much caution and/or ignorance. The field truly deserves to become a widespread academic field of interest, and a natural approach to tackle operational safety investments in industry.

Taking economic factors into consideration in the process of risk treatment and decision-making does not indicate that operational risk decisions will no longer take any emotional arguments into account, and that all decisions will be made by using and applying some kind of robotic financial models and approaches. On the contrary, moral arguments, such as equity,

Operational Safety Economics: A Practical Approach Focused on the Chemical and Process Industries,
First Edition. Genserik L.L. Reniers and H.R. Noël Van Erp.
© 2016 John Wiley & Sons, Ltd. Published 2016 by John Wiley & Sons, Ltd.

fairness, recklessness, acceptability, and others, can then play an even more important role in risk treatment and management than is the case today. By using economic theories, moral aspects can become an essential part of the decision-making process determining what risks to treat first and with what budgets. By doing so, risk management decisions can be defended by organizations in a more profound and balanced way and they may be more easily perceived as acceptable by authorities and the public.

In recent decades, organizations and authorities have struggled with the so-called "risk-based" decision-making approach. In risk-based decision-making, rational risks, determined by using best estimated probabilities and scenario-based consequences, were the cornerstone of risk prioritization, risk management, and budget allocation for risks. The approach, although rational and logical, has received a lot of criticism due to the lack of any emotional and economic factors. Therefore, so-called "risk orientation" is now the newest concept to guide decision-making with respect to risks. In risk-oriented decision-making, besides the classic rational parameters, moral and economic aspects are both taken into consideration when making recommendations to allocate safety budgets and to deal with the different types of risk. This book discusses a number of theories, models and ideas that may be employed to steer and further explore and develop the risk-oriented approach.

Chapter 1 provides an introduction to the field of risk management and risk thinking by managers, and argues that safety, or the avoidance of accidents, should be considered as an important business strategy for long-term profitability. Chapter 2 defines and discusses some essential concepts and terms that need to be well understood by managers before they can carry out an economic assessment with respect to operational safety. Chapter 3 provides the economic foundations for dealing with operational safety in organizations. The chapter is rather theoretical, but also provides some illustrative examples to help the reader understand the basic concepts of microeconomic theory applied in safety science. Chapter 4 discusses the different decision-making theories and links safety to decision-making. The chapter also treats moral aspects of safety and discusses some highly debated topics such as the value of statistical life. Chapter 5 offers the reader an extensive (but non-exhaustive) and comprehensive overview of costs and benefits of safety, and explains how to conduct a cost-benefit analysis with respect to operational safety. Chapter 6 provides the approach that can be employed to carry out cost-effectiveness analyses with respect to operational safety. Chapter 7 discusses Bayesian decision theory as an innovative way of dealing with operational safety economics. Finally, Chapter 8 elaborates on a variety of more advanced techniques that can be used by organizations to unambiguously advance operational safety within the company.

Applying the approaches, models, theories, and techniques explained in this book will take organizations to the next level of risk-oriented decision-making. As such, decisions are more defensible, more informed, more accurate, and more balanced, and companies will finally and truly consider operational safety as a domain to be profitable in the long term, right-fully as important as productivity and innovativeness. After all, as the renowned economist John Maynard Keynes famously put it, "It is better to be roughly right, than precisely wrong." Regretfully, currently safety economics is not well known at all in most companies and there-fore safety investment decisions often are guided by some kind of "gut feeling" and tend to be "roughly wrong," which is the worst possible situation. In some best-of-class companies, safety economics models and approaches are already used to some basic degree, and decisions in these companies tend to be "precisely wrong." Using this book, decisions should evolve toward being "roughly right" initially, and in the longer term toward being "precisely right." This book has been written by the authors – and should be used by its reader – with this conceptual thought in mind.

Index